£14-50.

Brazilian Economic Policy

Gian Singh Sahota

Brazilian Economic Policy

An Optimal Control Theory Analysis

PRAEGER SPECIAL STUDIES IN INTERNATIONAL ECONOMICS AND DEVELOPMENT

Praeger Publishers New York Washington London

Library of Congress Cataloging in Publication Data

Sahota, Gian S
Brazilian economic policy.

(Praeger special studies in international economics and development)
Bibliography: p
Includes index.
1. Brazil—Economic policy—Mathematical models.
I. Title.
HC187.S26 338.981 74-14047
ISBN 0-275-09690-4

PRAEGER PUBLISHERS
111 Fourth Avenue, New York, N.Y. 10003, U.S.A.
5, Cromwell Place, London SW7 2JL, England

Published in the United States of America in 1975
by Praeger Publishers, Inc.

Printed in the United States of America

To Kanta and Vivek

ACKNOWLEDGMENTS

This study is the result of several years of discontinuous research done at different places. Some preliminary work on this research was done while I was in Brazil: first as USAID-Vanderbilt visiting professor of economics at the Instituto de Pesquisas Economicas (IPE), University of São Paulo (1966-68), then as economic advisor to the Ministry of Finance (1967-69) and to the Comissao Interstadual da Bacia Paran-Uruguai "BACIA" (1968-69). Subsequent work was done while I was a visiting scholar at London School of Economics (summer 1969) and Harvard University (fall 1970), as Senior Simon Fellow at Manchester University (1971) and as and when other duties permitted at Vanderbilt University. In the development of this study I received help from numerous agencies and individuals. It is not possible to prepare an inventory of all of these; mention must, however, be made of the following agencies and individuals.

Among the agencies that, at one stage or another and in one way or another, contributed to the financing or advancement of this research are the Ministries of Finance and Planning (IPEA) of the government of Brazil; BACIA and ANPES, São Paulo; IPE and the Chair of Econometrics, University of São Paulo; USAID; the Graduate Program in Economic Development and the Graduate Center of Latin American Studies at Vanderbilt University; the London School of Economics; the Harvard Tax Program, Harvard University; Manchester University; and the Social Sciences Research Council. To all of these agencies I owe my thanks. Among the individuals from whom I benefited in the professional aspects of this study I wish to make acknowledgments to Antonio Delfim Neto, Affonso Celso Pastore, Carlos Antonio Rocca, Luiz de Freitas Bueno, Richard A. Musgrave, Carl S. Shoup, Arnold C. Harberger, David Laidler, Albert Fishlow, Albert Ando, Rudolph C. Blitz, Cliff J. Huang, and Yasuoki Takagi. Thanks are also due to Elionora Budniok and Frances Rich for typing the manuscript.

Some aspects of this study were discussed at various seminars, including the 1968 Evanston meetings, the 1971 European meetings, and the 1971 New Orleans meetings of the Econometric Society; those at the University of São Paulo (1969); at IPEA, Ministry of Planning and Coordination, Brazil (1969); at Sussex University (1971) and Manchester University (1971); and at Vanderbilt University (1972, 1974). Portions of it were published in Revista Brasileira da Economia, Economic Development and Cultural Change, and Econometrica.

CONTENTS

LIST OF TABLES

LIST OF FIGURES

THE APPROPRIATENESS OF THE THEORY OF TARGETS AND INSTRUMENTS FOR POLICY STUDIES

With the development of the mathematical theory of optimal control in the past decade, the theory of targets and instruments has assumed an added usefulness and elegance.[1] Aprioristic statements on the tendencies of the effects of various instruments on different objectives do not take us very far when several competing objectives and multiple-effect instruments are involved. Even when there is only one objective to achieve and there are several policy instruments, a quantification of the objective and an estimate of the quantitative effects of various instruments (even if approximate) are helpful for realistic economic policy making and planning. It is naturally worthwhile to have some quantitative estimate of how much has been achieved and how much is achievable (within a given time period). In general an economic policy based on models that have proved successful when statistically tested will be on a sounder footing than one based on mere intuition or "feeling." Quantitative-policy models will often reveal gross inconsistencies of the aprioristic models. "Propositions derived from aprioristic models have been proved to be dangerous primarily because so many aprioristic models are not realistic" (Tinbergen, 1956, p. 32).

Econometric techniques are by now sufficiently developed to yield fairly workable results, and high-powered computers have proved their potency in performing unbelievably complex feats, such as sending manned spacecrafts and computerized robots to the moon and bringing them back to earth with moon rock. In this age, therefore, it is no longer necessary to rely merely on the qualitatively-held notions about the reaction of the economy that were understandable before the twin breakthroughs. A simultaneous measurement of the quantitative effects of all targets and all instruments is now possible. Of course, a good deal of progress has still to be made with the formulation of models and (at least insofar as the developing countries are concerned) with the creation of trustworthy data.[2]

ECONOMETRIC MODELS AND QUANTITATIVE POLICY MODELS DISTINGUISHED

Econometric models that describe and predict the behavior of various market variables with respect to the relevant exogenous variables are now commonly used. Models of quantitative economic policy—variously called the models of targets and instruments, of ends and means, or of optimal control solutions—that describe and predict policy variables with respect to various objective variables are, however, still rare.[3] Until recently, the theoretical developments in this field (and their practical application by governments) were largely confined to the Scandinavian countries. The names of Jan Tinbergen and Henri Theil, and to a lesser extent Lief Johansen, Bent Hansen, and J. E. Meade, practically exhausted the list of the contributors to this theory as distinguished from econometrics proper. Recent American interest in this problem is reflected in the studies by Arrow and Kurz (1970), Poole (1970), Pindyck (1973), and Chow (1972a). The essential differences between a conventional econometric model and a quantitative economic policy model can be brought out by distinguishing two cornerstones of the latter, as in Theil (1965, p. 19):

(a) A formulation of either a social welfare function or simply individual objectives, representing the desires of the community or policy-makers;
(b) an econometric model describing the mechanism of the economy, in which policy instruments are explicitly introduced.

The "ultimate" end of economic policy can be stated to be the maximization of social welfare. This end is approached via several subordinate "ends," such as material welfare, equity, protection of persons and property, external security, and so on, that reflect the attitudes of policy makers and through them the desires of the community.[4] Since some of these subobjectives, too, cannot be easily quantified, economic policy usually treats what Tinbergen calls "derived" aims (Tinbergen, 1967, p. 39). These aims include improvements in allocation of resources, expansion of production, improvements in income distribution, and stability of employment and incomes, which aims are measurable and form a part of the set of endogenous variables in most econometric models. It is apparent that quantitative policy models include econometric models and more. Other differences will become clear from the following description of the theory of targets and instruments.

THEORY OF TARGETS AND INSTRUMENTS

In the preparation of a policy model, the first step is to formulate an econometric model to understand the mechanism of the system. Let the structural model be:

$$\sum_{j=1}^{J} \beta_{mj} Y_{jn} = \sum_{k=1}^{K} \alpha_{mk} X_{kn} + \sum_{i=1}^{I} \delta_{mi} \tau_{in} + U_{mn} \qquad \text{(i)}$$

where Y_j is a vector of J endogenous variables, including the "derived" objective variables, such as national income, price level, and tax revenue, as well as any other endogenous variables of the system; X_k is a vector of K exogenous variables and the lagged values, if any, of Y_j's (including those that are usually dumped in the constant of the regression, because of difficulties of measurement), which are not controllable by the policy maker; τ_i is a vector of I autonomous instrument variables (say, tax rates), directly determinable by policy makers; and U is a vector of error terms. The subscript m denotes the mth equation: $m = 1, \ldots, M$. For the model to be logically complete each endogenous variable Y_j must have an equation—i. e., $M = J$ must hold. The equations of the system are economically interdependent, but the linear ones are and must be algebraically independent. To emphasize the estimation aspect of the model, subscript n has been added to indicate the dimension of observations: $n = 1, \ldots, N$. Naturally, $N > J+K+I$, and N has no relationship with M. Model i represents the mechanism of our system, the cornerstone "b" mentioned above.

Denoting the estimated values of the endogenous variables by $\hat{Y}$ and omitting the n subscript, we have in matrix form:

$$B\hat{Y} = AX + D\tau \tag{ii}$$

The mechanism of our system having been econometrically determined in (ii), we turn to the next step in the policy model. Normally, not all Y's are the "derived" objectives for which targets may be set. We are now interested primarily in the target variables, say $\hat{Y}_1, \ldots, \hat{Y}_H$ instead of $\hat{Y}_1, \ldots, \hat{Y}_J$, $H < J$. We may denote the targeted-value variables by bars—i.e., $\bar{\hat{Y}}$ or simply $\bar{Y}$. The remaining $\hat{Y}_{H+1}, \ldots, \hat{Y}_J$ variables are what Tinbergen calls "irrelevant" variables (Tinbergen, 1952, p. 7), and can for simplicity be dropped. Thus we have:

$$\overline{BY} = AX + D\tau \tag{iii}$$

which can be solved for τ, to get:

$$\tau = D^{-1}\overline{BY} - D^{-1}AX$$

or

$$\tau = \pi\bar{Y} + \psi X \tag{iv}$$

In this form, the model indicates the values that must be given the vector of τ's to get the required values of the vector of $\bar{Y}$'s. They represent, in Tinbergen's terminology, the directives for economic policy, since they indicate how the political parameters have to be varied in relation to the changing data (Tinbergen, 1952, p. 28). Some of the $\hat{Y}$ arguments may have zero influence on some of the τ's in Equation iv, just as some of the X's and τ's may have zero influence on some of the $\hat{Y}$'s in Equation ii. Note that the logic of economic policy in Equation iv is an inversion of the logic of economic theory of Equation ii: the "given entities," the "data," of economic analysis—i.e., the τ's—become the "unknowns" of the policy analysis; the "unknowns" of economic theory, the Y's, become the data of economic policy. Observe also that the τ's may be determined not only by $\hat{Y}$'s but also by X's.

SETTING TARGETS

In setting the targets, two factors must be taken into consideration: a) the aspirations or desires of the community (certain targets might emerge simply from a state of equilibrium) and b) the possibility and feasibility of targets. The determination of the former is a political problem. The determination of the latter condition is an empirical problem and, in principle, objectively measurable. However, at least in developing countries, aspirations usually outrun what is actually possible; hence, an important task of a target-setting agency is to obtain measures of the achievability of various targets that are as accurate as possible, given various constraints (constitutional, institutional, political, social, economic, physical, and others) on instruments.[5]

Usually, one would start with initial conditions (the initial values of Y's, X's, and τ's), prognosticate the exogenous variables (X's), determine the gap between the targeted values of Y's and the predicted values of Y's from the model of Equation ii, and, finally, determine the level of τ's that would eliminate the gap as nearly as possible.

CHOOSING INSTRUMENTS

As just stated, the objectives, even if attainable in principle, might not be achievable in practice. A mathematical problem arises if the number of independent objectives is different from the number of distinct instruments available. As the system in Equation iv indicates, an exact solution requires that the number of instruments be equal to the number of targets—i.e., that $I = H$. In other words, the rank of matrix D must equal I. The system is unsolvable and all the targets cannot be achieved simultaneously if the number of instruments is less than the number of targets—i.e., if $I < H$. If $I > H$, there will be more than one solution; targets can be achieved simultaneously in more than one way.

In the policy analysis, it is always an advantage to have more instruments, more degrees of freedom, than the number of targets.

There are several reasons for this. Most instruments have constraints of one sort or another and might be usable only over a limited range of values. The effects of some of them may be small in relation to the magnitudes of goals. Thus in Equation iv, if the elements of D have low values, it will require very large values of τ to satisfy the equations. Even if there are no formal constraints, certain values of instruments might be inconsistent with the mechanism of the economic system—e.g., a profits tax of more than 100 percent. Sometimes the "irrelevant" variables left out enter through the back door by necessitating restrictions on some instruments, so that certain values of them are inadmissible on that account. Sometimes it may be found that two instruments, though legally distinct, are not economically distinct, so that in their economic effect they are really one instrument. In all such cases, a larger number of instruments will be useful to attain "a reconciliation by instrument differentiation" (Tinbergen, 1956, p. 137). Sometimes the targets might be incompatible, in which case the model will be inconsistent. It might be possible to avoid such incompatibility by an increase of instruments. A sufficiently large number of instruments might also attain a "distribution of pressure." For, first, the larger the number of instruments, the more amenable they are likely to be to "fairness" among those who bear the incidence of policy, and, second, the larger the number of policy instruments, the more "efficient" the instrument set, in the sense that the extent to which each instrument has to be changed will generally be smaller than otherwise (Tinbergen, 1952, p. 41). In the field of growth "the range of paths of optimal growth that can be achieved depends upon the range of instruments available to the government" (Arrow and Kurz, 1970, p. 119). It is because of such considerations that we have chosen almost all of the major taxes and major expenditure disaggregates as instruments. In the system described by Equation iv linear dependence of instruments will be a problem. In the optimal control theory model employed in this study even linear dependence is innocuous (see Chapter 12). In all, we analyze about a dozen instruments.

EFFICIENCY OF INSTRUMENTS

We have talked of the "efficiency" of an instrument above. This term needs some elaboration. In the context of quantitative economic policy "efficiency" is usually defined by the smallness of change in an instrument for the achievement of a given change in an objective in relation to institutional or other constraints (Johansen, 1965, p. 13). The effectiveness of an instrument for a particular objective can be measured only by holding the values of other relevant objectives constant, or at least by preventing them from changing to an undesirable direction. For those requirements to be possible, however, changes in other instruments are necessary. It makes no sense to try to measure the efficiency of an instrument as such in a mechanical fashion by taking partial derivatives, $\partial\tau_i/\partial Y_j$ (i.e., by holding the values of other instruments constant). Only rarely will an instrument be found to be a specialized one in the achievement of an objective, in the sense that negligible changes, if any, in other instruments might be called for; only in such a case will efficiency be easily measurable.

Instruments can also be influenced by exogenous variables, X: the so-called "data" of economic policy. If certain $\partial\tau_i/\partial X_k$ is positive and high, and if X_k has changed during the reference period, the required discretionary change in instrument τ_i to meet a target will be correspondingly less. In general, thus, built-in flexibility of taxes can be expected to add to the "controllability" of economic policy, where "controllability" is defined by Arrow and Kurz (1970, p. 120) as follows:

> A policy is said to be controllable by a given set of instruments if there exist values of the instruments, varying over time in general, which cause the private and government sectors together to realize that policy.

Furthermore, instruments should be stable—i.e., the values of an instrument "that achieve control converge to finite values" (Arrow and Kurz, 1970, pp. 120, 121).[6]

CONFLICT OF OBJECTIVES

A recurrent theme in public finance literature concerns the conflict of objectives. For example, full employment may conflict with price stability. A redistributive (progressive) tax may reduce the will, ability, and incentives of taxpayers to work, save, invest, take risks, and acquire skill and may thus retard growth. Revenue productivity of a tax may distort resource allocation. Very rarely if ever, however, will such prima facie conflicts be found to be inherent. As discussed above, many conflicts can, in principle, be mitigated by an adequate number of "controllable" instruments. To quote Shoup (1969, pp. 480-81):

> The conflicts usually cited arise not because the goals are inherently incompatible but because the instruments assumed available are too few, or their values too constrained, or their effectiveness zero or negative for certain ranges of their values. . . . The task of public finance research is therefore (a) to eliminate or reduce the conflicts of goals by inventing new public finance instruments, (b) to compute

or conjecture the range of values each of the new instruments would need to take, (c) to ascertain whether any of these ranges of values are unacceptably wide, and (d) if they are, to find out what goals outside the system those instrument values are incompatible with, since this is what "unacceptable" means.

Thus, the problem is not a trivial one of targets being contradictory; there is a serious problem of an inadequate set of instruments available for acquiring controllability. The targets are interdependent among themselves; instruments, too, are interdependent among themselves; most instruments influence most targets and are in turn influenced in their effectiveness by them. Tradeoffs within a set of targets and within a set of instruments will usually be imperative. This is the most important and interesting problem to tackle in the theory of targets and instruments.

SOCIAL WELFARE

The reference to tradeoffs, finally, brings us to the knottiest problem of the theory of quantitative economic policy: a weighting of targets to derive a measurement of the social welfare function—what Theil calls a "reference function." It implies such comparisons as the desirability or utility of a 1 percent increase in the rate of growth of national income against a decline by one percentage point in the Gini coefficient of income concentration. It involves a social welfare function:

$$W = W(Y_1^*, \ldots, Y_H^*) \qquad \text{(v)}$$

If the instruments may also enter the utility function of the community, the social welfare function assumes the implicit form:

$$W = W(Y_1^*, \ldots, Y_H^*; \tau_1, \ldots, \tau_I) \qquad \text{(vi)}$$

Note that in such a function there is no need to match the number of instruments with the number of objectives. There is only one objective, the maximization of W, within the existing constraints. The targets (Y^*) are flexible rather than fixed in absolute magnitudes. A convenient functional form to handle this problem follows from Theil's quadratic-function linear-constraints procedure, which involves a minimization of the sum of squared deviations from the desired target values subject to specified constraints on instruments (and objectives) and the constraints of the mechanism of the economic system (Theil, 1956, 1961, 1965).[7] The precise optimal policy model to be used is derived from

Pontryagin's Minimum Principle (Pontryagin et al., 1962) and is explained in Chapter 11. A computerized version of it is described in Appendix C. Here we describe only a brief conceptual form of the mathematical model employed for the purpose. The model consists basically of three (sets of) equations/constraints that in a very simplified form can be summarized as follows:

$$\text{I. Minimized cost} = \sum_{n=0}^{N} (Y_n - Y_n^*)'\Omega(Y_n - Y_n^*) \qquad \text{(vii)}$$

subject to the lower (L) and upper (U) boundary conditions on the m instrument variables:

$$\text{II. } \tau_m^L \leq \tau_m \leq \tau_m^U \qquad \text{(viii)}$$

and subject to the constraints of the mechanism of the economic system:

$$\text{III. } B\hat{Y} = AX + D\tau \qquad \text{(ii)}$$

where, as noted earlier, the lagged-value endogenous variables are included in X. The diagonal matrix Ω stands for a set of weights (to be assigned) for the objective variables. The subscript n stands for time period.

SIMULATION EXPERIMENTS VERSUS THE OPTIMAL CONTROL THEORY

In the context of planning or policy formulation, the use of econometric models has been confined largely to simulation. That is, experiments are made by specifying levels of policy instruments over the planning period to see what kinds of objective-variable path one ends up with. The analysis in Chapters 6 and 7 of this study falls in the instrument-oriented simulation approach of this nature. In the optimal policy solutions of Equations vii, viii, and ii, on the other hand, one specifies paths for the objective variables and solves for optimal paths of instrument variables. Control theory follows a goal-oriented approach; simulation follows an instrument-oriented approach. The former method starts with specified end-period values or the paths of the objectives and works out optimal sets of means to achieve or "track" them. The simulation method starts with a specified set of means and determines whatever ends can be expected to be achieved; it involves "decision rules." An example of a very simple decision rule is Milton Friedman's famous suggestion for a constant (3 percent or so) annual increase in money supply. Decision rules derived from simulated econometric models may not be that simple, however. In general (but not necessarily) decision rules tend to reduce government regimentation of an economy; control-theory solutions have to do with planning. In the final analysis, however, both approaches—one starting from the one end (the instrument end), the other starting from the other end (the objective end)—aim at an optimal solution in which objectives and instruments have to be solved simultaneously. Simulation gropes for the optimum through experiments; the control theory identifies it by optimizing solutions. Computationally, simulation is a laborious and time-consuming method; the optimal control model is in principle more efficient and should cost less computer time. The optimal policy model, in addition, has one distinct advantage over the simulation model: It yields tradeoffs among policy instruments and among objectives. This study concludes with an analysis of tradeoffs as derived from an application of the optimal control theory.

NOTES

1. The basis of the theory of optimal control is essentially Pontryagin's Minimum Principle (1962). Bellman's principle of optimality can be regarded as another (independently discovered) version of it. The theory has been developed essentially in relation to differential equations. An appropriate mechanical device for its solutions has been the analog computer. Its application to difference equations and practical economic problems as well as its solutions by digital computers may be said, for all practical purposes, to be a development of the 1970s. It has so far captured the interest of only a few mathematical economists, a notable one being Gregory Chow (1970, 1972a, 1972b, and 1973). A most recent application is that by Pindyck (1973).

2. Even for the developed countries, the model formulation and the accuracy of data get high weight. Klein (1960, p. 867) looks "toward improvements in precision of econometric judgments of the order of magnitude of 50 percent as a result of a better knowledge of the functioning of economic institutions, through the use of new measurements on variables, and through the use of more accurate data. In contrast [he] would expect marginal improvements of 5 or 10 percent through the use of more powerful methods of statistical inference."

3. Among the policy models that fall under the present definition those by Sandee and Pas for the Netherlands (1969); Krischen and Morissens for nine Western countries (1965), being a somewhat revised but abridged version of Krischen et al. (1964); May for Canada (1966); and Chow (1972b) and Pindyck (1973) for the United States can be mentioned.

4. The ultimate aims and derived objectives and the relationships among them are discussed by Krischen and Morissens (1965).

5. In this connection, without expressing our opinion one way or the other, we simply present an excerpt from Tinbergen (1956, pp. 22-23):

> The contributions made to our problem by most of present-day economists have been to declare the comparisons of different individuals' satisfaction an impossibility. This contribution is not very constructive, since it implies (i) that all feelings about social justice are meaningless, (ii) that the scientist cannot make any contribution, and consequently (iii) that it is left to others, who often lack a scientific education, to make such contributions. . . . Decisions are continuously being taken which imply such comparisons. . . .

6. The instability of an instrument has recently been the topic of discussion among several economists. In particular, see Holbrook (1972) and Chow (1973).

7. The main weakness of the Theil method is that it gives equal importance to over- and underachievement of targets. In the face of so many imponderables and unpredictable influences in economic policy and the probabilistic nature of the state parameters, Theil's conditions are not very realistic. In practice, targets are fixed to be achieved; it is normatively good to exceed the target values; it is bad to fall short of them. Otherwise, the normative notions of maximization or minimization of certain public goods will have no meaning. Yet it is a convenient and practical method in which the quantitative maximization of a preference function can be objectively tested. The application of activity analysis to such problems is still in its embryonic stage. This should be clear from such studies as those by May (1965), Chow (1973), Yunus (1971), and Pindyck (1973).

CHAPTER

2

AN OVERVIEW OF BRAZILIAN PUBLIC FINANCE

Major expansion of public activity took place in Brazil, as in most other countries, in the postwar period. While the ratio of tax revenues to GDP was about 13 percent in 1939 and approximately 15 percent in 1947, it rose to nearly 27 percent in 1968. During the period from 1947 to 1965 public investment increased at an annual rate of about 12 percent and public consumption at about 8 percent, while gross product rose at a rate of roughly 5.4 percent. Per capita gross product rose at about 2.5 percent per annum; per capita public expenditure rose at over 5.5 percent per annum.

The expansion of the public budget was due in part to the government's growing concern with economic development. The government's role was enlarged both in the direction of inducements to private producers as well as in direct production and financial activities. Typical instances of the inducement policies through fiscal and monetary means are its promotion of the economic development of backward areas, such as the development of the northeast and Amazon regions, through SUDENE, SPVEA, and other public organizations, and a host of autonomous and semiautonomous entities better known by their abbreviated names (e.g., DNOCS, DNOS, DNEF, DNPVN, CPCN, CNEN); the protection of coffee-growers from price fluctuations in the world markets; industrial development in general through Fundo de Reaparelhamento Economica, earmarked for the BNDE, and a resolute policy of import substitution; and so forth. As a result of the government's direct entry into the field of production and finance, the public sector has monopolistic or oligopolistic control by now in several branches of the economy. In socialist jargon, it is now estimated that the government holds the "commanding heights" of the Brazilian economy. Thus, the federal government owns about a half dozen large banks (that through direct and indirect means control about half of the banking business); five major steel companies (that give the government a direct control over nearly two-thirds of total steel production, railway traffic, and iron-ore extraction); giant electrical plants (that in combination with the state and local governments' electrical plants account for about half the total production of electrical energy); almost the whole of the petroleum industry, which meets 90 percent of Brazil's domestic consumption of petroleum; a substantial proportion of the petrochemical industry; and a host of other enterprises.* The available data for 1964 show that "mixed enterprises" (economia mista) account for about 16 percent of GNP.

Apart from the nontraditional activities referred to in the preceding paragraph, the government has also significantly enhanced its traditional activities of providing public services and the social and economic infrastructure. Brazil is the leading developing country in public transfer payments and compares favorably with the majority of the developing countries in the supply of "merit goods." An interesting aspect of Brazilian economic expansion is that while the tax revenue-to-income ratio has experienced a steadily rising trend during the reference period, the private saving-income ratio has not suffered a decline, except for a temporary sag in the late 1950s. On the whole, a rising trend is also visible in private saving. This may come as a surprise to students of public finance: even though in theory there are particular (though rare) taxes that may not only not discourage but even induce private domestic saving, most of the existing taxes all over the world are generally believed to be neutral at best and at worst to discourage private saving as a proportion of income. But the Brazilian experience, prima facie, appears to negate that impression.† Looked at from another angle, the ratio of taxes (including social security contributions) plus private saving to gross domestic product increased steadily from a little over one-fourth in 1950 to about two-fifths in 1968—although it must be noted, too, that although overall private (household plus corporate) savings have steadily increased, household savings have declined from 41 percent of the total in 1950 to 24 percent in 1958 (the latest year for which this breakdown is available). Private investment in real terms remained virtually stagnant at the absolute levels until the mid-1960s. As a result the percentage of private investment to GDP declined from about 12 percent in 1951 to less than 8 percent in 1965. Since then it has risen sharply: it rose to 10 percent of GDP in 1968 and 11 percent in 1969.‡ It appears that during

---

*Since 1968 the government has seemed to be reducing if not reversing the process of nationalization.

†However, see the discussion below.

‡The estimates for foreign private investment have recently become available. In 1971 foreign investment stood at U.S. $2,911 million (U.S. $1,790 million on account of new foreign investment and U.S. $1,121 million as foreign reinvestment). For 1972 foreign investment was estimated to be U.S. $3,404 million (U.S. $2,081 new investment

the noted stagnation period of private investment, incremental private savings of the private sector were drained off by public borrowing. By virtue of their higher credit, governments in general—but governments of developing countries in particular—are usually in a favorable position, vis-a-vis the private firms, to borrow from the public. Small firms find it especially difficult to borrow money at competitive terms when they have to compete with the government borrowers. In addition, the Brazilian government has also been using various kinds of compulsory deposit and forced saving schemes. For example, a 15 percent surcharge on corporation tax (since 1956), in the form of forced loans, is regarded by the corporations as not much different from tax. It may very well be, therefore, that the increased government investment has been at the cost of private investment (and increased public saving at the cost of household saving), at least in part. It is also possible that, due to high, chronic inflation and the probably redistributive effects of it, the kind of demand that private producers generally face has been partially choked off.

Leaving aside the possible tradeoff between public and private investments, it may be observed, nevertheless, that the overall rate of investment has remained (roughly) constant, at around 16-18 percent of GDP. In spite of the noted constancy, Brazil's investment commendably remains quite high for a developing country. Partly as a consequence of a high rate of investment, Brazil has enjoyed an average annual rate of growth of national income of approximately 6.8 percent during the 1950s and 6.1 percent during the 1960s, thus roughly doubling her GNP every decade. At this writing (1974) the Brazilian rate of growth is running between 9 and 11 percent per annum. The provisional figure for 1972 is 10.4 percent. As a result, despite a high rate of population increase, per capita income in 1967 prices increased steadily from U.S. $102 in 1952 to U.S. $183 in 1969 and U.S. $200 in 1972. By the standards of Asia, the United States, the United Kingdom, and many other countries unravaged by war, this is a very impressive rate of growth, though in the Latin American countries rates of growth of this magnitude are not regarded as spectacularly high.

An important feature of overall Brazilian public finance is a high level of treasury deficit. A close relationship between deficit finance and inflation is evident even from a cursory look at treasury's deficit and price levels (see Appendix A). The largest price increases occurred in the early 1960s; the deficits were also at their highest in those years. An econometric estimate widely cited in Brazil is the one by Delfim Neto et al. (1965), according to which a one-percent change

---

and U.S. $1,323 reinvestment); this was about two-thirds of 1 percent of that year's GNP, and nearly two-fifths came from the United States. These figures do not include foreign loans and financing. See Conjuntura (December 1973), p. 73.

in money supply (the single most important source of which is a treasury's deficit) causes a rise of two-thirds of 1 percent in prices in the same year. It is not surprising that the period of the so-called "corrective inflation"—from around the mid-1960s (Ellis, 1969), a steady corrective process as against the "shock treatment" of inflation in the preceding period—is associated with a concerted effort at reducing treasury's deficits, though as yet not the national budgetary deficits. This especially has been the case since 1967 when one of the authors of the cited econometric study, Neto, became the finance minister. The treasury deficit was, thus, reduced from 3.2 percent of GDP in 1964 to less than 1 percent in 1968, and there was actually a treasury surplus of 0.9 percent in 1969. This surplus resulted in part from the sale of indexed bonds, treasury bills, and other public borrowings, and in part from increased tax revenues.

A picture of the tax structure is presented in Appendix Table A.1. It is interesting to note that the two major taxes of Brazil, excise duties (imposto de consumo or consumption tax) and sales tax are levied on a value-added basis. Excise duties were converted from a turnover basis to a value-added basis in the 1958 reform. Most indirect taxes, including tariffs, were changed from specific rates to ad valorem rates. Most prices, as well as income tax brackets, were defined in terms of "multiples" of the minimum wage. Several other reforms were introduced. The conversion of the sales tax from a turnover to a value-added basis came in the 1966 tax reform, when "monetary correction" was applied to tax debts and firms' capital stock for depreciation and tax purposes. Several tax incentives for saving, investment, and exports were given. The excise tax and the sales tax remained the largest revenue-gatherers throughout the period, accounting for about 50 percent of the overall tax revenues of all levels of government. Although about 30 percent of the revenues from the excise tax comes from tobacco, this tax is quite general. The fuel tax grew from an almost insignificant percentage at the beginning of the reference period to 3.5 percent of GDP in 1968.

Brazil has a gift tax and an estate duty (both levied at progressive rates) but has practically no capital gains tax. The rural land tax is an insignificant part of the state revenues.

Unlike the developing countries in general, Brazil has a reasonably well-established social security system. Thus, a comparison of this system in about 40 countries done by Reviglio (1967) shows that in the year 1963 the ratio of social security receipts to GNP in Brazil was below only Uruguay and Chile. The social security expenditures were 26.7 percent of the central government's current expenditures (32 percent according to the revised estimates), again below only Uruguay and Chile. Only 6.5 percent of the Brazilian population, however, was covered, as against 40.6 percent and 22.9 percent, respectively, in Uruguay and Chile. Brazil falls roughly at the median point among the developing countries according to this measure. The rate for employees has been a maximum of 8 percent on wages and salaries; for employers it has been 14 percent; the total thus is 22 percent. Among the developing countries this is exceeded only by a few Latin American countries and some Arab countries. In 1963 (Reviglio's reference year) social security taxes formed approximately 12.2 percent of the total tax revenue of Brazil, which was the highest of all the countries studied by Reviglio having per capita income of U.S. $300 or less (Brazil's per capita income in that year was U.S. $156). Social security revenues have increased steadily since then, as indicated in Appendix Table A.1.

Another important feature of the Brazilian tax structure pertains to income taxes. For example, in the United States personal income taxes bring in more than twice as much revenue as corporation tax, but in Brazil the relationship is reverse: personal income tax revenues are only a third of corporation tax revenues. Another interesting feature can also be brought out by contrasting U.S. and Brazilian public finance, in relation to local finance. In the United States the tax revenue-to-GNP percentages of the federal, state, and local governments, respectively, were 2.6, 0.8, and 3.4 percent in 1902; 5.4, 4.4, and 4.6 percent in 1940; and 19.2, 4.2, and 6.8 percent in 1967 (Musgrave, 1971, Table 8-1). In Brazil the same ratios for the federal, state, and municipal governments were, respectively, 8.0, 3.7, and 1.3 percent in 1939; 10.0, 4.5, and 1.1 percent in 1950; and 18.1, 8.1, and 0.7 percent in 1967. Since municipal taxes are spent locally, the scope for interregional redistribution through the tax-expenditure process is much larger in Brazil than in the United States. Direct taxes, including social security taxes, have seldom risen above 50 percent of indirect taxes in Brazil. Of course, this is a typical feature of the developing countries, particularly those in the continental tradition. The U.S. direct-indirect tax ratio is the reverse of Brazil's.

It was noted above that Brazil has a large revenue budget. A closer analysis indicates that, in one sense, Brazilian accounting practices understate tax revenues and expenditures. For example, the item "other current receipts" in the federal budget has been negative, and increasingly so, since 1954, being -5 percent of GDP in 1968. These are made up of nontax revenues "net" of transfers for other than nonconsumption purposes (to certain public and mixed enterprises and nonprofit organizations). If the two sides of this category are shown separately, instead of one net series, they will inflate both tax revenue and expenditure levels. From a different view, however, the tax-revenue levels of Brazil seem to be high only in a technical sense. As will be detailed a few paragraphs below, the pricing policy of public monopolies, such as the publicly-owned merchant marine, railroads, and post office, is such that they incur large deficits that are met from the general revenue budget. In many other countries these services—e.g., the railways and the post office in India—earn profits that are generally passed on to the central budget. There is a strong case theoretically for the Brazilian practice of low public-utility pricing, so long as congestion does not start, and so long as adequate funds for the expansion of the service are provided from the central budget. Both conditions seem to be absent from the Brazilian public utility scene. The postal and (privately-owned) telephone services are woefully congested (the latter because of low prices fixed by the state), and the railways are believed to be inefficient and inadequate. At this point, however, we are not concerned with the pricing policy of Brazil's public services. The purpose is simply to point out that the rationale behind the pricing of public services and the levying of taxes on commodities and services is basically the same. The two measures cannot be dissociated under an efficient tax-price system. Therefore, it makes sense to measure overall tax revenues gross of profits/losses on public services, unless they are run with complete fiscal autonomy and are required to stand on their own.

On the whole, Brazil has a reasonably well-developed and well-balanced tax structure and has successfully exploited some of the most modern forms of taxation—e.g., value-added taxes. Yet Brazil is not free of some typical features of public finance in developing countries, such as a high ratio of deficit finance, a high ratio of indirect to direct taxes, a high proportion of earmarked revenues, a low share of local finance in the overall picture, and a low ratio of personal income tax to corporation tax.

The structure of expenditures is presented in Appendix A, Table A.1. Three alternative classifications are given: administrative, economic, and functional. An interesting feature of the administrative classification is the erosion of federal control over its revenues. We have just discussed a large category of federal nonconsumption transfers reported under "other receipts." Such transfers are usually earmarked items. The post office, the railroads, and the merchant marine and port authorities run perpetual operating deficits. For example, the operating deficits of the railroads increased steadily from 52.4 percent of their expenditures in 1957 to 72.7 percent in 1963; and they were 71.5 percent in 1964 (Abouchar, 1969, Table 2). In 1965 they exceeded the treasury deficit by about 13 percent and the revenues from import duties by about 40 percent. In the same year the post office was budgeted to incur a deficit of 94 percent of its expenditures (about half the size of the railroad deficit), and the merchant marine and port authorities were scheduled to incur a deficit nearly two-fifths that of the railroads (see Shoup, 1965, pp. 10-11). The combined 1965 deficits of these three publicly-owned services exceeded 1.5 percent of GDP. Secondly, there are transfers for consumers that, though handled to a large extent by social security and other autarkies, take away approximately 40 percent of the overall revenues of all governments and about 30 percent of federal revenues. Federal subsidies, which are ultimately passed on to consumers in the form of lower commodity prices, account for another 12 percent of federal revenues.

There are also a number of development items for which federal funds are earmarked.* Such earmarked funds introduce rigidity into Brazilian federal finance and reduce the government's fiscal flexibility for stabilization measures.

An important feature of the economic classification of expenditures is a rapid increase in transfer payments on the one side and a steady rise in the wages and salaries bill on the other. The expenditures on goods and services, on the other hand, show a gradual decline since 1960. It appears that since about 1960, in its preoccupation with checking inflation by (among other things) economizing on its consumption outlays, the government has adopted the least-resistance measures of cutting its outlays on goods and services rather than cutting the wage bill. This is a typical instance of the working of Parkinson's Law and the ratchet effect in bureaucratic employment.

The functional classification is only a partial one, as the data for several components, such as internal security, social welfare, and other services, are not available.† It may be observed that,

---

*These earmarked funds include the following: 8 percent of import duties to the National Port Fund; 4 percent of the excise duties to the Federal Electricity Fund; 100 percent of the automobile tax to Petrobras (an autonomous federal corporation); 10 percent of the stamp tax to education and health; 40 percent of the "single tax" on electric power to the Federal Electrification Fund; 100 percent of lighthouse duties to the Naval Fund; 70 percent of the race track tax to the breeding of horses; 100 percent of the income of the North East Bank for itself (the bank); 100 percent of the income of National Steel Co. to the University of Brasilia; 100 percent of the income of the National Motors Factory to the National Bank for Economic Development (BNDE); 100 percent of the income of São Francisco Hydroelectric Power Co., Cold Storage Co., National Railroad Co., and National Insurance Co. to BNDE; 100 percent of Petrobras income to Petrobras; 99 percent of additional income tax for economic development to railroads and 90 percent to highways; 15 percent of overall federal revenues and 20 percent of the state and local government revenues for education; 3 percent of overall federal revenues for the development of the Amazon region; and 1 percent of overall federal revenues for the São Francisco River Valley.

†The Brazilian sources do not report separate federal statistics for public education, health, and other functional classes as such. They are published by type of ministry and other judicial and legislative entities. For federal expenditures on health and education we used only the allocation to the respective ministries. Before 1954 only one ministry handled both health and education. The totals for these earlier years, therefore, were broken down into health and education categories on the basis of the ratios for the later years.

The state (and federal district) data are available by functional

except during the early 1960s, defense expenditure has remained, by and large, constant in terms of GNP. A significant reduction in defense expenditure was effected in the early 1960s, but the old level was restored soon after the revolutionary (military) government took over in 1964. The share of public education and health has steadily increased over the years. Education expenditure is now, roughly, even with defense expenditure.

---

classification. We could not obtain the figures for a couple of years in the late 1960s. These were, therefore, interpolated.

The municipal data are available up to 1959 by the required classification. Since the municipal share of public health and education is relatively small (about 10 percent of the total), we resorted to an extrapolation of these expenditures (based on the regressions for the earlier years) for the 1960s. In this connection, it may be mentioned that Article 169 of the Constitution requires that the federal government spend 10 percent (later revised to 12 percent) of its revenue on education and that the state and municipal units spend 20 percent.

The total of the three classifications reported in Table A.1 do not necessarily add up to an identical aggregate. In particular, the functional classification reports only a few of the categories that make up the overall budget.

Sources of data for most series, Conjuntura Economica (1969 and 1970 issues); for health and education, Anuario Estatistico do Brasil (various issues), and for defense, U.N., Statistical Yearbook (various issues).

CHAPTER

3

THE MECHANISM OF THE BRAZILIAN ECONOMY: AN ECONOMETRIC MODEL

The ultimate end of economic activity may be said to be utility or economic welfare. We have chosen for an intensive analysis four main components of the ultimate end—namely, economic growth, income distribution, economic stabilization (specifically, the stabilization of price level and foreign balances), and revenues as sources of the supply of public goods. The stabilization objective of high employment as such could not be analyzed due to lack of data.

The number of second-stage or intermediate objectives, such as investment, can be derived from the four stated objectives. For the sake of perspective we will denote the four first-stage objectives as the final objectives and all other intermediate objectives as the derived objectives. To have the mechanism of the Brazilian economy revealed, which is the first step in analyzing policy tradeoffs, quantitative relationships must be determined for both sets of objectives. That is, all objectives have to be treated as endogenous variables. For this purpose we specify an econometric model in which each of the final and derived objectives assumes an equation of its own. The equations are linked, as far as possible (simultaneously through feedbacks and causal flows), forwards or horizontally to the final objectives and backwards to policy instruments and exogenous variables. In subsequent policy models of this study the whole system is linked with the ultimate of social welfare. An important aim of the present econometric model is to determine the mechanism by which selected policy instruments influence the final objectives, directly as well as indirectly, through the intermediate objectives.

THE THEORETICAL MODEL

This chapter is devoted to a brief theoretical discussion of an aggregative econometric model of the Brazilian economy. The model is designed to study the short-run and long-run effects of various public policy instruments. It is based on annual data from 1950 through 1968, except where stated otherwise.

The specification of the model and of the individual equations is determined by three main considerations: the availability of data, the underlying theory, and special conditions and constraints of the Brazilian economy. The sources, definitions, and other explanations of the variable series are discussed in Appendix A. The relevant theoretical aspects of the model are built into the equation system and are the subject matter of this chapter.

Before we turn to these, a reference to the major constraints that have prevailed in the Brazilian economy during the reference period seems in order. Among these constraints the following may be mentioned:

- A high, sometimes runaway, inflation has been a dominant feature of the Brazilian economy throughout the reference period, hence, an equation for inflation and its feedback effects on the monetary and real sectors are duly recognized in the relevant sectors.
- Excess capacity coexisted side by side with high inflation during the 1960s, hence the specification of the Keynesian form of a demand model (along with the supply model), the inclusion of an excess-capacity equation, and capital stock corrected for under-capacity use.
- Restrictive monetary policy to curb inflation periodically led to acute shortages of working capital, which were regarded as a major source of excess capacity in the economy during the late 1960s. Due to the unavailability of data on the rate of interest, different proxy variables for credit rationing, etc., were tried in the model, including bank loans, excess reserve ratios, and the treasury's deficit finance. The monetary sector is treated as an important part of the model.
- Public expenditures (including transfer payments) invariably exceeded tax revenues, necessitating deficit finance. This is an added reason for including individual tax functions.
- Industrialization assisted by import substitution has been a laudable economic policy of the reference period. An analysis of the production function is included. The economic effects of import-substitution policy on industrialization are studied by analyzing real exchange rates for imports and exports.
- Brazil's situation has not remained an exception to the developing countries' ubiquitous problem of foreign balances, hence, export-import functions are duly studied.
- Several other special conditions and certain institutional constraints, such as a large proportion of earmarked tax revenues, and economic restraints, all of which were discussed in Chapter 2, need not be repeated here.

The structural model system consists broadly of four subsystems or sectors: a) the domestic real private sector, b) the foreign sector, c) the monetary sector, and d) the public sector. Each sector further consists of one or more subsectors. Hence, the definitional equations and identities are not collected together but are stated at appropriate places in different subsystems. The domestic real private sector consists of a set of demand equations for consumption and for the different components of investment, a set of supply equations, including two production functions and a capacity utilization equation, and two distribution functions for wage and nonwage incomes. Thus, both the demand conditions of the Keynesian-type model, meant primarily for short-run stabilization analysis, as well as neoclassical production and supply conditions for long-run growth analysis are included. Somewhat in the Harrodian tradition, the supply of capital is made an endogenous factor, and labor supply is treated as an exogenous variable. The short-run determination of factor prices, wage rates and interest rates, is ignored. The latter were ignored due to lack of data. The former were left out due to both a deficiency of data and the lack of relevance, since a minimum wage rate was fixed by the government throughout the reference period. Though there is no reason why an average wage rate should necessarily follow the minimum wage rate, the trend in the average wage rate nevertheless seems to be significantly associated with the minimum wage rate: both seem to have declined in absolute real terms since the early 1960s (see below).

Since foreign balances form an important aspect of the stabilization objective, and since tariff policy is analyzed as an important public policy instrument, export and import equations are treated as determined endogenously within the context of the entire system. The monetary sector consists of supply-and-demand equations for money, an equation for bank loans, and an equation for inflation (price changes). This sector includes most of the available monetary instruments, except the central bank rate. This rate was considered unimportant for the regression analysis as it has not changed more than twice during the period. Nor did an 8 percent p.a. nominal rate (until the monetary correction system initiated in the 1960s) carry much significance in face of high inflation. The monetary sector has several contemporaneous feedbacks with the rest of the simultaneous system.

The public sector is divided into two parts: taxation and expenditure. The taxation subsector has ten individual tax equations and one nontax revenue equation. While taxes have contemporaneous feedback effects with the overall system, expenditures are considered to be predetermined. Yet expenditures are not determined in a vacuum: they are made functions of the past values of the relevant endogenous variables. The expenditure sector thus forms a block whose variables wield contemporaneous influence upon the variables of the main system but are not in turn influenced by those variables. The expenditure variables may nevertheless have some interdependence within themselves in the form of substitution (competitive) and complementarity relationships. Moreover, they cannot be determined without reference to what is happening in the rest of the economy. Tax revenues, national income, etc., must lie in the background. Accordingly, the lagged values of endogenous variables seem to belong to this sector. Conventionally, however, the expenditure levels are fixed largely at the beginning of the year, taking the expected price increases into consideration. In a way, thus, we could have integrated this sector with the overall system as a block that comes before the rest of the system in a block-recursive model. Instead, we preferred to transfer this block to an appendix: the tests of substitution and complementarities among various expenditures and similar expenditure relations are, therefore, presented in Appendix B. Only public transfer payments from the disposable side of public revenues are treated as endogenously determined. The main system treats public expenditures (except transfers) as autonomous instrument variables. However, while studying the dynamic aspects of the model, when the lagged values of the endogenous variable also become endogenous in a set of difference equations, the public expenditures assume an endogenous posture.

The structural model is stated at this stage only in a general form, the purpose being simply to give a reader a compact view of the broad framework and to bring out certain theoretical underpinnings of various equations. The algebraic construction of the equations and the forms in which various variables enter different equations will be determined and reported at the estimation stage.

The quality of a model can be improved by putting as many restrictions on the parameters as are legitimately available from a priori information. Partly to avoid too many details and partly because the model has not been exactly specified at this stage, we have not put down these restrictions along with the structural framework. Those that are not obvious enough, however, will be discussed along with empirical results. Some others are taken up in the theoretical discussion of the individual autonomous sectors that follow the general specification of the model.

THE EQUATION SYSTEM

A. Domestic Real Private Sector

(1) $$Y' = C^p + C^g + W^g + I^{pl} + I^{pe} + I^g + I^{stk} - NRA$$

(2) $$C^p = f_2\left([Y^d]^e,\ M,\ \tau^{sle},\ W^g,\ C^g,\ \frac{\Delta P}{P},\ \omega_2\right)$$

(3) $$[Y^d]^e = [Y^d]^e_{-1} + \gamma\left(Y^d - [Y^d]^e_{-1}\right)$$

$$(4)\quad I^{pl} = f_4\left(Y,\ [(1-Z)K]_{-1},\ I_{-1}^{stk},\ X_{-1},\ \tau_{-1}^{c'}\left[\frac{Q-T^c}{K}\right]^e\right.$$

$$\left. I_{-1}^{g},\ BL,\ T^{dir},\ \text{Dummy 54},\ \text{Dummy 59},\ \omega_4\right)$$

$$(5)\quad [Q-T^c]^e = [Q-T^c]_{-1} + \delta\left([Q-T^c] - [Q-T^c]_{-1}^{e}\right)$$

$$(6)\quad I^{pe} = f_6(I_{-1}^{pe},\ Y'_{-1},\ Pop,\ t,\ U_6)$$

$$(7)\quad I = I^{pl} + I^{pe} + I^{g}$$

$$(8)\quad [I^{stk}]^* = f_8\left([Y'-I^{stk}],\ \Delta[Y'-I^{stk}],\ [1-z]_{-1},\ BL, \omega_8\right)$$

$$(9)\quad I^{stk} = I_{-1}^{stk} + \alpha\left([I^{stk}]^* - I_{-1}^{stk}\right)$$

$$(10)\quad K_t = \sum_{n=1950}^{t} I_n + K_{1949}\left(1 - \sum_{n=1950}^{t} \theta n\right)$$

$$(11)\quad Y^{na} = f_{11}\left(N,\ \left[\frac{\Sigma I}{\Sigma K}\right]_{-1},\ H,\ \frac{T^{ind}}{T^{dir}},\ t,\ K(1-Z),\ U_{11}\right)$$

$$(12)\quad Y^a = f_{12}(N,\ Fert,\ A,\ U_{12})$$

$$(13)\quad (1-Z) = f_{13}(g,\ I_{-1},\ BL,\ \tau^{inf},\ T,\ U_{13})$$

$$(14)\quad Q = f_{14}\left(Q_{-1},\ \frac{\Delta P}{P},\ e^{ex},\ \frac{\omega min}{y},\ Y',\ Y'_{-1},\ U_{14}\right)$$

$$(15)\quad W = f_{15}\left(Y',\ \frac{\omega min}{y},\ \frac{\Delta P}{P},\ U_{15}\right)$$

$$(16)\quad Y' = Y - NRA + T^{ind} - \text{Subsidies}$$

$$(17)\quad Y^d = Y' - (T^{dir} + T^{ind} + NTR) + TR$$

$$(18)\quad y = \frac{Y}{Pop}$$

$$(19)\quad g = 100\,\frac{\Delta Y'}{Y'} = 100\,\frac{Y'-Y'_{-1}}{Y'_{-1}}$$

$$(20)\quad Y = Y^a + Y^{na}$$

B. Foreign Sector

(21) $$Ex = f_{21}(e^{ex}, Y^{USA}, U_{21})$$

(22) $$I_m = f_{22}(e^{im}, Y', X_{-1}, (1-Z), I, U_{22})$$

(23) $$F = Ex - Im - NRA$$

Monetary Sector

(24) $${}^{d}M^{*} = f_{24}\left(\left[\frac{\Delta P}{P}\right]^{e}, Y', IL, r, \omega_{24}\right)$$

(25) $$\left(\frac{\Delta P}{P}\right)^{e} = \left(\frac{\Delta P}{P}\right)^{e}_{-1} + \sigma\left(\frac{\Delta P}{P} - \left(\frac{\Delta P}{P}\right)^{e}_{-1}\right)$$

(26) $$M = M_{-1} + K(M^{*} - M_{-1})$$

(27) $${}^{s}M^{n} = f_{27}(\tau^{inf}, Y', [\eta^{a}_{-1} - \eta^{c}], r, EO, U_{27})$$

(28) $${}^{d}M = {}^{s}M$$

(29) $$\frac{\Delta P}{P} = f_{29}(Y', M^{n}, P^{im}, \tau^{sle}, (1-Z), \omega min/y, U_{29})$$

(30) $$BL^{n} = f_{30}([\eta^{a}_{-1} - \eta^{c}], Y', \frac{\Delta P}{P} \frac{M_1}{M}, r, U_{30})$$

(31) $$P^{im} = \frac{P^{im,\,foreign}}{P^{domestic}} e^{im}$$

The Fiscal Sector

(32) $$T^{pi} = f_{32}(Y, Pop, \tau^{pi}, \frac{\Delta P}{P} Y^{na}/Y^{a}, U_{32})$$

(33) $$T^{c} = f_{33}(Q, \tau^{c'}, P, U_{33})$$

(34) $$T^{exc} = f_{34}(Y^{na}, \tau^{exc}, U_{34})$$

(35) $$T^{im} = f_{35}(Y', e^{im}, U_{35})$$

(36) $$T^{stp} = f_{36}(Y', Y^{na}/Y^{a}, U_{36})$$

$$(37)\quad T^{ss} = f_{37}(Y^{na}, \frac{\Delta P}{P}\, U_{37})$$

$$(38)\quad T^{fue} = f_{38}(Y', Y^{na}/Y^{a}, U_{38})$$

$$(39)\quad T^{sle} = f_{39}(Y', \tau^{sle}, Y^{na}/Y^{a}, U_{39})$$

$$(40)\quad T^{os} = f_{40}(Y^{a}, Pop, U_{40})$$

$$(41)\quad T^{mun} = f_{41}(Y', Pop, U_{41})$$

$$(42)\quad NTR = f_{42}(Y', U_{42})$$

$$(43)\quad Tr = f_{43}([1-Z], Y, EO, U_{43})$$

$$(44)\quad T^{disp} = T - Tr$$

$$(45)\quad T = \sum_{j=1}^{J} T_j$$

$$(46)\quad DEF = G + Tr - T - NTR$$

The variables are defined in Appendix A. The U's (and ω's) are the disturbance terms about which more will be said in the ensuing discussion. The model consists of 46 economically interdependent but algebraically independent structural equations for an identical number of endogenous variables. Of the 46 equations, 27 are stochastic relations; five are nonstochastic theoretical relations (i.e., Equations 3, 5, 9, 25 and 26); the remaining 14 are identities, definitions, or equilibrium conditions (namely, Equations 1, 7, 10, 16-20, 23, 28, 31, and 43-46). The nonstochastic equations pertain to those sectors that involve either the "expectation generating equations" or "partial adjustments processes" and will be eliminated in the process of translating the unobserved expected (those superscripted by e) or desired (those superscripted by *) values into observable values. The reduction procedure is discussed below. The left-hand superscripts stand for supply (s) and demand (d) functions. There are over 50 predetermined variables, including 17 lagged values of the endogenous variables, 13 policy parameters (tax rates, exchange rates, minimum wage rate, compulsory reserve ratio, etc.), and five (including alternative classifications, eight) public expenditures treated here as instrument variables. Additional details are given in Appendix A. Note that all tax collections, transfer payments, and investment of public enterprises are treated here as endogenous variables.

A BRIEF THEORETICAL DISCUSSION OF THE MAIN SECTORS

The salient features of the equations/sectors are briefly discussed below.

The Real Sector: The Demand Side

The Demand for Income (Equation 1)

This equation is conventionally defined as an identity. Carefully examined, however, the income equation represents a mixture of a definition, an equilibrium condition, and an institutional restraint.[1] It is a definitional equation since supply by definition equals demand. It can also be considered an equilibrium condition since quantity demanded and quantity supplied always adjust themselves to each other. Lastly, following Christ's logical argument, this equation can also be visualized to express the institutional facts of our economic system, that when buyers and sellers do not agree, the party that has the smaller (minimum) quantity to transact is the party that prevails (Christ, 1966, p. 67, footnote 17). For the operational purposes of this model, however, these distinctions are immaterial. An important thing to be noted is that frequently an error term is attached to this equation also. This would perhaps have been appropriate in the present model, too, so as to take account of statistical discrepancies and the errors introduced due to the deflation of different income components (C^p, I^p, W^g, C^g, etc.) by different but appropriate price series.[2] To keep with the more popular convention, however, no stochastic term is attached to Equation 1.

Private Consumption Sector (Equations 2 and 3)

This is one of the most important sectors of the demand side of the model whose feedback relationships with taxation and public expenditures must be duly recognized. The theory of consumption is one of the best-developed subjects in economics. The present equation for the consumption function is, as the reader will recognize, a standard specification embodying the permanent income hypothesis. The equation includes three state instruments (τ^{sle}, C^g, W^g) explicitly and several others through M, P, and Y implicitly influencing private consumption. Real money supply (M) is tested as a proxy for real balances or wealth, since no other wealth measure is available.[3] There are several reasons for testing price changes as an explanatory variable for consumption and for a few other equations. The rate of inflation has been high in Brazil and continues to be a topic of discussion and concern, in regard to its influence on consumption, investment, growth, and other categories. Its inclusion in estimation models is in fashion in Brazil. It can be an important factor in (and thus a proxy for) income redistribution, for which no time-series data are available, except for some information in the 1960 census and a survey bearing on the census about the same time (FGV, 1967). In the absence of a reliable money interest rate series, it may also reflect the influence of the monetary sector instruments on real variables. An independent effect of population is also recognized.

The dynamic aspects of consumption are incorporated by including the expected rather than the actual values of disposable income as an explanatory variable. It is assumed that consumer units do not (necessarily) adjust their consumption to the observed current income but rather to estimates of the current values of current income's unobserved components—namely, the permanent and the transitory components. This is done under the hypothesis that these two components affect consumer behavior differently. The permanent component of income can be defined as the present value of expected future incomes, i.e., wealth. The transitory component, as defined by Friedman, "is to be interpreted as reflecting all 'other' factors, factors that are likely to be treated by the unit affected as 'accidental' or 'chance' occurrences, though they may, from another point of view, be the predictable effect of specifiable forces, for example, cyclical fluctuations in economic activity" (Friedman, 1957, pp. 21-22). In adjusting its current consumption, the consumer unit is visualized as forecasting the permanent component of its future income over the subjectively predictable time horizon—i.e., as optimizing its consumption over time. In the application of this approach, it is assumed that consumer units base their forecasts on the past values of income and that, given the related knowledge, they try to optimize their forecasts. It is further assumed that the expectations are revised by some fraction of the difference between the current actual value of income and the previously expected value. Under these assumptions, we get the familiar "adaptive expectations" distributed-lag model, the precise derivation of which is briefly explained below.

Treating Equations 2 and 3 as constituting an autonomous sector and for simplicity's sake using only two variables, Equation 2 can be written as:

$$C^p = \alpha_0 + \alpha_1^M + \alpha_2 [Y^{d^e}]_{+1} + \omega_2 \qquad (2.1)$$

where $[Y^{d^e}]_{+1}$ is the expected future value of Y^d on which C^p depends in the current period. The forecasts are revised by a fraction γ of the gap between the realized actual value of Y^d and the previously expected value of it as in Equation 3, where $0 < \gamma \leq 1$. Assuming a geometrically declining lag scheme after Koyck (1954) and Nerlove (1958) and inserting Equation 3 in Equation 2.1, we get:

$$C^p = \alpha_0 + \alpha_1 M + \alpha_2(\gamma Y^d + (1-\gamma)[Y^d]^e) + \omega_2 \qquad (2.2)$$

Also from Equation 2. 1 we have:

$$[Y^d]^e_{+1} = \frac{1}{\alpha_2}[C^p - \alpha_0 - \alpha_1 M - \omega_2] \qquad (2.3)$$

Lagging Equation 2. 3 once and solving it simultaneously with 2. 2, we get:

$$C^p = \alpha_0 - (1-\gamma)\alpha_0 + (1-\gamma)C^p_{-1} + \alpha_2 \gamma Y^d +$$

$$\alpha_1 M - \alpha_1(1-\gamma)M_{-1} + \omega_2 - (1-\gamma)\omega_{2,-1} \qquad (2.4)$$

$$= a_0 + a_1 C^p_{-1} + a_2 Y^d + a_3 M + a_4 M_{-1} + U_2 \qquad (2.5)$$

where on a priori grounds:

$$0 < a_1 \leq 1$$

$$0 < \gamma \leq 1$$

$$0 < a_2 \leq \alpha_2 < 1 \qquad (2.6)$$

when the variables are measured in identical units. The prediction that a_1 is nonnegative and less than unity is based on the technical relationship in which $(1 - a_1)$ is a kind of relative weight and so positive where the weights applied to the current and all the past incomes add up to one. The coefficient a_2, being the short-run marginal propensity to consume, is positive and less than one.

The other predetermined variables of Equation 2 enter in the same form as M, M_{-1} in Equation 2. 4. This equation is what Nerlove (1958, p. 25) has called a "reduced equation," as distinguished from the "reduced-form equation" of a simultaneous equations model, which we shall also meet in the succeeding pages. It may be verified that if ω_2 is serially correlated, then U_2 will be serially uncorrelated, and vice versa. Other than that, the estimation of 2. 5 presents no special problem. As a matter of fact, it can be observed that the "reduced" Equation 2. 4 has no distributed lag but only a one-period lag.

It can be seen that in 2. 5, when the variables are measured linearly, $(1 - a_1)$ gives the coefficient of adjustment; a_2 is the short-run marginal propensity to consume; and if consumption were to settle down to a static level of $C^p = C^p_{-1}$ then $a_2/(1 - a)$ would give the long-run marginal propensity to consume (out of Y^d), which gives due weight to the permanent components of income.* The introduction of the lagged

*From a given disequilibrium state, 95 percent adjustment to the equilibrium state is reached in n periods, where n is obtained by solving for it in the relation $a_1^n \leq .05$.

value of the dependent variable on the right-hand side raises several problems of estimation that will be discussed below. Hence nothing is specified here as regards the distribution or serial correlation of U_2.

Some additional variables are included to test their influence on consumption. In particular, the sales-tax rate is entered in the presence as well as in the absence of the price-change variable to get an approximate idea of the shiftability of this tax, the extent to which it discourages consumption, and whether it does so, if at all, through price changes or independently. The substitution effect of private consumption with government consumption is tested by including C^g and W^g in the equation. Finally, price changes are introduced to see inflation's independent effect, if any, on consumption.

The Investment Sector (I^{pl}, I^{pe}, I^{stk}, I^g) (Equations 4-10)

Private Investment Equations 4 and 5. In comparison with the close agreement between theory and measurement of the consumption function, the gap between economic theory and the empirical estimation of the investment function remains wide.[4] The data limitations in Brazil put further restraints on the specification of this function. The interest cost of capital, for example, is not known. The series of rates of return (r) tried for some equations here is derived from stock exchange prices, which may reflect either the real rate of interest or the rate of return on investment or both (it may also represent neither). Moreover, the disaggregation used in this study is on a functional basis rather than on industrial or sectoral bases, and so the level of aggregation remains rather high. All we have done here is tried alternative explanatory variables and different combinations thereof.

A distributed lag model has been specified for this equation also. Though the resulting estimating equation is similar to that for the consumption function, the theoretical underpinnings are not necessarily the same. For instance, if it is assumed that investment is a function (in part) of expected rate of return—which in turn may be reflected by the current and past profits net of taxes ($Q - T^c$), or past sales ($Y' - I^{stk}$) or overall demand, or valuation of the firm's equity on stock exchange represented by our variable r—the equation is identical to Equation 2.5, both conceptually and operationally. Thus, if for keeping the derivation simpler the basic equation is specified as:

$$I^{pl} = \beta_0 + \beta_1[Q - T^c]^e + \beta_2 Y' + \omega_4 \tag{4.1}$$

and it is assumed, as in Equation 2, that forecasts about future rate of return, as represented here by rate of profits net of taxes—i. e., ($Q - T^c$) or ($Q - T^c)/K$—are revised by a fraction δ of the gap between the realized actual values of this variable and the previously expected values of it, the adaptive-expectations distributional-lag investment function, following the derivation used for the consumption function, is of the form:

$$I^{pl} = [\beta_0 - (1 - \delta)\beta_0] + (1 - \delta)I^{p}_{-1} + \delta\beta_1(Q - T^C) + \beta_2 Y' - \beta_2(1 - \delta)Y'_{-1} + \omega_4 - (1 - \delta)\omega_{4,-1} \quad (4.2)$$

$$= b_0 + b_1 I^{pl}_{-1} + b_2(Q - T^C) + b_3 Y' + b_4 Y'_{-1} + U_4 \quad (4.3)$$

where $0 < \delta \leq 1$. On the other hand, if we assume a "partial adjustment" scheme, the model is outwardly still similar to the form of Equation 2.5 but has basically different theoretical—and also estimation—implications. In order to derive such a relation, we assume that the desired level of capital, K^*, depends on output to be produced (Equation 4.4 below), but that the actual change in K is some function—say, ε of the distance between the desired K^* and the previously achieved level of K (Equation 4.5 below). We further assume, for simplicity, that θ' is an average annual rate of depreciation, so that the identity of Equation 4.6 below holds.

$$K^* = \beta_0 + \beta_1'(Q - T^C) + \beta_2' Y' + \omega_4' \quad (4.4)$$

$$K - K_{-1} = \varepsilon(K^* - K_{-1}) \quad (4.5)$$

$$K = I + (1 - \theta')K_{-1} \quad (4.6)$$

where $0 < \varepsilon \leq 1$ and $0 < \theta' < 1$. Inserting the value of K^* from 4.4 and K from 4.6 into 4.5 and solving for I, we get

$$I = \varepsilon\beta_0' - \varepsilon\beta_2'(Q - T^C) + \varepsilon\beta_2' Y' - (\varepsilon - \theta')K_{-1} + \varepsilon\omega_4' \quad (4.7)$$

$$= b_0' + b_2'(Q - T^C) + b_3' Y' + b_5' K_{-1} + \omega_4'' \quad (4.8)$$

It may be seen that, unlike the adaptive expectations model, if ω_4' of 4.4 is serially uncorrelated, so is ω_4'' of 4.7. This is a notable difference in the estimation problems between the two models. (More on this subsequently.) The value of θ' as used in national income accounts is between .028 and .05. Accordingly, as long as $\varepsilon > 0.05$ (that is, as long as the entrepreneurs desire to make up in one year more than 5 percent of the difference between desired and actual capital stock, as should certainly be the case), the coefficient $(\varepsilon - \theta')$ of K_{-1} can be expected to be negative but in no circumstances equal to unity. Accordingly, if the regression coefficient of K_{-1} in the above equation is not significantly negative, we do not have a partial adjustment model.

Normally, it is not possible to distinguish between an adaptive expectations and a partial adjustment model. For example, if the desired capital K^* is replaced by desired investment I^*, this model will be indistinguishable from the adaptive expectations model, like 4.3. In the present case, however, as we have seen above, the two are distinguishable.

It is also possible that we may have a combined "adaptive expectations" and "partial adjustment" model of investment. The reduced equation of such a model can be derived by assuming, for example, $(Q - T^c)$ of Equation 4.7 as its expected value $(Q - T^c)^e$, and following the earlier procedure to get:

$$I^{pl} = [\varepsilon\beta_0' - (1 - \delta)\varepsilon\beta_0'] + (1 - \delta)I_{-1}^{pl} + \varepsilon\beta_2'\delta(Q - T^c) + \varepsilon\beta_1'Y' - (1 - \delta)\varepsilon\beta_1'Y_{-1}' - (\varepsilon - \theta')K_{-1} + (1 - \delta)(\varepsilon - \theta')K_{-2} + [\varepsilon\omega_4 - (1 - \delta)\varepsilon\omega_{4,-1}] \quad (4.9)$$

$$= b_0'' + b_1''I_{-1}^{pl} + b_2''(Q - T^c) + b_3''Y' + b_4''Y_{-1}' + b_5''K_{-1} + b_6''K_{-2} + U_4 \quad (4.10)$$

which is the same as Equation 4.3, except that two new variables, K_{-1} and K_{-2}, have now been added. It may be verified that in this particular case the parameters are identifiables even in the combined model. Thus, $\delta = 1 - b_1''$, $\varepsilon = \theta - b_5''$, $\beta_2 = b_2''/\delta\varepsilon$, and so on. In the case where the coefficient of I_{-1} is not (significantly) positive, $\delta \cong 1$, and we do not have an expectations model. In the case where the coefficient of K_{-1} is not (significantly) negative, we do not have a partial adjustment model. In the case where both have the right signs and significantly so, we have a combined model. In the case where both have either wrong signs or values not significantly different from zero, we do not have a distributed-lag model but simply a contemporanoous relationship, even though some of the variables—e.g., I_{-1}^{stk} in Equation 4—might be lagged.

In addition to the variables that have appeared in the foregoing discussion, we tried several other variables. In particular, capacity-to-import at the beginning of the year (X_{-1}) is considered to be an important determinant of investment in Brazil (and in the underdeveloped countries in general). Net profits $(Q - T^c)$ are an indicator of future profitability and a source of investment funds (see, for example, Meyer and Kuh [1957]). We tried two alternative measures of it, namely r and Q. We also tried two forms of Q: a) the absolute amount of gross profits net of taxes $(Q - T^c)$, indicating the availability of investment funds and b) the net rate of return measure $(Q - T^c/K) - \theta'$, indicating the profitability of investment.[5] Capacity utilization, changes in investment in stocks, and bank loans (as an index of credit rationing) are among the other relevant explanatory variables tested in this equation. The corporation tax rate adjusted for the SUDENE tax incentive scheme is the only public instrument directly included. There were other spurts of tax incentives that could not be quantified. In 1954-55, for instance, the well-known Instruction of the Superintendency of Money and Credit (SUMOC 113) removed various bureaucratic obstacles to foreign investors; according to the estimates of some observers, this even amounted to a subsidy (see, for example, Gudin [1969, p. 8] and Gordon [1962]). While this gave a fillip to foreign investment by such giant concerns as Mercedes Benz, Volkswagen, Dunlop, and Platt, local enterprise seems to have grudged and suffered, especially due to lack of foreign exchange resulting from the collapse of coffee sales in 1954 and 1955. Likewise, the ad valorem tariff system introduced in 1958—which gave vast domestic monopoly power to Brazilian firms, instituted an automatic mechanism of monetary correction of fixed assets for depreciation and other purposes (Law 3470 of November 1958), and created similar other conditions favorable to investment—introduced another spurt of investment incentives. To take account of these nonquantifiable events, we introduced two dummies, one for 1954 and another for 1959. While the net effect of 1954 is not sure, the net effect of 1959 is expected to be positive.

The Other Investment Equations. The equations for investment in stocks (I^{stk}) are formulated in accordance with the same partial adjustments theory as is built into Equations 4.4 and 4.5, with the one exception that the "desired" stock is replaced by "desired" flow (changes) of investment in stocks and inventories. This change of variable measurement is necessitated by the fact that Brazilian data give only changes in stocks and not the absolute levels of stocks. The bank loans variable (BL) is also tested for this equation, though the causal flow is not clear. Public investment (I^g) is assumed to be mainly a predetermined category (a policy instrument; see above). Public enterprise investment (I^{pe}) is a kind of intermediate category not treated here but subject to the same dynamic behavior as private investment. There is usually a discretionary element of official policy intermixed with the economic motivations of the management of public enterprises. Yet some degree of autonomy and some standards for meeting demand as well as minimizing accounting losses invariably exist for these enterprises. The lagged value of the dependent variable here is meant to capture the past influences and does not necessarily indicate a distributed-lag relation.

The Real Sector: The Production and Supply Side (Equations 11-15)

The aggregate production functions (Equations 11 and 12) are the standard Cobb-Douglas type of formulations constrained by the availability of data. The functional-income-distribution equations, 14 and 15 (using the manufacturing sector as an index of relative factor-income shares), are formulated to capture the effects of long-run forces. The capacity-utilization equation, 13, also rightly belongs to this subsector.

Equation 14 describes profits (gross nonwage income) as a function of past profits, changes in general price level (reflecting, among other things, the income distribution effects of inflation), exchange rate for exports, minimum wage rate adjusted for productivity, and GNP. Equation 15 uses the same variables, except for e^{ex}, to describe wage income.

Equation 12 defines the nonagricultural production function in which, in addition to the traditional factors of production—capital (adjusted for capacity utilization) and labor—a few other variables assumed to represent technical change are included. Variable $(\Sigma I/\Sigma K)_{-1}$ is supposed to reflect embodied technical change under the vintage approach; E and H are used as labor-augmenting variables; and T^{ind}/T^{dir} is used simply to capture the effect, if any, of the overall tax structure on the incentives to work, effort, etc. The agricultural production function is a relatively simpler one (for lack of data), using only labor, fertilizer (as a proxy for all purchased inputs), and overall hectareage (as an index of fixed inputs) as the factors of production. This is a rather weak equation as the variables are not very trustworthy. For example, the series of hectareage is prepared simply by interpolation and extrapolation of the 1950 and 1960 census figures, as the annual series compiled by the ministry of agriculture are widely believed to be untrustworthy data. The same procedure is followed by the Berkeley Team study (Smith, 1969, p. 231). Labor input is, for lack of data, all existing labor gross of industrial labor. Lastly, fertilizer is measured as "apparent consumption," being the sum of domestic production and imports, as the actual consumption data are not available. Moreover, it is in gross tonnage and not in nutrient contents.

Capacity utilization is explained by the expansion in production and demand as indicated by the rate of economic growth; by the expansion of capacity as represented by past year's investment; by the inducement to utilize capacity as provided by the profit rate; and by constraints on working capital and demand in general, as indicated by changes in bank loans and the excess reserve ratio. Equations 7, 10, and 16-20 are definitions or "pure identities, equations that must hold for all time periods for whatever values the variables assume in that time period" (Klein and Goldberger, 1955, p. 31).

Since saving, as such, does not appear actively in the equations of the model (its long-run effects on growth being subsumed in the investment equation), no equation is specified for this variable. Similarly, no private saving-investment identity is included, as in annual observations such an identity becomes irrelevant in the presence of public saving and the massive movements of foreign private investment in Brazil. On the other side, even though changes in inventories are also a residual factor (given by GNP minus the aggregate demand components), it is treated as an endogenous variable with its own stochastic equation since it plays an active role in a number of equations, at least, through $(Y - I^{stk})$. Finally, while talking about the "residual categories," mention might also be made of the profit (nonwage) and wage equations of the manufacturing sector. In principle, nonwage income is a residual factor—that is, value added minus wage bill. In this model, however, both categories are treated as stochastic endogenous variables, and since value added (in manufacturing) as such does not appear in the model, this identity is also ignored. Similar reasons apply to any other identities that might appear to be relevant but are not specified since they play no independent role in the model.

The Foreign Trade Sector (Equations 20-22)

Equations 21 and 22 explain exports and imports. The discrepancy, if any, between the two is indicated by F and NRA in Equation 23. No balance or equilibrium condition is imposed. It may be recalled that the exchange rate for exports (e^{ex}) includes export premia and similar incentives, and the exchange rate for imports (e^{im}) incorporates protection measures in the form of tariffs and physical restrictions together (see Appendix A). Exports are made a function of price (e^{ex}) and foreign income for which the U.S. GNP is used as a proxy. For imports we try (e^{im}), Brazil's GNP, capacity to import at the beginning of the year, investment, and the index of capacity utilization. Excess capacity and high e^{im} are predicted to have negative effect on imports, while all of the remaining variables have a positive effect.

The Monetary Sector (Equations 24-31)

This sector consists of four stochastic equations: demand for money, supply of money, rate of inflation, and bank loans. The theoretical underpinnings in the formulation of the equations for the demand for money are (with minor differences) analogous to the private investment equations, and will not be repeated here. It should only be mentioned that changes in the desired stock of money M^* seem on a priori grounds in the context of Brazil to be a function of both expected price changes $\Delta P^e/P$ (the Cagan case [1956]) and expected real income changes Y^e (the Friedman case [1957]—or more appropriately the Zellner-Huang-Chau case [1965], since Friedman assumes instantaneous adjustment of the actual money balances to the desired money balances). If one specifies the model with two expected-value variables, however, the identification of the adjustment coefficients becomes much more difficult. Moreover, the permanent income hypothesis of the demand for money does not seem to have withstood its tests in the context of the developing countries of South America, possibly because their capital markets are even less developed than the economy as a whole (see, for example, Diz [1970], Pastore [1969], and Deaver [1970]). In view of this, only the expected price hypothesis of the demand for money has been incorporated into the present model. The tests of alternative specifications will be carried out, however.

Using the reduction procedure, illustrated earlier, for Equations 24-26, the "reduced" demand-for-money equation becomes:

$$M = f_{23}\left(M_{-1},\ M_{-2},\ \frac{\Delta P}{P},\ Y,\ Y_{-1},\ IL,\ IL_{-1},\ r,\ r_{-1},\ \omega_{23}\right) \qquad (24.1)$$

If, on the other hand, the demand for desired money balances is a function of measured price changes but the expected income changes, the "reduced" equation becomes:

$$M = f_{23}\left(M_{-1},\ M_{-2},\ \frac{\Delta P}{P},\ \left[\frac{\Delta P}{P}\right]_{-1},\ Y,\ IL,\ IL_{-1},\ r,\ r_{-1},\ \omega_{23}\right) \qquad (24.2)$$

Drawing upon our earlier explanations, it may be noted that if money balances are assumed to be always in equilibrium, the term M_{-2} drops out, whether the model conforms to Equation 24.1 or to 24.2. If, on the other hand, the demand for money depends upon measured income as well as measured price changes, in which case actual money balances approach equilibrium with a lag (the Bronfenbrenner and Meyer case [1960]), all the lagged terms in 24.1 and 24.2 drop out. In view of this, we shall use an inductive approach and leave it to the empirical verification of this study to determine which of these models best fits the Brazilian data.

In this connection, it is relevant to mention that when the coefficient of M_{-1} is nonzero and that of M_{-2} is zero, we have a first-order difference equation; when both coefficients are nonzero, we get a second-order difference equation. Depending upon the signs and magnitudes of these two coefficients, the path of expansion of the demand for money can be damped or explosive, monotonic or sinusoidal, or any combination of these. We will deal with these stability conditions when they arise in the empirical part. It is apparent that any equation having the lagged value(s) of its own endogenous variable among its set of explanatory variables will raise the indicated stability problem. This applies to the consumption, investment, inventory, profit, price, and money equations of the present model.

In addition to $\Delta P/P$ and Y, the demand for money also depends on the rate of interest and changes in the amount of foreign exchange that the holders of it offer for conversion into home currency. For instance, the government's purchases of the foreign currency that had flowed into the country as a result of large balance of payments surpluses and a slump-induced drop in imports in the years 1965 and 1966 led to a considerable currency expansion that was out of proportion with the planned currency issues. It is expected that changes in one or the other of the three alternative foreign-exchange variables available—namely, international liquidity, foreign surpluses on current account, and lagged import capacity—will capture the indicated component of demand for money.

The most important primary source of the supply of money is expected to be treasury deficits measured here by τ^{inf}. We do not have a reliable series of rates of interest as cost, and so we will try r as defined in this study. Excess reserve ratio has fluctuated considerably during the 1960s and is predicted to be an important source of secondary expansion of money supply. The bank loans variable is thrown in as a proxy for the government's policy of credit restrictions. Another likely candidate in explaining money supply is the ratio of currency held by the public to money supply.[6]

The price equation is an equally important relation of this sector in the context of Brazil. It links several policy instruments to the monetary and real sectors. For the dependent variable, both the GNP implicit deflator and the wholesale price series are tried. The first two explanatory variables of this equation—changes in income and money supply—are the standard ones. Note that while real money stock is used for all other equations of the model, for the purpose of price equation we employ nominal money stock. Relative import prices, sales tax, and real minimum wage rate corrected for changes in productivity (as measured by per capita income) are meant to capture the cost-push causes of inflation. However, while the relationships of prices with relative import prices and sales tax are expected to be positive, no such prediction seems to be warranted for the coefficient of the minimum wage in Brazil. For while prima facie a cost factor, low wages might reflect the efficiency of the worker in the developing countries, so as to more than cancel out the gain in reduced demand by a loss of real output.[7] At the same time the minimum wage can work as a proxy for several other prices that have been fixed as multiples of the minimum wage since 1960. Excess capacity and the accompanying unemployment are expected to exacerbate price increases for the same reason, while the restrictions of bank loans might lead to shortages of working capital causing excess capacity, as happened in the late 1960s. Both of these variables are, therefore, tested. The final explanatory variable is the currency-to-money ratio (M_1/M), which has experienced considerable fluctuations during the reference period. It is expected to be negatively correlated with the rate of inflation though, again, the direction of the causal flow is not clear.*

---

*The inclusion of a price equation is justified by the dynamic nature of the model. In static general equilibrium models of the text-

Bank loans are also treated as an endogenous variable. The excess reserve ratios of the banks, the growth rate, the rate of inflation (frequently entered multiplicatively with GNP, as Goldfeld does [1966], when proportionate changes rather than absolute levels of the variables are used), and the rate of interest are the main variables tested for this equation.

The Fiscal Sector (Equations 32-45)

In principle, revenue collections from a tax should be given by a technical relationship between the tax base and the tax yield, with the relationship provided by the tax rates (Griliches, 1968, p. 228), so that only the tax base needs to be estimated. In practice, however, due to various exemptions, differential rates, progressivity, and so on, such relationships are "actually very complicated and not expressible precisely in terms of two parameters in a linear equation" (Klein and Goldberger, 1955, p. 39). In the developing countries, more than in the developed countries, tax collections are not an automatic nonstochastic relationship based on tax rates and the taxable bases. They are affected by a host of factors, including changing administrative efficiency and tax effort, changing attitudes of the people toward tax payment, uneven and changing propensity of different groups of

---

book type—e.g., Modigliani (1944)—it is generally the case that the so-called Walras' Law prevails—that is, one of the market-clearing equations becomes redundant because of the budget identity. And it is the commodity price equation that is traditionally assumed to be redundant. This is not necessarily the case in dynamic general equilibrium systems of the present type, because, in the dynamic context, future incomes and expenditures are either cleared in the future markets, in which case the dynamic system reverts to an essentially static system, or future incomes and expenditures are expectations. Expectations, however, being subjective plans, are not cleared in the present market. But the budget constraints of individuals must be formulated in terms of expectations, so these constraints, unlike the case of static general equilibrium, cannot be used to make one of the current market-clearing conditions redundant. Moreover, in the present model, the general price changes are a function not merely of money but also of real sector expectations and exogenous cost factors. Although prices are not assumed to be inflexible, the so-called classical dichotomy between real and money sectors is absent from the present model. I am thankful to Albert Ando for clarifying the issues involved here through a private communication.

taxpayers to evade and avoid taxes, the increasing degree of the economy's commercialization, and so on. Accordingly, when one estimates a tax equation in a developing country, one does not merely estimate a summarized or average technical relationship, but some kind of a mixture of behavioral, institutional, and technical relationships. Moreover, even a summarized view of the responsiveness of different taxes is highly useful in itself; for example, the existence of hundreds of excise tax rates and complex personal income tax schedules inhibit a compact view of tax responsiveness.

The tax base for most modern taxes is income, expenditure or wealth, though there are some taxes whose base might not be expressed in terms of income or the value of wealth but in some physical units of wealth—for example, land tax, specific commodity taxes, and similar taxes. Certain taxes, such as the poll tax, which have no relation to income or wealth, are now of historical interest only.

Even when a tax is not directly related to income, the changes in its yield in comparison to changes in income are worth investigating for policy purposes. For a majority of taxes, however, a functional relationship between tax yield and income does exist. Accordingly, income is included as an explanatory variable for all taxes. However, appropriate income aggregates/components are used for different taxes: GDP for direct taxes in general, GNP for indirect taxes in general, agricultural income for some of the state taxes, profits for corporation tax, nonagricultural income for excise duties, the ratio of nonagricultural to agricultural income (in addition to absolute income) for fuel tax and a few other taxes, and so on. For income tax we tried per capita income as a substitute for GDP and population to get an alternative estimate of progressivity.

The most important variable in the context of the present study is, naturally, the series of the respective tax rates, where available. For income taxes we try alternatively the present rate, the past year's rate, and a two-year moving average. Population, too, has important influence on some tax yields, but its influence can be positive on some tax yields and negative on others. For example, the effect of population increase on personal income tax can be negative due to various deductions and exemptions (if population increase also implies an increase in family size or early marriages) and relatively lower per capita incomes, given an increase in GNP. Indirectly, it can cause an increase in land taxes: the pressure of population and the availability of labor can induce exploitation of the extensive margins of land, thereby bringing additional hectareage of land under the tax net. It can increase the yield from outlay taxes due to higher expenditures from given GNP, and so on. The population variable, therefore, is also tried for some of the taxes. The inflation variable, $\Delta P/P$, becomes relevant for some taxes for the reasons given earlier, when the variables are used in constant prices.

While the monetary effects of fiscal policy instruments are, in general, recognized both directly by introducing these instruments explicitly into the monetary equations and indirectly through other endogenous variables, the fiscal influence of monetary policy is incorporated only indirectly through income and other endogenous variables.

From the disposal side of the government's income, only transfers and subsidies (Tr) are treated as endogenous to our system. Excess capacity was selected as an index of unemployment necessitating increase in transfers. As an index of or proxy for various other causes of transfers and subsidies, we selected government's "foreign exchange operations" (EO). This category of government finance includes the financing of the minimum support price for coffee, which is the most important item in the general scheme of agricultural support prices. A major proportion of transfers are naturally tied to social security taxes (T^{ss}), but not necessarily to the current period's T^{ss}. Using current T^{ss} as an explanatory variable, though not tautological, would not be a very appropriate specification, perhaps.

The elasticities of all the taxes studied here with respect to income or its components can be predicted to be positive. For most taxes, this elasticity is expected to lie close to but on either side of unity, and rarely, if ever, to exceed a value of 2.

THE ESTIMATION PROCEDURE

There are two major sets of problems involved in the estimation of this model that must be attended to. One is concerned with the simultaneous equations problem of the full model, and the other relates to the estimation of the individual (in particular, the individual distributed-lag) equations of the model. We shall take up the problem of simultaneous estimation first.

It can be seen that the model is not segmentable, even if we neglect contemporaneous correlation among the error terms. It is interdependent, and is not even superficially recursive.*

---

*A model may be said to be superficially recursive when the endogenous variables from appropriately arranged lower-numbered equations may appear in the higher-numbered equations, but not vice versa—in other words, when the matrix of endogenous variables is triangular. The model is truly recursive when, in addition to the foregoing property, the disturbances are uncorrelated both contemporaneously and serially and appear either in the same equation or in the other equations of the system—i.e., when the covariance matrix is

Since the model is overidentified, empirical estimates are first made by using two-stage least squares (2SLS) under the assumption, to be relaxed later on, that the disturbances are contemporaneously uncorrelated.* The estimated values of the endogenous variables, which appear as explanatory variables in other equations, and their lagged values (in distributed-lag equations) are obtained in the first stage from the reduced-form equations. With a view to economize on the degrees of freedom in our 19-observation model, however, the method of principal components was employed to choose the exogenous variables for the first-stage estimates. It may be stated that the 2SLS estimators are consistent even when lagged endogenous variables are used (as in the present model).[8]

An important step in the present quantitative analysis is the calculation of impact multipliers or the reduced-form coefficients of linearly estimated or linearized equations. An easy way to obtain their values is simply to use the first-stage estimates of the endogenous variables in 2SLS estimation. The main constraint of doing so is that we have over 50 predetermined variables and only 19 observations, our first-stage reduced-form estimates being only the truncated forms of the overall estimation model. Accordingly, given our sample, the 2SLS estimates are supposed to use more information than the indirect least squares estimates of the reduced forms. They are, therefore, considered to be more efficient than the latter. The procedure used to obtain impact multipliers is to solve the 2SLS estimates of the structural equations to get the reduced-form coefficients. In a model such as the present one, in which all equations are not linear, however, the solved 2SLS method has the disadvantage of being computationally more complicated. Another general drawback of the reduced-form coefficients obtained from the solved 2SLS is that their standard errors cannot be calculated. Accordingly, while the confidence intervals of the structural coefficients are numerically known, the same are not known for the impact multipliers, which therefore have to be treated as if they were point estimators.†

---

diagonal. Of the 45 equations only two—namely, the export equation and the agricultural-production function—form autonomous blocks because of which the model can be said to be block-recursive. Since the remaining block consists of the entire model less two equations, the indicated block-recursiveness makes little difference to the estimation procedure or the identification problem.

*The equation in which the disturbances are contemporaneously correlated is estimated by 3SLS.

†An impact multiplier measures the change in an endogenous variable induced, in a given time-period, by a unit change in an exogenous variable that holds other exogenous variables constant but permits the interaction effects of other endogenous variables of the system to work themselves out. The reduced-form estimates of a linear model represent one-period impact multipliers. For a major work on impact multipliers, see Goldberger (1959).

Another basic problem in simultaneous estimation that is almost sure to be encountered in an aggregative model involving interdependent sectors such as the model of this study is the one that arises from contemporaneous correlation among the disturbances of different equations. The consideration of this problem will be better appreciated and more appropriately dealt with after we have discussed the issue of serially correlated disturbances and other problems that arise in connection with the distributed-lag equations. This is the second major problem of estimation to which we want to attend.

While the distributed-lag models (or at least some of them) embody, as we have seen above, strong theoretical propositions—e.g., people may adjust slowly, they may try to optimize forecasts, they may respond differently to the (unobserved) permanent and (observed) transitory components of a change in an economic category, and so on—these models have some weaknesses in estimation. One major weakness appears when the error terms are serially correlated. We have seen above that in the adaptive-expectations model, even when an equation is correctly specified, so that the original disturbances are not serially correlated, the disturbances acquire serial correlation in the transformed form. Ordinarily, however, serial correlation may be present in the true disturbances due to the likelihood of specification errors. It is well known that the application of the ordinary least squares in the presence of serially correlated disturbances biases the coefficients. The biases become more serious in distributed-lag models. In those models the lagged endogenous variable, being a weighted average of past disturbances, becomes correlated with the current disturbance term and thereby creates a simultaneous equations problem. Hence, the least squares estimates of such an equation will be neither efficient nor consistent. Moreover, the lagged endogenous variable will usually have a strong tendency to pick up some of the effects of the (left-out) variables that cause serial correlation. For instance, if a least-squares regression is fitted to Equation 2.4, the $\omega_{2,-1}$ will be the left-out variable whose effect will be picked up by C^p_{-1}. "There is no question that in most cases the addition of the lagged dependent variable to the regression will reduce the serial correlation of the residuals" (Griliches, 1961). The calculated serial correlation coefficient in such an equation will tend to be zero, even though the true one is nonzero. The Durbin-Watson statistic will be biased towards its desirable value of 2.[9] Hence, we do not even have a reliable test of serial correlation in this case. With a view to overcoming this problem we employed a corrective method, which will be explained after we have discussed some of the other weaknesses of this model.

The lagged endogenous variable tends not only to pick up part of the serial correlation but may also pick up trends in the data.[10] Moreover, in the case where, for various reasons, the lag variables may be collinear with the zero-lag exogenous variables—e.g., partly because of serial correlation in the residuals and partly because of measurement errors in the zero-lag exogenous variables—the lag variable may be stronger and may "steal" from the other included variables. The effect of all these biases usually is that the coefficient of the lag variable, like the average lag, is overestimated, and the adjustment coefficient is underestimated. Worse still, when the other included variables do not account for a very large fraction of the variation in the current dependent variable, the computed coefficient of the lagged dependent variable is likely to acquire nonzero (usually positive) and significant values even though the true value of it may be zero. The bias will be reduced as the explanatory variables included in the equation and the data series become larger.*

Even when the disturbances are not autocorrelated, a distributed lag model can have certain other weaknesses. For instance, the "standard" least squares estimates of the coefficients in a distributed-lag model with no serial correlation of errors are consistent and asymptotically efficient, but they are not unbiased in small samples (see Mann and Wald [1943]). It is to be noted that the lagged endogenous variables appear not only in the "basic" model but also in the reduced-form model.

The weaknesses discussed above appear in the distributed-lag models whether the model is estimated as a single (reduced) equation or as part of a system of several interdependent equations. In the case of an interdependent simultaneous-equations model, such as the present one, moreover, we meet the additional problem of the existence of contemporaneous correlation among errors. While serial correlation of disturbances is due largely to either the specification errors or the reduction procedure of the distributed-lag equations, contemporaneous correlation of the disturbances—at least in the macromodels of national-income categories such as the present one—may be embedded in our economic system. A pip in the disturbance of a major tax function, for example, is not unlikely to be associated with contemporaneous undulations in the disturbances of the consumption, investment, and income equations. It may, therefore, be unrealistic to assume no correlation among the disturbance terms of different equations (without proper tests). When disturbances are contemporaneously correlated, even the 2SLS method may fail to give consistent estimates of the coefficients.

In this study, two measures were adopted to overcome these weaknesses. First, particular attention was paid to the specification of the model both in terms of the explanatory variables, the data permitting (and additional lag terms thereof were tried in the regression equations), and in terms of the algebraic forms used. The second remedial measure pertains to the estimation of the model in presence of both serially and contemporaneously correlated disturbances. We shall first explain the procedure used to tackle serial correlation.

---

*Objections might also be raised against the artificial imposition of a geometrically declining distributed-lag scheme. This assumption is made for the convenience of estimation.

The basic problem in estimating an equation with autocorrelated disturbances can be approached by visualizing a model in which we have an additional endogenous variable (of serial correlation) that is unobservable. It is well known that the single-equation least-squares estimation of such a model will not yield consistent estimators. The aim, therefore, is to make or approximate a simultaneous estimation of both the basic endogenous variable, which is observable, and the serial-correlation variable, which is unobservable. With this aim, several methods have been suggested in recent years to get consistent estimators in the presence of autocorrelated disturbances. For instance, one can specify a serial-correlation equation, take the reduced form of the basic equation and the serial-correlation equation (which will result in the introduction of additional lag terms), and estimate the basic coefficients and the serial-correlation coefficient simultaneously by using nonlinear regression procedures (as suggested, for example, by Griliches [1967]) or Aitken's generalized least squares (Aitken, 1934), or follow Malinvaud's suggestion to identify the coefficients from a linear estimation of a nonlinear regression (Malinvaud, 1966, p. 469). Taylor and Wilson (1964) have developed a "three-pass method," which is superior to the ordinary least squares but probably not to some of the other estimation techniques. This method is also not considered in general to result in consistent estimators (see, e.g., the Monte Carlo results of Wallis [1967] and remarks by Griliches [1967, footnote 38]). The instrumental variable—or, more formally, the 2SLS method—(with appropriately-chosen instrumental variables) is generally considered to yield consistent, and probably also asymptotically very efficient, results (see Sargan [1958], Malinvaud [1966, p. 12], and remarks by Griliches [1963, pp. 22-26], among others). In this study, for the distributed-lag equations, in the first instance we employed the 2SLS method. For all such equations, and when serial correlation is observed in the nonlag equations that cannot be removed by better specification of the functions due either to lack of relevant data or to other reasons, we followed the procedure recommended by Wallis (1967) and employed the Cochrane-Orcutt transformation procedure (1949).[11] The combination of the procedure employed here is briefly illustrated below, using the consumption function of Equation 2.5 as an illustration. It is applicable both to the equations of our basic model as well as those of the reduced-form model, since the disturbances of both models can be serially correlated.

In the 2SLS there are essentially two steps: in the first step, Equation 2.5 is estimated by omitting C^p_{-1} from the equation. That is, we estimate:

$$C^p = a_0 + a_2 Y^d + a_3 M + a_4 M_{-1} + U_2 \qquad (2.7)$$

where, for simplicity, we neglect the problem of simultaneity among Y^d, C^p, and M. In this round of computations the quality of the estimates of the individual coefficients is of secondary importance. The important thing is to get the best possible estimate of C^p. In the second step, the computed values of C^p (to be denoted by a caret, as $\hat{C}^p$) from the first step are lagged once to get $\hat{C}^p_{-1}$, which series is then used as an instrumental variable for $\hat{C}^p_{-1}$, and a new regression is run. That is:

$$C^p = a_0 + a_1 \hat{C}^p_{-1} + a_2 Y^d + a_3 M + a_4 M_{-1} + U_2 \qquad (2.8)$$

The 2SLS procedure does not necessarily remove serial correlation in errors; it only makes the Durbin-Watson statistic a valid test. The problem of serial correlation is a problem of specification. The Cochrane-Orcutt transformation is presumed to remove the serial-correlation bias. The procedure as recommended by Wallis (1967) consists of three steps. The first step is to estimate Equation 2. 5, not by omitting C^p_{-1}, as in the preceding paragraph, but by replacing it by the lagged values of the other independent variables, such as Y^d_{-1}, as instrumental variables. In the second step, the first-order serial correlation coefficient, ρ, is calculated from the regression residuals of the first step as follows:

$$\rho = \frac{\frac{1}{n-1} \Sigma U_2^* \cdot U_{2,-1}^*}{\frac{1}{n} \Sigma U_2^{*2}} - \frac{\xi + 1}{n} \qquad (2.9)$$

where n stands for the number of observations, each U_2 indicates the error term obtained from the estimates of the first step, and ξ denotes the number of independent variables in the regression. Apparently, the covariance terms in the numerator are one less than the variance terms in the denominator. The last term, i. e., $(\xi + 1)/n$, is a correction term for small sample bias as suggested by Wallis (1967, p. 559). In the last step the Cochrane-Orcutt transformation procedure, which consists in applying ordinary least squares to the transformed variables, is applied as follows:

$$C^p - \rho C^p_{-1} = a_0 + a_1(C^p_{-1} - \rho C^p_{-2}) + a_2(Y - \rho Y^d_{-1}) + a_3(M - \rho M_{-1}) + a_4(M_{-1} - \rho M_{-2}) + U_2' \qquad (2.10)$$

If one is not sure a priori whether a model is a distributed-lag or a nonlag, serial-correlation model (which indicates that the equation is not well specified and that some of the relevant variables are left out), a simple test can be made by including the lagged value(s) of the exogenous variable(s) also (see Griliches [1967, p. 34]). If the serial-correlation model is a more appropriate one, the coefficient(s) of the lagged exogenous variable(s) should equal the negative of the product of the coefficient of the current value of the respective variable times that of the lagged endogenous variable. We tried such alternative specifications in this study. These remarks are particularly relevant to the investment, inventory, money, profits, and consumption equations, which may be subject to distributed lags or serial correlation or both. The tax equations, on the other hand, are not presumed to be subject to distributed lags, but may depict serial correlation of errors.

As already mentioned, the 2SLS method gives consistent estimators. An additional merit of this method, as pointed out by Griliches (1967, p. 41), is that its consistency does not depend upon particular assumptions about the structure of the disturbances, though it requires the exogenous variables to be truly exogenous. In the lagged model, computed by ordinary least squares instead of 2SLS, the near-zero value of the calculated ρ, which is usual, is no evidence of the absence of serial correlation of residuals, as the serial correlation might be picked up largely by the lagged endogenous variable. It may be seen that when $\rho = 0$, Relation 2.10 reduces itself to Equation 2.5; when $\rho = 1$, we have a first-differences model. Apparently, it must be a very inadequately specified equation whose error terms yield a value of one for ρ. According to Watson and Hannon, "[It] appears that the first difference transformation may lead to very low efficiencies unless $\rho = 1$ (Watson and Hannon, 1956, p. 441). Hence, as is emphasized by Griliches, "[A] blanket application of the first differences transformation is not recommended" (Griliches, 1963, p. 79).

To obtain consistent estimates of the coefficients in the presence of contemporaneous correlation of the disturbances one has to apply, in one form or another, Aitken's generalized least squares to the whole set of equations to estimate all parameters simultaneously. In this method the variances and covariances of the disturbances have to be explicitly utilized like any other independent variables. But since they are not known, Aitken's method is not workable. Hence, various alternatives have been suggested, in one way or another utilizing the variances and covariances calculated, to begin with, from one-at-a-time least-squares estimates. One such method is that of three-stage least squares (3SLS) as developed by Zellner and Theil (1962). Operationally, this technique consists of an analysis of the covariance matrix of the residuals obtained from 2SLS estimates. In a way, this method has similar approach as the one suggested above for overcoming serial correlation: in the latter method one has to analyze the variances of the individual equations; in the former the whole covariance matrix of disturbances has to be simultaneously analyzed. In both cases the residuals are obtained from two-stage least squares. Three-stage least squares are, in general, considered to yield more efficient estimators than two-stage least squares. This technique has been successfully used by Zellner and Theil (1962) and others in estimates of simultaneous equations. A similar technique to get efficient estimates of coefficients in a "seemingly unrelated" set of (demand) equations has been used by Zellner (1962). This method is more efficient than ordinary least squares and "the gain in efficiency," according to Zellner, "can be quite large if 'independent' variables in different equations are highly correlated" (Zellner, 1962, p. 348). This technique is considered to be of general application. An alternative method is to apply the maximum-likelihood principle. For the application of this method, the assumption of multivariate normal distribution of disturbances is essential.

It may be noted that this method "collapses" to yield single-equation least-squares estimates either when the disturbance terms of different equations have zero covariances or when the independent variables used in different equations are identical even if disturbances across equations are correlated (Zellner, 1962, p. 351). Since, in the reduced forms, the independent variables are the same in all equations of the model, this method reduces itself to ordinary least squares. The same qualification applies to those tax (T_j) equations of the structural model of this study, which have the same independent variables. The generalized least-squares method and its practical variants, therefore, will strictly apply only to those structural equations of the model that do not have common explanatory variables.

This estimation technique removes only contemporaneous correlation among disturbances; the problem becomes more complex when the disturbances are both contemporaneously and serially correlated. Here a practical approach may be, first, to remove serial correlation from the individual equations in a way suggested in Equations 2.7-2.10 and then, as a second round, to take measures to remove contemporaneous correlation. This is the procedure employed, for example, by Parks (1967), who combines the conventional transformation method to remove serial correlation with Zellner's estimation technique for "seemingly related" equations models to remove contemporaneous correlation, thus generalizing Zellner's procedure.

In this study an attempt was made to obtain efficient estimates by trying to eliminate both contemporaneous and serial correlation, where present. Unfortunately, the sample series available is not large enough to permit significant improvements in estimation by the methods reviewed above. Insofar as serial interdependence of disturbances is concerned, it was removed as carefully as possible by the procedure discussed in Equations 2.7-2.10. As to the contemporaneous interdependence among the disturbances, we separated those equations among whose disturbance terms the covariances, as calculated from the application of the relations of the type of Equations 2.7-2.10, exceeded about 0.20, with the following further qualifications for the tax equations: Because most tax equations have a large majority of variables in common and because of the shortage of degrees of freedom, for the purpose of the covariation flow from the tax sector to the nontax equations only a mongrel tax function for aggregate T was used. For the reverse flow, each individual tax equation was treated independently of the remaining tax equations, and 3SLS estimates were applied, where required, to one tax equation at a time combined with the rest of the (nontax) system. In summary, the 2SLS method is used to remove simultaneous equations bias; the Cochrane-Orcutt transformation to remove the bias arising from the serial correlation of disturbances; the 3SLS technique to eliminate the possible bias due to contemporaneous correlation of errors.

Before concluding this chapter, the difference between automatic revenue responsiveness (directly relevant to the revenue mobilization objective) and built-in flexibility of income (directly relevant to the stabilization objective) should be made more explicit. For this purpose, let us assume an extremely simplified version of the income determination model, say:

$$Y = C + I \tag{ix}$$

$$C = a_0 + a_1(Y - T), \quad a_1 < 1 \tag{x}$$

$$T = \alpha_0 + \alpha_1 Y, \quad \alpha_1 < 1 \tag{xi}$$

The reduced form, expressed in first differences, becomes:

$$\Delta Y = \frac{\Delta I}{1 - a_1 + a_1 \alpha_1} \tag{xii}$$

Denote the left-hand term as ΔY^b when $\alpha_1 = 0$ (i.e., change in income when tax yield does not vary with income) and ΔY when $\alpha_1 > 0$ (i.e., changes in income when the taxes concerned vary with income). Apparently, for a given ΔI, $\Delta Y \leq \Delta Y^b$. The marginal automatic responsiveness of tax yields with respect to income is:

$$\alpha_1 = \frac{\Delta T}{\Delta Y} \tag{xiii}$$

and built-in flexibility, defined as the fraction of changes in income that is prevented because of the existence of automatic tax responsiveness, according to the famous Musgrave-Miller formula (e.g., Musgrave [1959, pp. 505-12]), is:

$$z = 1 - \frac{\Delta Y}{\Delta Y^b} \tag{xiv}$$

$$= \frac{a_1 \alpha_1}{1 - a_1 + a_1 \alpha_1} \tag{xv}$$

It can be easily verified from xii that the larger the absolute values of the marginal propensity to consume (a_1) and the marginal tax yield (α_1) with respect to income, the larger will be the value of built-in flexibility (z).

The Musgrave-Miller formula (xiv) is extended in two directions. The first extension is a kind of generalization of the authors' single-tax-economy formula to a multitax-economy formula that follows from our macro model. It can be easily shown that the Musgrave-Miller formula, which in its simple form given in xiv above has been widely used for empirical measurements (see, e.g., Pechman [1956]), is always an overestimate of built-in flexibility. To see this, replace xi above by xi. a-xi. c, as follows:

$$T_1 = \alpha_0 + \alpha_1 Y \qquad \text{(xi. a)}$$

$$T_2 = \beta_0 + \beta_1 Y, \quad \beta_1 < 1 \qquad \text{(xi. b)}$$

$$T = T_1 + T_2 \qquad \text{(xi. c)}$$

Then the single-tax equation estimates of built-in flexibility are:

$$\bar{z}_1 = \frac{a_1 \alpha_1}{1 - a_1 + a_1 \alpha_1} \quad \text{for } T_1 \qquad \text{(xvi)}$$

$$\bar{z}_2 = \frac{a_1 \beta_1}{1 - a_1 + a_1 \beta_1} \quad \text{for } T_2 \qquad \text{(xvii)}$$

The joint built-in flexibility measure in simultaneous estimation is given by:

$$z = \frac{a_1 \alpha_1 + a_1 \beta_1}{1 - a_1 + a_1 \alpha_1 + a_1 \beta_1} \qquad \text{(xviii)}$$

It may be verified that:

$$\frac{a_1 \alpha_1 + a_1 \beta_1}{1 - a_1 + a_1 \alpha_1 + a_1 \beta_1} \leq \frac{a_1 \alpha_1}{1 - a_1 + a_1 \alpha_1} + \frac{a_1 \beta_1}{1 - a_1 + a_1 \beta_1} = (\bar{z}_1 + \bar{z}_2) \qquad \text{(xix)}$$

and in general $z \leq \bar{z} = \sum_j \bar{z}_j$.

It is thus inductively clear that in the presence of more than one tax related to income, the Musgrave-Miller measure of the built-in flexibility of a single tax will always overstate its value. Additional taxes and other leakages will have similar effect. Extended interdependencies complicate the relation, but the inequality of xix is almost always true.

The foregoing formulae are extremely simplified versions of the relations used in this study where there are several a_1's (the marginal propensities to consume, invest, etc.) and several marginal tax rates and other leakages (α's, β's, etc.). In the reduction procedure, moreover, the a_1, α, and β, too, get weighted by other structural coefficients.

It may be noted that the indicated bias in the individual tax flexibilities has nothing to do with the econometric estimates of a's and α's and other coefficients as such. The present bias is inherent in the partial nature of Formulae xvi-xvii, irrespective of the presence of absence of bias in the estimated marginal propensities to spend (a's) and marginal tax rates (α, β).

In order to get more realistic measures of built-in flexibilities of the individual taxes, the $\bar{z}$ measures as given by relations such as Formulae xvi-xvii must be corrected for their upward bias. This is done in the present study, first, by calculating the uncorrected $\bar{z}_j$ in accordance with the relations of the kind of xvi-xvii, and then by adjusting them proportionately downward, the coefficient of proportionality being provided by the ratio $z/\bar{z}$. Thus, the corrected built-in flexibility, which can be called z_j^*, is given by:

$$z_j^* = (z/\bar{z})\bar{z}_j \qquad \text{(xx)}$$

The second extension relates to the basis of comparison of built-in flexibility. In the Musgrave-Miller formula the change in equilibrium income with variable tax payments is compared with the change in equilibrium income when tax payments are held constant in absolute value. It is apparent that tax payments, especially the tax payments that are related to income or expenditure, are almost never held constant in the face of changing income. Moreover, in the developing countries, the changes in tax payments as a proportion of income are of greater significance than the mere existing levels of tax collections. The current public expenditures are generally geared to the current revenues (at the minimum). The built-in tax flexibility loses its operational significance, if the leakages from the flow of private incomes are reinjected concurrently by the public sector. If the marginal tax yield is significantly higher than the average tax yield then even if the previous year's proportions of government expenditures to income rather than the previous year's absolute values are maintained in the current year, there will be some lag of expenditures to revenues. Accordingly, it would be of interest in the context of the developing countries to compare the change in equilibrium income with income-elastic tax payments (ΔY) and the changes when tax collections change in proportion to income—i.e., when marginal and average tax yields are the same. In other words, what is being suggested here is a proportionate income-tax benchmark instead of the lump-sum tax benchmark of the Musgrave-Miller formula. Denoting the suggested change in equilibrium income by $\Delta\bar{Y}$, and expressing the redefined coefficient of built-in flexibility by z, we get:

$$z' = 1 - \frac{\Delta Y}{\Delta \bar{Y}} = \frac{a_1(\alpha_1 - \bar{\alpha}_1) + a_1(\beta_1 - \bar{\beta}_1)}{1 - a_1 + a_1(\alpha_1 + \beta_1)} \qquad \text{(xxi)}$$

where $\bar{\alpha}_1$ and $\bar{\beta}_1$ stand for average tax yields. As long as, for a given tax structure, $\alpha_1 > \bar{\alpha}_1$ and/or $\beta_1 > \bar{\beta}_1$, the ratio of the tax payments to income will rise and the increase in equilibrium money income prevented by the tax concerned will be greater than the increase in income prevented by the average tax yield, and vice versa. To distinguish the flexibility as given by this formula from that given by the Musgrave-Miller formula of Relation xii, we shall refer to z' as the "differential built-in flexibility" and that computed from Relation xviii—i.e., z—as the "absolute built-in flexibility."

NOTES

1. In connection with the discussion of Equation 1, see the discussion in Christ (1966, p. 67, footnote 17, and p. 189, footnote 16).

2. Unusual though it might appear, this practice of introducing a stochastic term in the income definition is not without precedent. See, for example, Smith (1963, p. 707). See also the suggestive remarks of Christ (1866, p. 587).

3. Klein and Goldberger (1955) and several other studies try similar proxy measures for wealth. The measure M in this study, however, differs from the Klein-Goldberger measure in that their measure includes currency, bank deposits, saving and loan shares, and U.S. government bonds, while our measure includes only currency and demand deposits. The Ando-Goldfeld study (1968) also uses a measure identical to ours.

4. For a significant advance in narrowing this gap in recent years, see Jorgenson (1965). See also Eisner (1965).

5. This distinction is emphasized by Johansen (1966, p. 463).

6. Wogart (1969) finds this behavioral variable to take precedence over the policy variables of reserve ratios and such in explaining the secondary expansion of the Brazilian money supply during 1947-66. He finds it related negatively with inflation and positively with growth. See also Delfim Neto (1966).

7. Minimum wage does not, of course, imply corresponding changes in average wage, which is a more relevant variable in this context. The available information, however, suggests that the observed decline in the real minimum wage rate during the mid- and late 1960s was also accompanied to some extent by a change in the same direction of the average wage rate. See, e.g., "Salarios no Gunabara," Conjuntura Economica, 24, no. 2 (1970), for the 1964-68 period.

8. For a proof of consistency of 2SLS when there are no lagged endogenous variables, see Basmann (1957). The proof of consistency in the presence of lagged endogenous variables is given by Christ (1966, pp. 438-40).

9. The Durbin-Watson test applies strictly to nonautoregressive models. The test is given in Durbin and Watson (1950).

10. The relations for various biases in lag models are derived by Griliches (1961 and 1967). The bias of the coefficients of the lagged endogenous variable will usually be upward as in most economic time series serial correlation is positive. "Disturbances in econometric equations almost never appear to have negative serial correlation" (Christ, 1966, p. 529).

11. For an autocorrelated disturbance case, the Cochrane and Orcutt procedure is very close to Aitken's generalized least squares, almost the only difference being that in the former procedure we lost one extra degree of freedom. See Johnston (1963, p. 186).

CHAPTER

4

SUBSTANTIVE RESULTS

The estimated equations are discussed individually in this chapter. Although a number of different variables and alternative regression forms were tried for various equations, in general only one or two, usually the better or the typical ones, are reported. Where of interest, the results of the nonreported regressions are also discussed. In general, the linear forms gave slightly (but not overly) better statistical fits, in terms of an overview of the significance levels of the coefficients and the appropriately calculated standard errors of estimate, than the logarithmic equations, while for a few equations the latter forms produced better statistical results. For some of the equations, in particular the monetary-sector equations, the proportionate-difference forms gave somewhat superior results. In many equations linear and nonlinear forms of the variables are intermingled. Where a majority of the coefficients had very high standard errors, the shorter regressions were tried by dropping out the variables with nonsignificant coefficients. Some of the variables were, however, retained for analytical purposes, even though they could not acquire acceptable significance levels. The major statistical problem in this respect arises because of the very high intercorrelations of the national-income components and those variables that have trend effects, such as population and labor force.

The regressions are numbered to correspond with the equation numbers of the theoretical model of Chapter 3. The estimation method used is indicated at the end of each equation by the following symbols: LS (least squares), 2SLS (two-stage least squares), COT (Cochrane-Orcutt transformation), 3SLS (three-stage least squares), and RF (reduced form). The numbers in parentheses below each estimated parameter are the t values of the respective coefficients. (The sample mean values of some of the variables are given in the last row of Appendix A, table A. 1, which may be used to calculate the elasticity coefficients from the marginal coefficients of the linear regressions.)

The coefficients of multiple determination as reported in this study have been corrected for the degrees of freedom used up, denoted by $\bar{R}^2$. The Durbin-Watson statistic (d) is also reported: it is a measure of serial correlation of errors. In the case of no serial correlation, d converges to a value of 2 (Durbin and Watson, 1950). In general, all distributed-lag equations and all of those nonlag equations that had a $d < 1.50$ were (re)estimated by the application of the Cochrane-Orcutt transformation. As is well known, however, for a first-difference equation or the COT type of transformed-variable equations, the $\bar{R}^2$ is always lower in comparison to the corresponding absolute value forms. In those cases in the present study where the $\bar{R}^2$ in the ordinary regression was rather low, the $\bar{R}^2$ in the corresponding COT equation tended to vanish (this applied, in particular, to personal income tax equations; for such equations, therefore, the COT was not used).

The 3SLS estimates were carried out for all those equations among whose disturbances the covariances exceeded 0.20.* The logarithms were taken to the base e—i.e., natural logs—as these make the elasticity interpretation and its calculation in semilog equations easier than in the case of the common logs.

THE DEMAND SECTOR

The Consumption Function

To give the reader a taste of the relevance of the high-powered econometric techniques employed here and also a picture of the typical differences among the alternatively-estimated equations, a set of four to five regressions computed by alternative techniques, each for consumption and investment functions, is given in Table 1. The five regressions reported for the consumption function are log 3SLS, linear LS, linear 2SLS, linear 3SLS, and linear COT-3SLS. As multicollinearity

---

*The covariances having values of 0.2 or higher among the disturbances of the endogenous variables, using a mongrel function for T in place of individual T_j's, are as follows: between C^p and Q = .5191; between I^{pl} and Y' = .4389; between I^{pl} and Ex = .4701; between I^{pl} and Im = .4682; between 1-Z and Y' = .4483; between Q and Im = .5237; and between Ex and Im = .5073. Usually, whenever an intercorrelation exceeded 0.2, the addition of the small sample bias to it raised its value close to 0.5. The high values of intercorrelations noted above, on the other hand, did not give a value of ρ much above 0.5. Because of these reasons and for the convenience of calculation, for all such equations, a value of ρ of 0.5 was employed.

tends to magnify standard errors, a 10 percent level of probability is, in general, used as a test of significance.

It may be seen that the coefficients of the lagged value of C^p and of Y^d are significantly different from zero in the LS, 2SLS, and 3SLS estimates, evidently confirming our hypothesis of the adaptive-expectations model. The lagged values of the exogenous variables were dropped, largely because of multicollinearity. The linear regressions give superior results in comparison to the (LS and 2SLS) log regressions. This can be judged by calculating the long-run elasticity of consumption from the 3SLS log regression, which comes to 1.33 (0.64/[1 - 0.52]) in Regression 2.5.1. The sample mean of C^p/Y^d is 0.88, which gives a marginal propensity to consume (MPC) of 1.17. Since MPC is unlikely to exceed unity, the coefficient of $\ln C_p^{-1}$ apparently robs from the other coefficients (refer to the theoretical discussion in Chapter 3). On the other hand, in a consumption equation with a significantly positive intercept term, a long-run MPC of 0.75 (0.636/[1 - 0.154] from Regression 2.5.3) is a very plausible value. It may be observed that while in 2SLS and 3SLS the coefficient of C_{-1}^p declines substantially (and pari passu the adjustment coefficient γ goes up), in comparison with that given by LS it still retains its statistical significance, though only at the 10 percent level instead of 1 percent in the LS estimates.

The coefficient of lagged consumption is positive and significant in 2SLS but almost evaporates in COT and 3SLS estimates of Regression 2.5.5. The warnings of Griliches that in the case of serial correlation the lagged series of the endogenous variable may "rob," especially when the variables are measured with errors, seem to be well founded (Griliches, 1967). Not until the transformation is applied does this variable in the present equation cede its spoil. The coefficient of Y^d regains its share and is now almost exactly equal to that found in 2SLS of Equation 2.5.3.

The result has implications for economic theory. Prima facie, the permanent-income hypothesis is not supported. Nor is the influence of expected prices on consumption borne out. Yet this second conclusion seems a hasty one: for example, the period of expectations might not be a year. In an inflation economy, where real (fixed) incomes can be changed significantly from month to month, even from week to week, consumption can be geared to the expected income of the future two to three months or three to four quarters. There is no way of testing this hypothesis, however, since quarterly/monthly income-consumption data are not available.

The coefficient of private consumption with respect to public consumption is negative, as expected, and manages to acquire significance in some of the alternative regressions, in particular in the log regressions. It stops short of the conventional significance levels in some of the reported regressions, presumably because of multicollinearity. This result indicates that the consumption of publicly supplied goods is a close substitute for private consumption. The marginal substitution rate is -1.56 in the linear regression and comes to about -1.8 from the elasticity measures of the log regressions. These values give the impression that the rate of substitution is rather high. But this impression is immediately corrected when one realizes that the public consumption expenditure is about half of tax revenues. The computed value is, as such, consistent with the coefficient of the marginal propensity to consume, noted in the preceding paragraph.

The coefficient of sales tax (τ^{sle}) is not significant and changes signs in alternative regressions. The maintained hypothesis here is that sales taxes are passed on to the consumer in higher prices and thereby discourage consumption. The present regressions, however, are not a real test of this hypothesis, because sales taxes are supposed to work through prices, and ordinarily prices need not even enter the present relation, which uses real-quantity variables. Nevertheless, we do test the rate of inflation to discern its indirect effects through income redistribution and such. For the same reason τ^{sle} is also tested in the presence as well as in the absence of a price variable. If the effects of τ^{sle} worked through $\Delta P/P$ the coefficient of τ^{sle} in the presence of $\Delta P/P$ should lose its significance in comparison with its effect in the absence of $\Delta P/P$.[1] Nothing of the sort seems to have happened. The coefficient of $\Delta P/P$ is also negative, but not significant. In any case, whether or not sales taxes are passed on in prices and whether or not they influence consumption must be tested at the micro level. The present attempt seems to be nothing more than an academic exercise.

The coefficient of M is not significant at all. Hence, the real balance or the wealth effect on consumption, as may be represented by real money balances, appears to be nonexistent in Brazil.

The Private Investment Function

The statistical fit for the private investment function is not as good as that for consumption. Considering the unpredictability of much of the behavior of investors, however, the fit is pretty good: we have succeeded in explaining over 80 percent of the variability of investment. The coefficient of lagged capital stock, $K(1-Z)_{-1}$, does not acquire significance. Hence, the "partial adjustments" model of Equations 4.4-4.10 is not supported. Some support is found for the "adaptive expectations" hypothesis of Equations 4, 5, and 4.1-4.3. That is, investment is positively related to its past value (I^p_{-1}); current sales $(Y'-I^{stk})_{-1}$; rate of return, profitability, and the availability of internal funds ($[Q-T^c]/K$ and its alternative measures); and changes in bank loans ($\Delta BL/BL$). It is negatively related to excess capacity (Z); corporation tax rate (τ^c both lagged and current); inflation ($\Delta P/P$); and the change in sales $\Delta(Y-I^{stk})$. As between the rate-of-return variable ($[Q-T^c]/K$) and the value of sales ($Y'-I^{stk}$), the regressions provide little basis for choice. The sales variable seems to throw the profit variable out. Our measure of profitability, however, is an extremely crude one (both Q and K are subject to measurement errors), while the sales variable is, undoubtedly, as trustworthy a variable as any other income category of the national accounts. Moreover, what is termed "profits" here is the entire category of nonwage income. It will, therefore, not be justified to interpret the results as supporting the sales-maximization to the rejection of the profit-maximization hypothesis. Both variables in our interpretation are indexes of profit maximization.

The standard errors of the coefficients of I^p_{-1} and ($Y'-I^{stk}$) become magnified due to multicollinearity between them. One indication of this is provided by a comparison of the LS Regression 4.3.2 and the 3SLS Regression 4.3.3. Although the scales tilt in favor of the ($Y-I^{stk}$) in 3SLS, partly because of the removal of simultaneous-equations bias, multicollinearity might also have something to do with it. Yet the long-run coefficients, 0.233 in Equation 4.3.2 and 0.184 in 4.3.3, remain surprisingly close to each other. Thus, while the short-run coefficients tend to remain indeterminate, this does not appear to be the case with the long-run coefficient. As between profitability ($[Q-T^c]$ net of depreciation), availability of internal funds ($[Q-T^c]$ gross of depreciation), and rate of return ($[Q-T^c]/K$), there is little to choose. All three variable forms acquire, by and large, equivalent levels of statistical significance.

The conclusions of the foregoing paragraph hold as long as proper attention has not been given to serial correlation. Since the d statistic is not a relevant test of serial correlation in a distributed-lag equation, there is no way to tell how far the estimates are biased. The COT results of Regression 4.3.4, indeed, reveal a similar picture as encountered in the consumption equation: the forces lying behind serial correlation are collected in the coefficient of the lagged endogenous variable. Because of the removal of the trend component, the COT method is also likely to reduce the indeterminacy caused by multicollinearity. The coefficient of the lagged endogenous variable declines from a high of 0.64 in the LS estimates through 0.48 in the simple 3SLS estimates to a miserable low of 0.11. The adjustment coefficient, thus, seems to be extremely large, accounting for 89 percent of the adjustment in one year. The indication of a shorter-than-one-year period of adjustment is as apparent here as for consumption.

The most important of the explanatory variables, next to sales, perhaps, is excess capacity in the beginning of the year. The subtler the estimation technique, the more significant the coefficient of this variable. Relative to all other variables, it is also the most "robust" in different regression forms. A reduction of one percentage point in excess capacity induces a fixed private investment of about Cr$70 million (in 1963 prices). This is about 7 percent of the mean investment of the reference period. Excess capacity, thus, not only reduces the present rate of growth but also attenuates the future productive capacity and, to that extent, the future rate of growth. The causes of excess capacity, thus, are pertinent factors in understanding the rate of growth, to which we will return.

The coefficient of τ^{inf} is not significant, except at the 10 percent level in the LS regression. Nevertheless, its persistently negative sign should be marked. For, on the other side, τ^{inf} is positively related to money supply, and the latter variable is an important determinant of inflation (see below). Our results, thus, are somewhat at variance with those suggested (on a rather weak basis) by Behrman and Klein (1970). Further research at a disaggregate level is in order (see, however, the discussion of the excess-capacity equation below).

The remaining coefficients may be disposed of summarily. The tax rate variable ($\tau^{c'}$) has a negative coefficient, but it is not significant. There is weak evidence that bank loans are positively associated with fixed investment, though the causal flow is not clear in this case. The residuals of this variable are contemporaneously correlated with those of the export, import, and GNP variables. Since GNP appears as an explanatory variable of this equation, only the former two series of residuals were used for 3SLS.

Public enterprise investment (I^{pe}) seems related to hardly any variable other than lagged GNP. From this result it appears that public investment might as well be treated as a predetermined variable along with other public expenditures.

Changes in inventory investment as such do not seem to be related to any conceivable variable, whether entered in logs, absolute values, or first differences. The fit is very poor, and none of the coefficients acquire significance. It may accordingly be treated as a residual factor. Moreover, since this investment does not (directly) involve any policy instruments, it need concern us no more in this model.

The Consumption Function

(2.5.1)
$$\ln C^p = \underset{(.8)}{1.2} + \underset{(2.05)}{.52} \ln \hat{C}^p_{-1} + \underset{(3.01)}{.64} \ln \hat{Y}^d - \underset{(-.38)}{.0042}\tau^{sle} - \underset{(-.23)}{.37} \ln (C^g + W^g)$$
$$\bar{R}^2 = 0.956$$
$$d = 1.85 \qquad \text{(3SLS)}$$

(2.5.2)
$$C^p = \underset{(1.72)}{2148} + \underset{(2.24)}{.568} C^p_{-1} + \underset{(.75)}{.3} Y^d - \underset{(-1.0)}{118.7}\, \tau^{sle} - \underset{(-0.52)}{1.49} W^g - \underset{(-0.49)}{0.72} C^g$$
$$\bar{R}^2 = 0.950$$
$$d = 2.12 \qquad \text{(LS)}$$

(2.5.3) $$C^p = \underset{(5.4)}{2842} + \underset{(1.81)}{.154\,\hat{C}^p_{-1}} + \underset{(3.30)}{.636\,\hat{Y}^d} - \underset{(-.08)}{6.78\,\tau^{sle}} - \underset{(-1.54)}{1.56\,(C^g + W^g)}$$

$\bar{R}^2 = 0.948$
$d = 1.53$ (2SLS)

(2.5.4) $$C^p = \underset{(2.8)}{2199} + \underset{(1.74)}{.378\,\hat{C}^p_{-1}} + \underset{(2.37)}{.501\,\hat{Y}^d} + \underset{(.43)}{4.35\,\tau^{sle}} - \underset{(-1.44)}{1.47\,(C^g + W^g)}$$

$\bar{R}^2 = 0.948$
$d = 2.24$ (3SLS)

(2.5.5) $$(C^p - \rho C^p_{-1}) = \underset{(3.3)}{1874} - \underset{(-.23)}{.06\,(\hat{C}^p_{-1} - \rho\hat{C}^p_{-2})} + \underset{(3.10)}{.646\,(\hat{Y}^d - \rho\hat{Y}^d_{-1})} + \underset{(.01)}{1.36\,(\tau^{sle} - \rho\tau^{sle}_{-1})} - \underset{(-1.01)}{1.11\,[(C^g + W^g) - \rho(C^g - W^g)_{-1}]}$$

$\bar{R}^2 = 0.817$
$d = 2.17$ (COT-3SLS)

Private Investment Function

(4.3.1) $$\ln I^{pl} = \underset{(-.8)}{-.369} + \underset{(3.45)}{1.07\ln \hat{I}^{pl}_{-1}} + \underset{(1.22)}{.04\,\frac{\hat{Q}-\hat{T}^c}{K}} - \underset{(-1.28)}{.116\ln \Delta(\hat{Y}' - \hat{I}^{stk})} + \underset{(1.07)}{.037\,(1-\hat{Z})} + \underset{(1.58)}{.008\,\frac{\Delta \hat{BL}}{BL}}$$

$\bar{R}^2 = 0.766$
$d = 2.76$ (3SLS)

(4.3.2) $$I^{pl} = \underset{(-1.3)}{-335.2} + \underset{(2.05)}{.64\,I^{pl}_{-1}} + \underset{(.83)}{.084\,(Y'-I^{stk})} + \underset{(1.62)}{36.2\,(1-Z)_{-1}} - \underset{(-1.12)}{22.25\,\tau^{c'}_{-1}} - \underset{(-1.72)}{47.9\,\tau^{inf}}$$

$\bar{R}^2 = 0.803$
$d = 2.56$ (LS)

$$(4.3.3)\qquad I^{pl} = \underset{(-3.4)}{-719.4} + \underset{(1.63)}{.483}\,\hat{I}_{-1} + \underset{(4.02)}{.095}\,(\hat{Y}-\hat{I}^{stk}) + \underset{(3.40)}{7.02}\,(1-Z)_{-1}$$

$$- \underset{(-1.39)}{38.9}\,\tau^{inf} + \underset{(1.25)}{4.57}\,\frac{\Delta \hat{BL}}{BL}$$

$$\bar{R}^2 = 0.822$$
$$d = 2.28 \qquad (3SLS)$$

$$(4.3.4)\qquad (I^{pl} - \rho I^{pl}) = \underset{(-2.6)}{-346.9} + \underset{(1.35)}{.105}\,(\hat{I}^{pl}_{-1} - \rho \hat{I}^{pl}_{-2})$$

$$+ \underset{(4.07)}{.119}\,([\hat{Y}'-\hat{I}^{stk}] - \rho[\hat{Y}'-I^{stk}]_{-1})$$

$$+ \underset{(2.7)}{68.9}\,([1-\hat{Z}]_{-1} - \rho[1-\hat{Z}]_{-2}) - \underset{(-.84)}{24.9}\,(\tau^{inf} - \rho\tau^{inf}_{-1})$$

$$+ \underset{(1.06)}{3.6}\left(\frac{\Delta \hat{BL}}{BL} - \rho\left[\frac{\Delta \hat{BL}}{BL}\right]_{-1}\right)$$

$$\bar{R}^2 = 0.533$$
$$d = 2.09 \qquad (COT\text{-}3SLS)$$

Public Enterprise Investment Function

$$(6)\qquad \ln I^{pe} = \underset{(-2.9)}{-17.0} - \underset{(-.19)}{.0103}\,\tau^{def} + \underset{(2.9)}{2.24}\ln Y'_{-1} + \underset{(1.26)}{.3}\ln \hat{I}^{pe}_{-1}$$

$$\bar{R}^2 = 0.920$$
$$d = 1.53 \qquad (2SLS)$$

Inventory Function

$$(8)\qquad I^{stk} = \underset{(-.4)}{-36.6} + \underset{(1.28)}{.35}\,I^{stk}_{-1} + \underset{(1.55)}{.0265}\,(\hat{Y}'-\hat{I}^{stk}) - .145\,\Delta(\hat{Y}'-I^{stk})_{-1}$$

$$\bar{R}^2 = 0.254$$
$$d = 2.31 \qquad (2SLS)$$

Agricultural Production Function

$$(12)\qquad \ln Y^{a} = \underset{(-1.02)}{-.98} + \underset{(7.48)}{1.525}\ln N - \underset{(-1.17)}{.137}\ln Fert$$

$$\bar{R}^2 = 0.899$$
$$d = 1.75 \qquad (LS)$$

Nonagricultural Production Function

$$(11.1)\quad \ln Y^{na} = \underset{(-.8)}{-1.27} + \underset{(1.94)}{.041}(\Sigma I/\Sigma K)_{-1} + \underset{(1.51)}{.266}\ln N + \underset{(3.58)}{.793}\ln(1-Z)K$$

$\bar{R}^2 = 0.976$
d = 1.61 (LS)

Overall Production Function

$$(11.2)\quad \ln Y = \underset{(-1.54)}{-1.39} + \underset{(2.43)}{1.24}\ln N + \underset{(1.72)}{.337}\ln(1-\hat{Z})K + \underset{(3.12)}{.025}(\Sigma I/\Sigma K)_{-1}$$

$$+ \underset{(.46)}{.026}\ln H^g$$

$\bar{R}^2 = 0.995$
d = 1.85 (3SLS)

Capacity-Utilization Function

$$(13.1)\quad \ln(1-Z) = 5.12 + .0029\hat{g} + .0086\,\tau^{inf} - .079\ln \hat{I}_{-1}$$

$\bar{R}^2 = 0.820$
d = 1.37 (3SLS)

$$(13.2)\quad (1-Z) - \rho(1-Z)_{-1} = \underset{(-1.01)}{-.886} + \underset{(3.77)}{.366}(\hat{g}-\rho\hat{g}_{-1}) + \underset{(1.26)}{.285}(\tau^{inf}-\rho\tau^{inf}_{-1})$$

$$- \underset{(-1.15)}{.0011}(I_{-1}-\rho I_{-2})$$

$\bar{R}^2 = 0.382$
d = 2.73 (COT-3SLS)

Nonwage Income Function

$$(14.1)\quad \ln Q = \underset{(-1.02)}{-3.63} + \underset{(3.02)}{.853}\ln\hat{Q}_{-1} - \underset{(-1.37)}{.0035}\,e^{ex} + \underset{(2.07)}{.373}\ln\frac{w^{min}}{y}$$

$$+ \underset{(.62)}{.354}\ln\hat{Y}'$$

$\bar{R}^2 = 0.974$
d = 2.17 (3SLS)

$$(14.2)\quad (Q-\rho Q_{-1}) = \underset{(-1.4)}{-412} + \underset{(.02)}{.006}(\hat{Q}_{-1}-\rho\hat{Q}_{-2}) + \underset{(.79)}{3.56}\,(e^{ex}-\rho e^{ex}_{-1})$$

$$- \underset{(-.94)}{2.74}\left(\frac{w^{min}}{y} - \rho\left[\frac{w^{min}}{y}\right]_{-1}\right)$$

$$+ \underset{(3.71)}{.282}\,([\hat{Y}'-\hat{I}^{stk}] - \rho[\hat{Y}-\hat{I}^{stk}]_{-1})$$

$\bar{R}^2 = 0.899$
$d = 2.11$ (COT-3SLS)

Wage Bill Function

$$(15)\quad W = \underset{(1.75)}{-281.2} + \underset{(8.8)}{.0759}Y' + \underset{(1.50)}{1.76}\frac{\Delta P}{P} - \underset{(-3.8)}{3.77}\frac{w^{min}}{y}$$

$\bar{R}^2 = 0.912$
$d = 1.52$ (LS)

Export Function

$$(21)\quad \ln Ex = \underset{(.3)}{.275} + \underset{(1.64)}{.0446}\,e^{ex} + \underset{(5.41)}{.944}\ln GNP^{US}$$

$\bar{R}^2 = 0.815$
$d = 2.55$ (3SLS)

Import Function

$$(22)\quad \ln I_m = \underset{(-.25)}{-1.21} - \underset{(-1.95)}{.0032}\,e^{im} - \underset{(-.54)}{.262}\ln X_{-1} - \underset{(-.80)}{.022}(1-\hat{Z}) + \underset{(4.49)}{1.04}\ln\hat{I}$$

$\bar{R}^2 = 0.774$
$d = 2.94$ (3SLS)

Money Demand Equation (Annual Data, 1950-68)

Note that all proportionate differences like $\Delta M/M$ were entered in percentages to reduce rounding errors. The results reported below are the reconverted values—i.e., the relevant variables (other than intercepts and the variables that were not multiplied by 100 before being entered in the regression)—are now divided by 100.

$$(24.1)\quad \frac{\Delta M}{M} = \underset{(-1.04)}{-.109} - \underset{(-.064)}{.0145}\left(\frac{\Delta M}{M}\right)_{-1} + \underset{(.09)}{.06\,g} - \underset{(-2.94)}{.164}\left(\frac{\Delta P}{P}\right)_{-1}$$

$$+ \underset{(1.91)}{.000375\,X_{-1}} + \underset{(3.07)}{.1386}\frac{\Delta IL}{IL}$$

$\bar{R}^2 = 0.336$
$d = 1.61$ (LS)

$$(24.2)\quad \frac{\Delta M}{M} = \underset{(-1.02)}{-.089} + \underset{(.09)}{.053\,g} - \underset{(-1.88)}{1.58}\,\Delta\left(\frac{\Delta P}{P}\right)_{-1} + \underset{(1.90)}{.00035\,X_{-1}} + \underset{(3.12)}{.1407}\left(\frac{\Delta IL}{IL}\right)$$

$\bar{R}^2 = 0.325$
$d = 1.64$ (LS)

Money Demand Function (Monthly Data, January 1969-June 1970)

$$(24.3)\quad M = \underset{(3.8)}{3.9} - \underset{(-.96)}{.239\,M_{-1}} - \underset{(1.0)}{.238\,M_{-2}} - \underset{(-3.22)}{.049\,r} - \underset{(-.67)}{.00044\,r_{-1}}$$

$$- \underset{(-2.4)}{.0107}\frac{\Delta P^w}{P^w} + \underset{(2.96)}{.0498\,t}$$

$\bar{R}^2 = 0.773$
$d = 2.11$ (LS)

Money Supply Function (Nine Observations, 1960-68)

$$(27.1)\quad \frac{\Delta M}{M} = \underset{(3.7)}{1.174} + \underset{(3.09)}{.118\,\tau^{inf}} + \underset{(1.86)}{.0239\,g} - \underset{(-3.89)}{572.3}\frac{M_1}{M} - \underset{(-1.22)}{.0126\,r} - .19\frac{\Delta P}{P}$$

$\bar{R}^2 = 0.713$
(M deflated by GNP deflator) $d = 3.08$ (LS)

$$(27.2)\quad \frac{\Delta M}{M} = \underset{(3.1)}{.556} + \underset{(.32)}{.00713\,\tau^{inf}} + \underset{(3.32)}{.0249\,G} - \underset{(-3.49)}{299.7}\frac{M_1}{M} - \underset{(-.74)}{.00444\,r}$$

$$+ \underset{(.84)}{.097}\frac{\Delta P^w}{P^w}$$

(M deflated by $\bar{R}^2 = 0.816$
wholesale price level) $d = 2.54$ (LS)

$$(27.3)\quad \frac{\Delta M}{M} = \underset{(2.5)}{.83} + \underset{(2.02)}{.132\,\tau^{inf}} - \underset{(-.53)}{.009\,r} - \underset{(-2.18)}{5.24}\frac{M_1}{M} - \underset{(-2.05)}{.462}\frac{\Delta P^w}{P^w}$$

$\bar{R}^2 = 0.394$
(M deflated by GNP deflator) $d = 2.13$ (LS)

Inflation Function (Annual Data 1950-68)

$$(29.1)\quad \frac{\Delta P^w}{P^w} = \underset{(-1.04}{-.0996} + \underset{(4.29)}{.64}\left(\frac{\Delta P^w}{P^w}\right)_{-1} + \underset{(.38)}{.19}\left(\frac{\Delta M}{M}\right)_{-1} + \underset{(2.45)}{.304}\frac{\Delta P^{im}}{P^{im}}$$

$\bar{R}^2 = 0.673$
$d = 1.56$ (LS)

$$(29.2)\quad \frac{\Delta P^w}{P^w} = \underset{(-1.07)}{-.0767} + \underset{(2.29)}{.362}\left(\frac{\Delta P^w}{P^w}\right)_{-1} + \underset{(2.59)}{.516}\left(\frac{\Delta M^n}{M^n}\right) + \underset{(1.03)}{.125}\frac{\Delta P^{im}}{P^{im}}$$

$\bar{R}^2 = 0.771$
$d = 2.17$ (LS)

$$(29.3)\quad \frac{\Delta P^w}{P^w} = \underset{(-1.0)}{-.085} + \underset{(2.16)}{.585}\left(\frac{\Delta P^w}{P^w}\right)_{-1} + \underset{(1.79)}{.051}\left(\frac{\Delta M^n}{M^n}\right)_{-1} + \underset{(2.28)}{.291}\frac{\Delta P^{im}}{P^{im}}$$

$\bar{R}^2 = 0.668$
$d = 1.62$ (LS)

$$(29.4)\quad \frac{\Delta P^w}{P^w} = \underset{(-1.4)}{-.108} - \underset{(-.25)}{.087}\left(\frac{\Delta P^w}{P^w}\right)_{-1} + \underset{(2.95)}{.63}\frac{\Delta M^n}{M^n} + \underset{(1.23)}{.312}\left(\frac{\Delta M^n}{M^n}\right)_{-1}$$

$$- \underset{(-.18)}{1.627\,g} + \underset{(1.02)}{.180}\left(\frac{\Delta M^n}{M^n}\right)_{-2} + \underset{(.45)}{.06}\frac{\Delta P^{im}}{P^{im}}$$

$\bar{R}^2 = 0.772$
$d = 1.77$ (LS)

Inflation Function (Monthly Data, January 1969-June 1970)

$$(29.5)\quad P^w = \underset{(5.5)}{151.0} + \underset{(11)}{.72}\,P^w_{-1} + \underset{(.23)}{2.1}\,M^n + \underset{(2.83)}{21.2}\,M^n_{-1} - \underset{(-5.3)}{132.8}\,e^{im}$$

$\bar{R}^2 = 0.997$
$d = 1.19$ (LS)

$$(29.6)\quad P^w = \underset{(5.1)}{148.6} + \underset{(9.88)}{.701}\,P^w_{-1} + \underset{(1.07)}{10.2}\,M^n + \underset{(2.63)}{20.5}\,M^n_{-2} - \underset{(-5.2)}{136.3}\,e^{im}$$

$\bar{R}^2 = 0.997$
$d = 2.23$ (LS)

Bank Loans Function (Nine Observations)

$$(30)\quad \frac{\Delta BL}{BL} = \underset{(1.23)}{-.127} + \underset{(2.18)}{.0407\,g} - \underset{(1.66)}{2.778}\,(\eta^a_{-1} - \eta^c)$$

$\bar{R}^2 = 0.275$
$d = 2.47$ (LS)

Fiscal Sector Equations

Personal Income Tax

(32.1) $$T^{pi} = \underset{(-1.6)}{-39.0} + \underset{(2.56)}{5.64\,\tau^{pi}_{-1}} + \underset{(5.52)}{.0069\,Y}$$

$\bar{R}^2 = 0.613$
$d = 0.97$ (LS)

(32.2) $$(T^{pi} - \rho T^{pi}_{-1}) = \underset{(-.6)}{-14.3} + \underset{(.90)}{2.55}\,(\tau^{pi} - \rho\tau^{pi}_{-1}) - \underset{(-.33)}{.064}\left(\frac{\Delta P}{P} - \rho\frac{\Delta P}{P_{-1}}\right)$$
$$+ \underset{(6.3)}{.529}(y - \rho y_{-1})$$

$\bar{R}^2 = 0.128$
$d = 1.24$ (LS-COT)

Corporation Tax

(33.1) $$T^{c} = \underset{(.45)}{19.8} + \underset{(2.98)}{.0456\,\hat{Q}} + \underset{(.91)}{2.67\,\tau^{c'}_{-1}}$$

$\bar{R}^2 = 0.874$
$d = 1.83$ (2SLS)

(33.2) $$T^{c} = \underset{(-.2)}{-9.9} + \underset{(1.71)}{.013\,\hat{Y}} + \underset{(.65)}{1.83\,\tau^{c'}_{-1}} + \underset{(1.06)}{.12\,T^{ss}} - \underset{(-.30)}{.129}\,\frac{\hat{\Delta P}}{P}$$

$\bar{R}^2 = 0.909$
$d = 1.87$ (2SLS)

Excise Duties

(34) $$(T^{exc} - \rho T^{exc}_{-1}) = \underset{(-7.2)}{-221} + \underset{(4.59)}{39.7}\,(\tau^{exc} - \rho\tau^{exc}_{-1})$$
$$+ \underset{(2.08)}{.051}\,[(Y - I^{stk}) - \rho(Y - I^{stk})_{-1}]$$

$\bar{R}^2 = 0.920$
$d = 1.63$ (LS-COT)

Import Duties

(35) $$T^{im} = \underset{(-1.2)}{-23.2} + \underset{(4.44)}{.011\,\hat{Y}'} - \underset{(-.08)}{.019\,e^{im}}$$

$\bar{R}^2 = 0.543$
$d = 0.94$ (2SLS)

Stamp Duties

$$(36)\quad \ln T^{stp} = \underset{(-7.1)}{-7.2} - \underset{(-.34)}{.029}\frac{Y^{na}}{Y^{a}} + \underset{(8.8)}{1.30}\ln Y' - \underset{(-.81)}{.07}(T^{ind}/T^{dir})$$

$$\bar{R}^2 = 0.967$$
$$d = 2.42 \qquad \text{(LS)}$$

Social Security Tax

$$(37)\quad (T^{ss} - \rho T^{ss}_{-1}) = \underset{(-6.5)}{-132} + \underset{(12.9)}{.0915}(Y^{na} - \rho Y^{na}_{-1}) - \underset{(-3.5)}{1.84}\left(\frac{\Delta P}{P} - \rho\frac{\Delta P}{P_{-1}}\right)$$

$$\bar{R}^2 = 0.904$$
$$d = 1.55 \qquad \text{(COT)}$$

Fuel Tax

$$(38)\quad T^{fue} = \underset{(-2.3)}{-145.7} + \underset{(3.94)}{.025}\hat{Y} + \underset{(5.6)}{16.2}\frac{Y^{na}}{Y^{a}}$$

$$\bar{R}^2 = 0.840$$
$$d = 1.25 \qquad \text{(2SLS)}$$

Sales Tax

$$(39)\quad (T^{sle} - \rho T^{sle}_{-1}) = \underset{(-6.6)}{-306} - \underset{(-1.0)}{.0198}(\hat{Y}' - \rho\hat{Y}'_{1}) + \underset{(5.32)}{86.7}(\tau^{sle} - \rho\tau^{sle})$$

$$+ \underset{(1.48)}{64.0}\left(\frac{Y^{na}}{Y^{a}} - \rho\left[\frac{Y^{na}}{Y^{a}}\right]_{-1}\right)$$

$$\bar{R}^2 = 0.938$$
$$d = 1.99 \qquad \text{(2SLS)}$$

Other State Taxes

$$(40)\quad \ln T^{os} = \underset{(3.8)}{11.74} - \underset{(-2.17)}{.84}\ln Y' + \underset{(3.3)}{.56}\text{ dummy }54 + \underset{(1.51)}{.3}\text{ dummy }59$$

$$\bar{R}^2 = 0.318$$
$$d = 1.85 \qquad \text{(LS)}$$

Municipal Taxes

$$(41)\quad T^{mun} = \underset{(1.6)}{21.8} + \underset{(6.44)}{.0105}\hat{Y}'$$

$$\bar{R}^2 = 0.691$$
$$d = 1.82 \qquad \text{(2SLS)}$$

Public Transfers

$$\text{(43)} \quad \ln Tr = \underset{(4.5)}{2.03} + \underset{(5.9)}{.74} \ln T^{ss} + \underset{(1.26)}{.2837}(1-Z)K$$

$$\bar{R}^2 = 0.988$$
$$d = 1.56 \qquad \text{(LS)}$$

THE SUPPLY SECTOR

The Production Function for Agricultural Income

This is a rather disappointing regression. There is an extremely high intercorrelation between labor (N) and crop acreage (A); accordingly, only one of these inputs was entered at a time. The variables of this function, as noted earlier, are very rough approximations of their real values. An astonishing result of this regression is the negative sign (of fertilizer's coefficient) though it is not significant. Since the fertilizer variable was supposed to be a surrogate for all purchased inputs that are likely to be the carriers of new techniques, the result appears to be rather implausible. Several explanations can be offered. Among these, one is the possibility of an incorrect measure of the variable. First of all, it is defined as "apparent consumption" (domestic production plus imports), since the actual-use data are not available; and, besides, it is measured in tonnage rather than in nutrient content. Second, it is strongly trended and is also collinear with N; due to this and since N is the more important factor (insofar as the proportion of its contribution to output is concerned), N may be stealing from fertilizer. Third, the main emphasis in Brazil has been on industrialization: the fiscal, import, and monetary policies have all been geared to favor industrialization, so agriculture has, on the whole, been a neglected sector, and technical change, if there is any at all, must be very feeble in this sector. Even the prevailing professional opinion in Brazil, as reflected by some of the influential members of Fundaçaõ Getulio Vargas, has been that technical change, in any case, is not going to benefit the Brazilian farmers (one of their arguments is that the low elasticity of demand for agricultural products is chiefly responsible for technology's lack of influence; see, for example, Paiva [1967]). Programs of irrigation and similar agricultural projects have been conspicuous in their absence from the Brazilian plans. In such a situation a negative coefficient for fertilizer does not seem a surprising result.

In any case, since none of the explanatory variables of the agricultural production function are endogenous or even lagged endogenous, this function is a segmented one. Agricultural income may, thus, be legitimately assumed to be an exogenous variable of our system. For further analysis, however, it was made nonsegmented by virtue of Y^a being a component of Y.

The Production Function for the Nonagricultural Sector

The production function for the nonagricultural sector is a reasonably good one. The relative coefficients of labor and capital (which in a Cobb-Douglas function also stand for factor shares) are not inconsistent with the observed average factor shares, though prima facie the labor share seems to be rather low. The sum of coefficients (1.06) exceeds unity, but not significantly so. There is, thus, as expected, little evidence of scale economies. On the other side, the coefficient of embodied technical change is 0.041, giving a 4 percent per annum rate of technical change—an unbelievably high rate.

None of the variables other than the conventional factors of production acquires significant coefficients. The signs of the trend factor and the tax structure (T^{ind}/T^{dir} or T^{c}/T^{pi}) are negative, but t values are extremely low. The coefficient of public health expenditure, presumed to be a proxy for investment in man, is positive but not significant.

With a view to making a rough check on the relative magnitudes of the nonagricultural production function, because of what has been said about the agricultural production function, and because the inputs used are the aggregates for the overall economy, an overall aggregate production function was also estimated and is reported in Regression 11.2. It may be observed that the coefficient of embodied technical change is somewhat lower but still large from the standards observed in the advanced countries. The coefficient of labor, as one may expect, is a kind of mean of the agricultural and nonagricultural production coefficients. Since the coefficients change significantly between different regressions, it is difficult to put much confidence in them. The inputs must be broken down by sectors for future research.

The Capacity Utilization Function

Capacity utilization is related positively to current growth and current deficit finance and negatively to lagged investment. In alternative regressions it gets a negative and significant coefficient with respect to overall taxation. The COT has not improved the statistical fit: while straight 3SLS estimates indicate a positive serial correlation of errors, strangely enough the COT-3SLS estimates display a negative serial correlation. Since the $\bar{R}^2$ falls to a low of 0.38, we shall use the results of the straight 3SLS estimates.

It was seen earlier that inflationary finance is negatively related to fixed private investment; we now note that capacity utilization is positively and significantly related to inflationary finance (τ^{inf}). It is perhaps this dual role of inflation that keeps the controversy still unsettled regarding the favorable or unfavorable role of inflation in Brazil. The results of this study indicate that, apart from the generally-held view that inflation encourages less productive rather than more productive fixed investment, inflation may retard the overall rate of fixed investment and, hence, long-run growth. On the other side, once inflation is current, any reduction in the means of payment thwarts capacity utilization and therefore growth. In order for existing capacity to be kept in normal full use, inflation must continue. The "shock treatments" of the early and middle 1960s undoubtedly bear out this conclusion: the "corrective inflation" of the late 1960s—by which the government attempted not to throttle the operation of industry by major cuts in its working capital (for example) but to correct the causes of inflation gradually by other means—is still in its experimental stage, although it seems so far to have succeeded remarkably in raising the rate of growth and simultaneously lowering the rate of inflation. The combined results of the investment function and the capacity-utilization function lead to the conclusion that inflation does not encourage long-run growth but that once it is ushered in, it can retard short-run growth if attempts are made to exterminate it.

The high significance level and the "robustness" of the coefficient of I_{-1}, in the (1-Z) equation indicates that, by itself, the capacity-creating role of investment in Brazil tends to overpower its demand-generating role.

A surprising result of the capacity-utilization equation is that the bank loans variable has a negative coefficient in alternative algebraic forms, and significantly so, whether it is entered along with τ^{inf} or in place of it. A possible explanation of this wrong sign might be that the causal flow is the reverse of what has been specified in this equation and comes from the demand side; the higher the level of capacity utilization, the higher the availability of internal funds to the firms, and the lower, therefore, the need for bank loans. In either case, our result is inconsistent with the observed experience of at least the late 1960s, when a shortage of working capital was widely regarded as the major cause of excess capacity. It was as a result of this diagnosis that in 1967-68 the new Finance Minister Delfim Neto eased credit and gave tax holidays to the firms; those measures appear to have contributed to reversing the slump and doubling the rate of growth to around 9 percent or so in the course of one or two years.

THE DISTRIBUTION SECTOR

The Profit Equation

The nonwage or profit equation has similar magnitudes and significance levels in all forms of regression from LS to straight 3SLS, but they alter significantly in the transformed 3SLS regression. An interesting change to be noted is that while in the nontransformed regression (14.1) the coefficient of real wages is positive and significant, indicating the common influence of a third variable such as growth or the trend on both wage and nonwage income, in the transformed regression the same coefficient acquires the right (negative) sign, though not significantly so. The only variable that retains and in fact increases in significance is the sales or GNP variable. No influence of past events seems to linger on except through current GNP.

The Wage Income Equation

The wage bill equation (for the manufacturing sector) has a marginal coefficient of 0.0759 against an average (W/GP) of 0.0563, indicating an improvement in the manufacturing wage bill as the economy grows. This might or might not mean an improvement in the economic lot of workers in relation to that of nonwage earners, since an increase in wage labor from transfers of agricultural labor to industry is an essential phenomenon of economic development. A rough pencil work may proceed as follows: First, the share of the secondary sector in GNP was approximately 20 percent in 1950 and rose to approximately 22 percent in 1968; second, assuming a matching transfer of labor from the other sectors to the secondary sector, the marginal wage share must be at least 10 percent higher than the average share in the receiving sector, if the base shares are to maintain; since this share has, in fact, been 35 percent (0.0759/0.0563) higher, economic development must be improving the wage share in national income.[2]

The negative coefficient of the wage bill (W) with respect to minimum wage relative to per capita income (w_{min}/y) calls for a comment, since ordinarily one would expect a positive relationship. The first impression is that the coefficient has the wrong sign (which is statistically significant). It must be recalled, however, that this is an equilibrium equation. A possible explanation on the labor supply side is that in their efforts to recapture their real income by moonlighting (which is widespread in Brazil at all levels of earnings), by women's labor, etc. (supply shift), working families more than recover the previous levels. The low-income families unable to afford to send their children to school may instead be compelled to let them become wage labor. This result obtained under the falling real wage rate may not hold symmetrically under a rising real wage rate; such dichotomous results are not entirely uncommon. A possible explanation from the demand (for labor) side is that the marginal rate of substitution between labor and nonlabor inputs is rather high, so a small change in the wage rate leads to large changes in employment. Finally, the wage bill includes the salaries of skilled labor (including foreign engineers in foreign owned firms), which are believed to be complementary with capital. Since the indicated responses are a consequence of distorted factor prices, the result does not negate the a priori conclusion that the welfare of workers has been impaired.

Surprisingly enough, the sign of the coefficient of price changes is also negative, though not significantly so. There is no evidence that inflation reduces the wage share from the income-source side (in distinction to the income-use side). Allowance must be made, however, for the labor transfers discussed above.

THE FOREIGN SECTOR

The Export Function

We have talked about the block-recursive nature of our model. The export equation forms a segmented block that comes before the remaining equations since it is determined entirely outside our system by an instrument variable, e^{ex} (defined as so many Brazilian cruzeiros per U.S. dollar), and foreign income levels. It is interesting to note that, if the coefficient of e^{ex} is stretched to be significant at the 10 percent level, an improvement in the real terms of trade leads to an increase of foreign earnings. The price elasticity of Brazilian exports—chiefly coffee—cannot be very elastic. On the other hand, the income elasticity of demand for Brazilian exports (again, mainly coffee) is nearly one.

The Import Function

Imports are positively related to GNP, but when investment and GNP are both included, investment throws GNP out; thus, the main determinant of imports seems to be investment. Even though an increasing proportion of capital equipment requirements of Brazil have been met from domestic production of the heavy industries, the import content of Brazilian investment, especially in the form of raw materials, remains high. According to our estimates (not all reported here) the import content of Brazilian investment is more than 46 percent.

Excess capacity and reduction in capacity to import do not seem to lead to the economizing of imports. A higher exchange rate for imports seems to discourage imports, though the coefficient is not significant.

THE MONETARY SECTOR

The monetary sector variables are measured in proportionate changes. These measures seem to give better statistical results and make better sense for this sector than do absolute values or first differences.

Demand Function for Money

The fit of the demand function for money is a rather disappointing one. The variability explained in alternative regressions hardly ever exceeds 0.35. The $\bar{R}^2$ for the regression reported here, though significantly different from zero at the 1 percent level, is only 0.336. The lagged endogenous variable fails to acquire acceptable significance levels. The techniques of 2SLS, 3SLS, COT, etc., were not applied, therefore, as, due to the poor fit, the variables could not withstand the burden of these tests.[3] In the COT regression, for instance, the $\bar{R}^2$ completely evaporates. Likewise, the hypotheses for the partial adjustment and the adaptive expectation processes cannot be properly tested. As will be seen below, however, cogent evidence of the lagged effects does emerge from the reduced forms of the decomposed monetary sector. In the single-equation estimates, therefore, the entire emphasis was placed on getting the best fit, without regard to the said twin hypotheses.

Partly because of the poor fit for the demand function, partly taking cues from the earlier results for the consumption and investment functions, which created doubts about the relevance of one-year periods for the dynamic analysis, and partly due to data constraints, three alternative samples were tried for the monetary sector. The first sample consists of the annual data for the reference period of this study (1950-68). The second sample is for the subperiod of the 1960s (1960-68), because the data for a few of the monetary variables, such as the reserve ratios and exchange operations, could not be procured for the earlier years. The third sample was assembled simply to test the shorter periods for the lag analysis, and consists of monthly data from January 1969 to June 1960 (16 observations). For the monthly sample, the data for money supply (M), price level (P), stock prices (r), exchange rate for imports (e^{im}), and bank loans are available from

Conjuntura Economica 24, no. 10 (1970). No index of output by month is available. A trend variable (t) was used as a proxy for output growth.

When variables are used in first differences or proportionate differences, the R^2 is always lower than when they are used in absolute values. On this basis and since the coefficients have expected signs, we are reluctant to reject the regressions of Equations 24.1 and 24.2. For whatever it may be worth, the annual sample for the demand function indicates a negative relationship with lagged prices and positive relationships with capacity to import and international liquidity. All three coefficients are consistent with the predicted signs. The monthly sample indicates a similar negative relationship with present and lagged price level. The trend coefficient is positive, as expected. The coefficient of r is negative (and significant), indicating the portfolio effect. The coefficient of lagged money supply is negative but significant in neither sample. The fit is much better in the monthly sample than in other samples.

A good deal of work on the demand for money in Brazil has been done by Pastore, who has also assembled monthly data on real rates of interest. Two of the typical regressions from Pastore's estimates based on monthly data are reproduced below in the terminology of the present study (Pastore, 1969, p. 112):

$$M/P = 22.2 + \underset{(5.8)}{1.05\,Y'} - \underset{(-4.3)}{3.86\,(\Delta P/P)^e} + \underset{(4.6)}{0.48\,(M/P)_{-1}}$$

$R^2 = 0.870$, $d = 1.35$
Reference period 1954-66, monthly data

$$M/P = -.04 + \underset{(4.6)}{.008\,Y'} - \underset{(-3.4)}{.004\,r} + \underset{(9.3)}{.795\,(M/P)_{-1}}$$

$R^2 = 0.882$, $d = 2.04$
Number of monthly observations = 43

where r stands for the real rate of interest. It may be mentioned that the actual rate of change of prices (the basis of the expected rate of change of prices) has exceeded the short-term money rate of interest almost throughout the reference period used by Pastore. This is prima facie evidence that there is strong credit rationing and that the short-run money market is not functioning in the usual way. This is presumably the reason why Pastore preferred to use the real rate of interest instead of the nominal rate of interest, the latter being the relevant price involving choice between money and short-term securities, both of whose values are fixed in terms of money, provided the short-term money rate of interest is greater than the expected rate of change of prices. In any case, the fit obtained by Pastore is superior to the one obtained for this equation in the present estimates: one reason for this, of course, is that we employ proportionate differences, while Pastore uses absolute-value variables. Nevertheless, the coefficients of the two equations are consistent with each other and have the expected signs.

Transparently, the monthly observations give a better fit. Unfortunately, the simultaneous solution of the monthly demand functions when the rest of the system is computed from annual data, especially when the equations include various lagged relationships, creates problems. Hence, for better or worse, Equations 24. 1 and 24. 2 will be used for reduced-form analysis.

The Money Supply Function

In contrast to the demand function, the supply function is a cheering one. The fit is good and most coefficients are significantly different from zero at the conventional levels. The money supply measures, alternatively deflated by the wholesale price level and the GNP deflator, are tested. Both measures of the price level, P and P^w, are also tested. Money supply is positively related to treasury deficit and growth. The coefficient of r has the wrong sign but not significantly so. The coefficient of $\Delta P/P$ shows somewhat unstable behavior. It is negative in the GNP-deflated money supply regression and positive in the wholesale-price-deflated money supply regression with identical sets of explanatory variables. It is negative and significantly so only in Regression 27. 3, in which money deflated by the GNP deflator and the variable $\Delta P^w/P^w$ are used.

The Price Equation

The first regression (29. 1) is simply a replication of the Delfim Neto et al. (1966) relation to see whether it still holds. It does, and the past year's changes in prices and current changes in import prices explain about two-thirds of the current price variation. The lagged real money variable $(\Delta M/M)_{-1}$ is thrown in for a purpose: it is not significant at all in this regression. The second regression (29. 2) substitutes nominal money $(\Delta M^n/M^n)$ for the real money $(\Delta M/M)$ variable, for it does not have to invoke the classical dichotomy: after all, it is the nominal and not the real quantity of money that expresses the price level; it is now highly significant. The import-price variable $(\Delta P^{im}/P^{im})$ loses much of its explanatory power to $\Delta M^n/M^n$; the former is now not significantly different from zero. Whatever influence relative import prices have on the domestic price level seems to work through the money supply. It thus appears that the cost-push effects of higher relative prices of imports will perhaps not be translated into higher domestic prices unless the money supply makes it possible to do so. This is clear enough from the third equation (29.3) in which $\Delta M^n/M^n$ is replaced by $(\Delta M^n/M^n)_{-1}$: the $\Delta P^{im}/P^{im}$ variable comes back into its own. Note that the coefficient of $(\Delta M^n/M^n)_{-1}$ also retains its significance, but both its magnitude and the significance level are now considerably reduced, while those of $(\Delta P/P)_{-1}$ are correspondingly increased. This is clearly due to multicollinearity between $(\Delta P/P)_{-1}$ and $(\Delta M/M)_{-1}$. In the final regression reported for this equation (29.4), we experimented with three-year lag values of the money supply: namely, $\Delta M^n/M^n$, $(\Delta M^n/M^n)_{-1}$, and $(\Delta M^n/M^n)_{-2}$. Now only the money supply variables retain their significance levels. (The significance levels of the lagged values are low, probably due to multicollinearity among the three consecutive-year values of M^n.) In view of these results, it is difficult to imagine what scope there remains for the structural and cost-push factors to explain Brazilian inflation.

The monthly regressions (29.5 and 29.6) tell a similar story. It is interesting to note that prices are affected not only by the past year's money supply but also by past months' money supply. The lag effects appear to be dispersed all along the line emanating from the past month's money supply to the past 36 months' money supply and possibly even longer.

Having found indications for both monthly and annual lags, a natural next step was to try quarterly lags (a sort of mean lag). For this purpose, the monthly data series for money and price variables were aggregated into quarterly series, and the annual series of GNP were broken down into quarterly values using Harberger's procedure, which consists of expressing the changes in the annual average values of national income as algebraic functions of the unknown quarter-to-quarter changes, the variance of which is then minimized subject to the constraint that the series generated be consistent with each of the known year-to-year changes in the annual average income (Harberger, 1963, p. 249). The money and price series are available in International Financial Statistics (International Monetary Fund, annual). Several lags were tried. The following equation with 32 quarterly observations (1961-68) is reported:

$$\begin{aligned}\frac{\Delta P^w}{P^w} = \underset{(.03)}{.028} + \underset{(5.8)}{.4136}(\Delta P^w/P^w)_{-1} - \underset{(-.5)}{.08}(\Delta Y'/Y') + \underset{(2.5)}{.131}(\Delta M/M) + \\ \underset{(4.2)}{.198}(\Delta M/M)_{-1} + \underset{(3.5)}{.123}(\Delta M/M)_{-2} + \underset{(1.6)}{.073}(\Delta M/M)_{-3} + \\ \underset{(2.7)}{.117}(\Delta M/M)_{-4} + \underset{(1.5)}{.086}(\Delta M/M)_{-5} \qquad R^2 = 0.803,\ d = 1.92\end{aligned}$$

The sum of the coefficients of $\Delta M/M$ over 15 months is $0.728 < 1$, which means tests of additional lags are in order, although they were not carried out in this study. Silva has tried similar relations with larger samples, longer lags, and with alternative model specifications; he gets, in general, similar results (see Silva [1973]).

The Bank Loans Equation

The final equation of the monetary sector, that for bank loans, is given in Regression 30. It has the right positive relationship with rate of growth. The relationship with excess reserve ratios, however, has the wrong sign, but not significantly so. That the coefficient of exchange operations acquires significance in neither the money supply function nor the bank loans function is somewhat surprising. Its influences are perhaps cancelled out, to some extent, through the aggregation of the monthly data into annual series and of several categories of exchange operations into an overall one tried here.

The results of the foregoing equations of the monetary sector can now be collected. These are best visualized from the linearized reduced forms of the decomposed monetary sector, given below. (The process of linearization will be explained in Chapter 5.) The dynamic properties of the reduced forms of the entire system will be analyzed in Chapters 7 and 8.

$$(47)\quad \dot{M} = .81\dot{M}_{-1} + .0382\dot{M}_{-2} + .0167\dot{M}_{-3} + 1.408\dot{P}_{-1} - .2205\dot{P}_{-2} - .17\dot{P}_{-3} - .8298\dot{P}^{im} + .7012\dot{P}^{im}_{-1} + .5447\dot{X}_{-1} + .7002\dot{IL} - .88\dot{IL}_{-1} + 3.63\dot{g} + 183.4\dot{\tau}^{inf} - 12.5\dot{r} - 1601.0(\dot{M}_1/M)$$

$$(48)\quad \dot{P} = .417\dot{P}_{-1} + .13\dot{P}_{-2} + .1001\dot{P}_{-3} + .038\dot{M}_{-1} - .023\dot{M}_{-2} + .01\dot{M}_{-3} + .486\dot{P}^{im} - .415\dot{P}^{im}_{-1} + .0136\dot{X}_{-1} + .0174\dot{IL} - .022\dot{IL}_{-1} - 1.536\dot{g} + 4.58\dot{\tau}^{inf} - .312\dot{r} - 40.1(\dot{M}_1/M)$$

Here the dots stand for total derivatives. The reduced forms 47 and 48 were solved from Equations 24.1, 27.3, and 29.4, taking all coefficients reported there.

It may be seen that neither the price nor the money equation conforms to any simple one-period lag relation or the distributed-lag processes with geometrically declining weights, as hypothesized in Chapter 3. Both are difference equations of the third order in more than one variable. Further analysis of these difference equations takes us into the dynamic properties of these equations, which is the subject matter of Chapter 8. The foregoing discussion will, therefore, be continued there.

THE FISCAL SECTOR

Taxes

The estimation of tax equations creates very few theoretical and statistical problems in comparison with the private sector equations. The statistical fits are quite good, and most (though not all) of the relevant variables have significant coefficients. The coefficients of tax collections with respect to their respective tax rates are all positive and significant, with the exception of import duties. The coefficient of e^{im} has a negative sign for import duties, but not a significant one. This, however, is not an implausible result; it only indicates that, since higher tariffs and a higher exchange rate (more cruzeiros per dollar) mean higher import prices, the price elasticity of imports is probably extremely high. Import duties, in any case, are nowadays mostly levied to restrict imports rather than raise revenues. Numerous exemptions further curtail yield from tariffs. Two regressions are reported for T^{im}, which indicates that investment is a more important explanatory variable here; it throws out GNP.

For personal income tax we tried alternatively (along with other variables) per capita income, profits, and GNP. Per capita income gives a better index of the progressivity of this tax, especially since GNP and population cannot be entered together due to a strong multicollinearity between them. The disturbances of this variable remain serially correlated even when all the relevant explanatory variables (y, τ^{pi}, τ^{pi}_{-1}, $\Delta P/P$) have been entered. And unlike the economic variables in general, the serial correlation is negative, the d statistic in the log equation being 3.86. When, on the other hand, COT is applied, The $\bar{R}^2$ vanishes. In view of this the regression analysis of this tax remains so far incomplete and perhaps inconclusive. For such cases and for other reasons to be noted at appropriate places, an additional sample was assembled and will be briefly analyzed in Chapter 5. It might be mentioned that price change ($\Delta P/P$) and the composition of income (as measured by Y^{na}/Y^{a}) were also tried for this tax but were not found to give significant coefficients.

For some of the taxes for which the tax rate series could not be assembled we tried alternative instrument measures that could serve as indexes/proxies for the tax rates. For fuel tax collections, for instance, τ^{exc} was tried (but was not found significant). The indirect/direct tax ratio was inserted in the T^{stp} equation; it acquires a negative sign but is not significant. The sales tax rate (calculated simply as T^{sle}/adjusted Y) is not an entirely satisfactory measure (see Appendix A). It is difficult to say how much of the influence of income changes is picked up by this variable. A regression of T^{sle} with τ^{sle}, Y, and Y^{na}/Y^{a} is reported. The regression without τ^{sle}

also gives a good fit. The coefficient of Y is significant in the absence of T^{sle} but not in the presence of T^{sle}. The coefficient of the composition of income (Y^{na}/Y^{a}) has the expected sign for sales tax, but not significantly so. No additional information is provided by the substitution of Y^{na} for Y. The state taxes other than sales tax—i.e., T^{os}—indicate a negative relation with GNP. The $\bar{R}^2$, too, is only 0.32. This is possibly due to the repeal of several fees and prices and nuisance taxes and their replacement by a sales tax, and due to the predominance of property (e.g., rural land) taxes in T^{os}, which are not related to income. The land taxes almost everywhere are notoriously regressive. Land taxes are in need of a drastic reform in Brazil, especially when high growth, rapidly increasing population, highly skewed distribution of land holdings, the ownership of a sizeable proportion of agricultural land by foreigners, and so on, coupled with high inflation make land a lucrative source of capital gains. However, while the yield from land taxes can have zero elasticity with respect to income, there is no reason for its becoming negative. The income coefficient must be picking up nonincome effects. Hence, the coefficient of income for T^{os} can be ignored.

Stamp duties indicate a significant response to the rising share of nonagricultural income in the economy. Social security taxes show significant relationships with nonagricultural income (a positive one) and inflation (a negative one). Municipal taxes give a significant positive relationship with income in the presence as well as in the absence of the municipal tax rate variable, calculated simply as T^{mun}/Y'. No reasonable relationship could be obtained for nontax revenues. Hence, this category was treated as an exogenous variable.

Several variables were tried for public transfers. Eventually only T^{ss} emerges as a significant variable for this regression. Capacity utilization weighted by capital stock gives the right (+) sign, but not significantly so.

Public Expenditures

A set of public expenditure equations is given in Appendix B, Table B.1. These are not of particular interest at this stage, as they are not influenced by contemporaneous influences, and so are treated in this study as exogenously determined. However, when the dynamic influences are considered (when, therefore, the predetermined lagged values of the endogenous variables, which are included among the explanatory variables of these equations, also become endogenous), the public expenditures, by implication, will also become endogenous variables. What might be of interest at this stage are their elasticities with respect to lagged GNP and, for some purposes, their differential rates of growth over time. The elasticities are 0.84 for E^g, 0.48 for H^g, and 1.54 for D (as calculated from the linear multiple regressions of Table B.1). The simple geometric time rates of growth are given in Table B.2 of Appendix B.

It will be seen that the major expansion has doubtless occurred in transfer payments at almost all levels of government, though the bulk of transfers has naturally been handled by the autarquias. Two of the other noteworthy features of public expenditures may be underlined. First, the expansion of public expenditure has been more labor intensive than material intensive, at all levels of government. The most impressive of such changes is in the federal expenditures: a 3 percent per annum rise in wages and salaries against only a 0.5 percent per annum increase in goods and services. The same percentages are, respectively, 4 and 2 percent for the autarquias, about 3.3 and 1.2 percent for the state governments, and about 4 and 2.4 percent for the municipal governments. Various factors can be pinned down for these differentials: organizational measures, economy drives, material-saving technological improvements, relatively higher productivity growth in the private sector (and thereby lower prices and costs of materials to the government), scale economies, and so on. A substantially wider gap between the rates of growth of the two categories at the federal level suggests, however, that the most important of all causes might be the economy drive to combat inflation, the major onus of which traditionally falls on the central authority. It is usually more difficult for a government to reduce (in Brazil, the expansion of) its personnel staff than to cut down (the expansion of) material cost. It may be remarked that the time rates of growth per se do not give the causes of growth; they merely measure the end result. The causes are suggested by the multiple regressions. It will be seen from Table B.1 that C^g and W^g are highly competitive expenditure categories. The W^g expenditure is more consistently related to government's disposable income (T_{-1}^{disp}), while C^g fails to indicate such relationships (see Regressions B.1, B.2). The latter category is more closely related to deficit finance (Regression B.2). This is added evidence that the ax of disinflationary measures falls more effectively on C^g than on W^g.

Another feature of public expenditure that emerges from the present analysis is that the autarquias, the state governments, and the municipal governments have systematically added to the public saving at an almost identical rate of approximately 4 percent per annum. The federal government has been making a negative contribution. Likewise, the federal government is also a laggard in the investment field. It is possible that the federal government has increasingly passed on its erstwhile saving and investment functions to its autarquias. If such is the case, then it also explains, in part, the observed sharp reduction in the relative share of its expenditure on goods and services, since the expenditures of the public autarquias, in comparison with those of the general government, are likely to be more material- (e.g., investment) than labor-intensive.

NOTES

1. For an extensive exploitation of this method to explain the effects of monetary variables on growth, see Wallich (1970).

2. See Chapter 3, Note 7, above. The data on average wage rate is hard to come by. Apart from the evidence that the average wage rate is trailing the falling minimum real wage rate available in Conjuntura Economica, 24, no. 2 (1970), certain indirect calculations also support the general impression to the same effect. See, for example, Maneschi (1971).

3. The least-squares estimates are likely to give more acceptable results whenever the variables are highly multicollinear or the fit is poor due to specification errors. There are indications, for instance, "that full information estimates are very sensitive to specification error and likely to be seriously affected by its presence; it is not yet known if the three-stage-least-squares method is equally liable to be affected by specification error" (Johnston, 1963, p. 268). Similarly, in presence of multicollinearity the ordinary least squares prove to be the most "robust" (again, see Johnston [1963, p. 274]).

CHAPTER

5

THE RESPONSIVENESS OF TAX YIELDS

For the resource-mobilization objective of this study the measures of the responsiveness of tax yields are directly relevant. The main concern in the developing countries has, indeed, been more with the tax elasticities than with built-in flexibilities of income.[1] The latter measures have been of primary interest in the developed countries.[2] Of course, one measure follows from the other. The main emphasis on the one measure rather than the other in the two worlds signifies the developing countries' primary concern with resource mobilization through taxation and the developed countries' primary concern with stabilization. In the present chapter we analyze the responsiveness of taxes with respect to income. The responsiveness of income with respect to taxes or the built-in flexibility of income will be taken up in Chapter 7.

SOME SUPPLEMENTARY REGRESSION ANALYSIS: A COVARIANCE MODEL

Before turning to the analysis of tax responsiveness, however, some additional regression analysis for taxes is needed for several reasons. First, the presence of serial correlations in some of the tax equations left the regression analysis of a few incomplete in Chapter 4—e.g., in the case of the personal income tax. One way to get around this problem is to use the cross-section data. Second, due to the somewhat low significance levels of a few of the important coefficients of taxes and such in Chapter 4, the tests of the "robustness" or at least of the approximate orders of magnitude of the relevant coefficients seem desirable. Alternative samples may provide such tests. Third, the tax rate series have been assembled and used for the present purpose for the first time in Brazil. It would be presumptuous to claim that the accuracy of these series cannot be improved, but a margin of error should be allowed here. Alternative tests of the automatic responsiveness of taxes, if they can be devised, are therefore warranted. It is hoped that a covariance analysis of a combined time-series and cross-section sample, which is proposed here, will provide the indicated alternative estimates. Last, for a few of the taxes the tax rate series could not be assembled: for those taxes all that could be done in Chapter 4 was to estimate their buoyancies rather than automatic or built-in responses. An attempt will be made in this chapter to net out the built-in components by carrying out a kind of covariance analysis.

The new sample referred to in the preceding paragraph consists of 12 cross-sections (1949-60) for 21 states for those taxes for which state-level data are available. These data are pooled, and the regressions are run by introducing dummies for the years and the states. Formally, the following model was fitted to the taxes concerned:

$$(49)\quad T_{jit} = Y_{it}^{s_{j1}}\left(\frac{Y^{ind}}{Y}\right)_{it}^{s_{j2}} \exp\left(s_{j00} + \sum_{i=2}^{m} s_{j0i}D_i + \sum_{t=2}^{n} s_{j0t}D_t + U'_{jit}\right)$$

where "exp" stands for the exponential value (e = 2.71828). The subscripts j, i, and t stand, respectively, for tax j, state i, and year t. Although this is not a straightforward covariance method, the use of the multidimensional dummies, nevertheless, provides a standard covariance analysis.[3] This equation, therefore, is frequently referred to in this study as a covariance-regression model. The D_i and D_t are the dummy variables for i states and t years, respectively, which take the value of unity for the ith state and tth year as the case may be and zero otherwise. The coefficients s_{j1} and s_{j2} are the slope elasticities, and s_{j00}, s_{j0i}, s_{j0t} are the general constant, the state correlates, and the year intercepts, respectively. The U'_{jit} are the error terms that are assumed to have zero mean and constant variance, as well as to be uncorrelated with their past values. Since a general constant term—i.e., s_{j00} (whose dummy variable takes the value of unity for all observations)—has been included, one of the states and the first year of the sample series have been omitted from the set of dummy variables to avoid singularity. This equation will be estimated by ordinary least squares.

The main purpose of using this model is to separate the effects of discretionary or other autonomous measures. It is expected that the year intercepts will pick up the effects of changes in tax rates, improvements in the administrative efficiency of tax collection, changes in taxpayer behavior due to changes in the popularity of the government, say, and similar autonomous changes. The real explanatory variables, particularly the income variable, will thus reflect the built-in components of the given tax, net of discretionary changes.

Likewise, the state correlates are supposed to pick up similar differences among the states. In addition to this main conceptual merit, the covariance-regression method also has certain statistical merits.[4] For instance, it is supposed to reduce simultaneous-equations bias. (For a proof, see Hoch [1967].) Since cross-section observations are combined with time-series observations, the typical tendency of most economic time series to move together may be mitigated, possibly reducing multicollinearity.* It is expected that abstracting from the peculiarities of individual classes of the sample would increase the quality and the acceptability of the coefficients. Finally, having obtained the individual state and year effects, there is scope to explain the computed differences by a second-stage analysis by regressing, say, the estimated state correlates on the relevant variables from the states.

A major demerit of the present method, as pointed out by Griliches (1965) is that the introduction of year and state dummies can reduce the available variance of the independent variables, as this procedure leaves one only with the "within" (instead of "between" and "within") state and year variance components. This may cause larger standard errors and, in the case where the variables are measured with errors, a downward bias in the estimated coefficients. Thus, it may reduce the significance levels as well as the magnitudes of the coefficients. While interpreting the results of this method, therefore, this possibility will be kept in mind.

A drawback of the sample of this section in relation to the main sample of the present study is that the data at the state level are rather scarce and, as a result, there is not much scope for applying some of the rather high-powered econometric techniques that were employed in preceding chapters. In any case, the purpose of this model is not to supplant the earlier analysis but to supplement it. The regression results for the southern group of states are presented in Table 1. The results for the northern group of states are also discussed in the text but to economize on space are not reported in full

---

*A conventional method of skirting multicollinearity by the use of pooled time-series and cross-section samples has been to make prior estimates of one or more of the collinear coefficients separately from cross-sections data and to estimate the remainder of the collinear coefficients separately from the time-series data. For example, in the consumption function studies, often the income elasticity of demand is separately measured from the cross-section data and the price elasticity alone is estimated from the time-series data. It is assumed that by doing this the implications of multicollinearity between income and price are avoided (see, for example, Klein [1962, pp. 62-64]). The method used here, on the other hand, estimates the coefficients concurrently and purports to reduce multicollinearity by introducing variations in observations out of the common trend.

in the tables. The divisions of the states into two regions is discussed in the footnotes to Table 1.

The role of the composition-of-income variable warrants a brief discussion. At the aggregate level, this variable was represented by Y^{na}/Y^{a}. At the state level the variable Y^{ind}/Y (the ratio of industrial income to GNP) was considered to be a better index of the income composition. One way of looking at this variable is to regard the variable of real income as divided into two parts: the level and the composition of income. It is hypothesized that the yields of different taxes, and indeed even the patterns of income and expenditure, may differ not only according to different levels of income but also to different compositions of it. For example, the consumption (excise) taxes are levied mostly on industrial commodities. The company income tax is a profit tax that is levied on corporations, which are mostly industrial in nature. Agricultural and industrial incomes are subject to somewhat different income tax rules. Recently agricultural commodities have been exempted even from the sales tax. Tariffs are levied mainly on industry-based goods and very little on agriculture-based goods, as usually it is the manufacturing industry that gets protection while agriculture remains, by and large, unprotected (this is the case at least in Brazil).

Of course, if the changes in the composition of income have a fixed short-run as well as medium-run relationship with changes in the level of income, the separation of the two influences may not be of much operational use. Our model has important short-run characteristics and uses and has to be employed for annual projections. In the short run, however, the composition and the level of income are not likely to depict a strictly fixed relationship. Even if they did, it would still be worthwhile to study the separate influences of the two variables. Moreover, as will be seen when we analyze the coefficients of these variables, in the present sample at least there are certain statistical advantages in including the Y^{ind} variable along with Y. The inclusion of the former, for instance, raised the magnitudes as well as the significance levels of the latter in a majority of the regression equations of this study. We shall return to this point.

The coefficients of income from several alternative specifications of the regression equations are reported. For instance, Table 1 reports the marginal tax yields both from log equations and linear equations. The coefficients of Y with and without dummies and with and without Y^{ind}/Y are also reported.

Although the main focus of the present covariance-regression analysis is on estimating the automatic responses of individual taxes with respect to changes in income (and its composition), it will also be of interest to have a brief look at the state and year effects that are supposed to be separated in this analysis.

TABLE 1

Automatic Responses of Different Taxes as Computed from Pooled Time-Series and Cross-Section Data, Southern States, Brazil, 1949-60

	Federal Tax					
	T^{pi}	T^{c}	T^{exc}	T^{stp}	T^{sle}	T^{mun}
	(1)	(2)	(3)	(4)	(5)	(6)
Constant	-3.682	-2.814	-2.459	-3.054	-.447	-1.167
ln Y'	.736	.653	.677	.688	.661	-.086
	(3.69)	(3.64)	(3.83)	(6.21)	(5.59)	(-.50
ln Y^{ind}/Y	.824	.744	.774	.635	.668	-.163
	(4.27)	(4.29)	(4.53)	(5.92)	(5.36)	(-.91)
Sergipe	.324	.256	.289	-.911	-1.342	-.568
	(2.98)	(2.62)	(3.00)	(-9.98)	(-18.51)	(-.91)
Bahia	1.244	1.344	1.329	.352	-.169	1.838
	(4.02)	(4.82)	(4.85)	(2.05)	(-.86)	(6.90)
M. Gerais	1.157	1.392	1.337	.527	-.629	2.248
	(2.24)	(2.99)	(2.93)	(1.84)	(-2.08)	(5.20)
Rie de Janeiro	1.236	1.393	1.357	-.254	-.251	1.817
	(2.84)	(3.56)	(3.52)	(-1.05)	(-.95)	(4.80)
Guanabara	2.343	2.58	2.522	2.402	.275	1.882
	(3.74)	(4.58)	(4.55)	(6.91)	(.73)	(3.51)
São Paulo	2.290	2.63	2.538	.997	.621	4.331
	(2.70)	(3.43)	(3.38)	(2.11)	(1.22)	(5.95)
Paraná	.980	1.089	1.072	.651	.229	1.355
	(2.97)	(3.67)	(3.67)	(3.55)	(1.16)	(4.78)
S. Catarana	.771	.861	.846	-.425	-.536	.845
	(2.70)	(3.35)	(3.34)	(-2.68)	(-3.05)	(3.36)
Rio G. Sul	1.907	2.088	2.048	.452	.295	2.224
	(3.84)	(4.66)	(4.65)	(1.64)	(.99)	(5.22)
1950	.048	.027	.020	.028	.104	.040
	(0.76)	(0.48)	(.36)	(.80)	(2.24)	(.59)
1951	.098	.134	.125	.214	.257	.005
	(1.54)	(2.36)	(2.23)	(6.08)	(5.53)	(.07)
1952	.137	.121	.121	.309	.248	.141
	(2.08)	(2.05)	(2.08)	(8.47)	(5.20)	(2.06)
1953	.080	.074	.076	.277	.211	.079
	(1.15)	(1.18)	(1.23)	(7.18)	(4.44)	(1.17)
1954	.136	.135	.135	.291	.318	.143
	(1.91)	(2.11)	(2.15)	(7.39)	(6.33)	(1.99)
1955	.092	.056	.068	.418	.371	.365
	(1.13)	(.77)	(.95)	(9.27)	(6.70)	(4.60)
1956	.139	.106	.120	.452	.398	.382
	(1.85)	(1.46)	(1.69)	(10.13)	(7.23)	(4.84)
1957	.289	.244	.260	.447	.523	.551
	(3.31)	(3.11)	(3.36)	(9.23)	(8.89)	(6.55)
1958	.265	.277	.274	.436	.545	.557
	(2.68)	(3.12)	(3.13)	(7.94)	(8.27)	(5.91)
1959	.231	.298	.277	.484	.621	.527
	(2.22)	(3.18)	(3.00)	(8.36)	(9.16)	(5.43)
1960	.349	.465	.428	.539	.724	.617
	(2.97)	(4.39)	(4.12)	(8.26)	(9.55)	(5.69)
R^2	.995	.996	.996	.998	.995	.987
Marginal yield 1	.009	.016	.025	.008	.0364	-.0009
	(3.58)	(3.54)	(3.88)	(6.15)	(5.26)	(-.47)
Without dummies and Y^{ind}/Y						
ln Y'	1.563	1.458	1.471	1.399	1.152	.971
R^2	.923	.935	.912	.895	.906	.856

NOTE TO TABLE 1

Entries in parentheses are the t values obtained by taking the ratios of the respective coefficients to their standard errors, giving two-tail t tests of the coefficients against zero. The t values of the dummy coefficients are computed similarly and are reported as they came out in the computer output. However, since the dummy coefficients make sense in relation to one another, they should be tested against one another and not against zero. The reported t values of the dummy coefficients, therefore, are not entirely relevant. In order to make the suggested test, the reported t values have to be divided into the respective dummy coefficients to get the standard errors. The t test is then given by the following formula:

$$t = \frac{s_{j01} - s_{j02}}{(\mathrm{Var}_{j01} + \mathrm{Var}_{j02} - 2\,\mathrm{Cov}[s_{j01}, s_{j02}])}$$

with twice the degrees of freedom of the regression concerned. Here, s_{j01} and s_{j02} are any two dummy coefficients of a given regression. Since the covariance terms are not reported in this study, approximate t ratios may be obtained by taking the sum of the variances only and ignoring the covariance terms of the denominator.

The new variable Y^{ind} of this table stands for income originating in the industrial sector.

The 21 states of the country were divided into two groups to identify broadly homogeneous regions. The relatively less developed and less commercialized states of the north, the northeast, and the middle west of Brazil were grouped as one set of states and are interchangeably referred to as "the northern states" or "the less-developed region." The relatively more commercialized, coastal, or more developed states of the east and the south of Brazil are put in the second group and are referred to as "the southern states" or "the more-developed states." The state of Espírito Santo was included in both groups for the following reasons: Since one of the state dummies has to be omitted to prevent the matrix from becoming singular, and since the state for which the dummy variable is suppressed becomes the base from which the other state correlates have to be measured, Espírito Santo was considered to provide a common base of comparison for both sets of states; moreover, it is in some ways (e. g., on the basis of per capita income) a median state of Brazil. The number of observations in the pooled cross-section and time-series data is 144 for the northern states and 120 for the southern states. The detailed regression results for northern states are not reported here, but their summary results are given in later tables.

The state totals of income tax series were divided into personal- and corporate-tax series on the basis of the yearly national ratios for the corresponding series. This is not an unobjectionable measure, but alternative procedures would have been equally arbitrary.

THE YEAR EFFECT

The main discretionary measures are expected to appear in the year intercepts rather than state intercepts, particularly in the case of the federal taxes, which are, by and large, uniformly levied all over the country. Three most important discretionary effects that are supposed to be picked up by the year intercepts are changes in the efficiency of tax collection, changes in the public's compliance (i. e., tax avoidance and tax evasion), and changes in tax rates and revision of the tax bases over the years. In the procedure adopted to net out the automatic responses of personal income tax the third influence is accounted for. Since, in the regressions for which the dummy coefficients are reported, the dependent variables (T_j) are entered in (natural) logs, if the year-dummy coefficients are multiplied by 100, they will measure percentage changes in tax collections over the base period. It may, thus, be seen that the discretionary changes of the sample period have caused significant declines only in the yields of import duties. The year coefficients of the sales tax also acquire negative signs when the regression is linear, but they are not significant. In no case are the changes smooth like a trend. The changes have been, by and large, continually upward, for the consumption tax in the southern states (but not in the northern states), for stamp duties in both groups of states, for federal direct taxes in southern states, and for state sales taxes in all states. In the rest of the tax categories the changes have been intermittent and unsystematic. As compared with the base periods, the overall federal tax collections were, due to discretionary changes, higher by about 32 percent in the southern states and about 22 percent in the northern states; the overall state tax collections were higher by about 38 percent in the southern states and 42 percent in the northern states; the overall local tax collections were higher by about 63 percent in both groups of states. Thus, roughly, from one-third to two-fifths of the increase in tax collections in Brazil can be attributed to the discretionary measures. It is probable that these effects would be picked up by the income variables in the buoyancy estimates.

THE STATE EFFECT

It will be seen that most of the state intercepts are significantly different from zero. Since these intercepts are dispersed between negative and positive signs, their significance levels, when they are tested against one another (which is the relevant test), rise in most cases. But they also decline in some others. The state-dummy coefficients may be interpreted as follows: when the differences between two dummy coefficients are multiplied by 100, they give percentage differences in tax collections between the respective states. For example, taking an extreme case, for T^{stp} in the southern states group the state coefficient for Sergipe in -0.911 with a t value reported for a test against zero of -9.98 and standard error of about 0.09. The state coefficient for the same tax for Guanabara is 2.402 with a standard error of 0.034. The covariance between the two coefficients is approximately 0.012. The stamp tax collections not explained by income and the composition of income, thus, as (2.402 - [-0.911])100 = 331 percent higher in Guanabara than in Sergipe, with the t ratio for this difference being 29.0, which is significant at the .05 percent level. Incidentally, similar (though much lower) differences between these two states exist for nearly all taxes. This is, however, an extreme case: Sergipe is on the one extreme with the lowest tax yield for a given income base, and Guanabara is on the other extreme with almost the highest tax yield for a comparable base. Significant differences also exist among several other states. In the case of state taxes, Espírito Santo (represented by the general constant term) seems to be doing better than all other states in the southern region and most states in the northern region. In regard to local taxes, São Paulo among the southern states and Pernambuco among the northern states appear to come out at the top. In the case of personal-income-tax collections, holding the effects of the levels and the composition of income constant, we can rank the states in a probable descending order of tax collections as follows: Guanabara, São Paulo, Rio Grande do Sul, Bahia, Rio de Janeiro, Minas Gerais, Paraná, Santa Catarina, Sergipe, and Espírito Santo, among the southern group of states; and Pernambuco, Pará, Ceará, Alagoas, Piauí, Rio Grande do Norte, Amazonas, Goiás, Maranhão, Mato Grosso, and Espírito Santo (repeated) among the northern states. Similar probable ranking in the cases of the other taxes may be read from the table. The emergence of states like São Paulo, Guanabara, and Rio de Janeiro at the top in tax collections for a given income base is, in part, explainable in terms of the traditional tendency of administrations to concentrate their tax-collection efforts on the most lucrative regions and big tax-paying groups. The cost of enforcing tax collection efforts on widely dispersed, small, and relatively poor taxpayers is usually disproportionately high. The inequity in tax collections, however, is evident, and perhaps unavoidable. This kind of inequity is usually not visible and in the consciousness of the legislators and the tax administration. The transfer of the (potentially or actually) collected tax revenues from the relatively well-to-do south to the poor north through the official government policy, on the other hand, is a widely-known fact.

Looking at T^{sle}, it may be noted that while the year-dummy coefficients are practically unaffected by the omission of the Y^{ind}/Y variable from the regression, the state-dummy coefficients increase almost without exception both in magnitudes and in statistical significance. This is a typical result. The same tendency persists in the regressions of almost all other taxes (not reported here to economize on space). This result gives a strong suggestion that the composition-of-income variable belongs to the regression, and that when it is not included, its positive effect is picked up by the state dummies. Since many states differ in their composition of income, this is, indeed, what one would expect to be the result.

ELASTICITIES AND MARGINAL TAX YIELDS

The elasticity measures give a quick view of the magnitudes of percentage changes of the tax yields in relation to income. They are useful measures for visualizing the unit-free relationships among the variables. In the cases of individual taxes, however, it is sometimes more meaningful to measure the changes in tax collections in terms of a cruzeiro change in income and other relevant variables. Moreover, it is the marginal tax yields that are to be utilized in calculating built-in flexibilities. Accordingly, marginal tax yields are also calculated. Where these yields are calculated from nonlinearly estimated equations they are evaluated at the mean sample values. Hence, these are denoted "mean" marginal tax yields.

The elasticities and marginal yields of taxes with respect to income are presented in Table 2. It will be seen that the corresponding coefficients of all those taxes for which the tax rate series have been used in the regressions of the macro model are closer to each other in the two models than the rest of the taxes. In general, the corresponding coefficients of the covariance model are almost consistently lower than those of the macro model; but they are less far below for direct taxes (mainly because the effects of their tax rate changes have largely been held constant) than the indirect taxes (for most of which tax rate changes could not be held constant). Indeed, the differentials of the corresponding pairs of the two sets of coefficients are narrower in these two models than those observed in the alternative algebraic (log vs. linear) forms of the same model (compare the primed and nonprimed coefficients in column 2 of Table 2). Of course, the buoyancy elasticities (i. e., those based on gross tax-yield changes including the effects of discretionary measures and such) are much higher than the built-in elasticities of either model. (The buoyancy estimates for the state-level data are given in the bottom lines of Table 1.) The marginal values naturally depict a similar picture except that discernible differentials exist between the corresponding coefficients in the two regions of the state-level sample.

The built-in elasticities as measured by Y alone (when Y^{ind}/Y is also in the regression) may understate the overall effects of income on tax yields, as the changes in the levels of income might not, and in the developing countries usually do not, come without some changes

TABLE 2

"Mean" Marginal Tax Yields and Tax Elasticities with Respect to Income

Tax	Elasticity		Marginal Yield	
	Main Model	Covariance Model	Main Model	Covariance Model
(1)	(2)	(3)	(4)	(5)
T^{pi}	.89e (±.33)	.736	.0069	.009s .008n
T^{c}	.65e (±.60)	.635	.013	.061s .019n
T^{exc}	.86e' (±.59)	.677	.020	.025s .026n
T^{im}	.85e (±.39) 1.58e' (±.86)	..	.011 ..	..
T^{stp}	1.22b (±.30)	.688	.012	.011s .005n
T^{ss}	3.54b (±.46)c 2.03b' (±.20)c	..	.092 ..	..
T^{fue}	2.21b (±1.06) 1.77b' (±1.03)	..	.025 ..	..
T^{sle}	1.62b (±.34)c	.661	.030	.0364s .0254n
T^{mun}	.96b (±.303) .71b' (±.20)	0.00	.0105 ..	0.00 0.00
Mongrel T	1.034e'(±.4)	..	.319	..

Note: The symbols in the body of the table stand for the following: e = built-in elasticity; b = buoyancy elasticity; the primed e and b denote the values from the log equations; and e and b without primes denote the values from the linear equations. All coefficients are with respect to GNP, except for those for T^{pi} and T^{c}, which are with respect to GDP, and those marked by the symbol c (specifically T^{ss} and T^{sle}), which are with respect to nonagricultural income. The first entry of column 5, marked by s, is for the southern states, and the second entry, marked by n, is for the northern states. The "±" values in parentheses in column 2 stand for "± 2 standard errors."

In the cases of the log equations, the marginal tax yields were computed by taking the products of the computed elasticities and the corresponding (geometric) mean values over the sample period. Hence, they are denoted as "mean" marginal yields. This is also the interpretation of the coefficients of the linear equations. The marginal yield of a given tax with respect to income is in cruzeiros, per cruzeiro change of income per year. The marginal tax yields with respect to other variables have similar interpretation.

in the composition of income. As a rule, built-in changes in tax yields should include all changes dependent upon income, whether they are caused by changes in the levels of income, in its composition, or its geographic shifts. The composition of income changes relatively slowly. (Note that this variable is entered in the regressions of Table 1 as the log of the ratio of industrial income to overall income.) The coefficients of this variable are highly significant in the relatively developed and fast-growing southern states, but are scarcely significant in the nonindustrial and (the erstwhile) stagnant states of the northern region. On the other hand, when this variable is left out, the coefficients of Y become very low and statistically insignificant for all taxes in the southern states, while the cceefficients of Y in the northern states remain practically unaffected both in magnitudes and significance levels. For one thing, the intercorrelation between Y and Y^{ind}/Y is not very high: it is 0.229 in the log form and 0.154 in the linear form for the northern states, and 0.522 in the log form and 0.470 in the linear form for the southern states. No problem of multicollinearity between these variables, therefore, can be expected to raise its ugly head. For the northern states, the interpretation is simple: when the coefficient of a variable is not significantly different from zero and is not highly correlated with the other variables, its omission (which under the stated conditions is not unjustified) is unlikely to alter the coefficients of the "included" variables significantly. The case of the northern states is interesting. When both variables are in the regression, all of the Y^{ind}/Y coefficients are significantly different from zero at the .5 percent level, and all of the Y coefficients, except for T^{im}, are also significant at the .5 percent level. On the other hand, when the variable Y^{ind}/Y is left out, none but one coefficient of Y (the exception being for T^{stp}) remains significant, even at the 10 percent level. A possible explanation of this interesting twinship between these two variables is mainly statistical: When one of these two variables is left out, the unexplained variance is inordinately large, increasing the standard error of the coefficient of the included variable. On the other side, as discussed in the course of the specification of the model, the inclusion of the dummy variables—a score of them here—removes the "between" variations and leaves only the "within" variations to be explained by the real variables—the solitary variable Y here. When these dummy variables are left free to pick up the effects of the left-out variable Y^{ind}/Y, they tend to become strong enough even to rob the included variable Y. The inclusion of the complementary real variable (Y^{ind}/Y here) seems to lend strong support to the other partner, eliminating the indicated weakness of estimates. Another explanation may be more econometric: If the left-out variable belongs to the regression and is correlated with one of the included variables, then if the indicated included variable changes more rapidly than the left-out variable, the coefficient of the former is likely to be underestimated (see Griliches [1957]). Thus insofar as the covariance model is concerned, the regressions that include the variables Y, Y^{ind}/Y, and the dummies (reported in detail in the tables under reference) appear to be decidedly the best.

Accordingly, we return to consider the aggregate effects of changes in the level and composition of income. We have just seen that they cannot be computed by omitting one of them. On the other hand, there is no way of aggregating the two variables themselves without first computing their coefficients, as they are measured in different units. After their separate coefficients are known, however, we can aggregate the variables by using the computed coefficients as their (geometric) weights. That amounts to inflating the base—i.e., income—by adding to it a weighted change in the composition of income. But that still does not give us a composite built-in elasticity that we can compare with the alternative measures. An approximate measure of the aggregate elasticity may be to add them by using the relative changes in the two variables as the weights. During the reference period 1949-60 of the covariance model, for instance, real income (Y) increased by about 84 percent and its composition (Y^{ind}/Y) rose by about 15 percent. In other words, on an average, a one percent increase in income was associated, roughly, with a .18 percent rise in Y^{ind}/Y. This is, however, an average measure: the association may not hold for individual years. Nevertheless, assuming the observed average relationship to hold, we may get an approximate measure of the aggregate elasticities of Y and Y^{ind}/Y by using the (simple) weights of 100 and 18, respectively. This was done, and the calculations, along with others to be compared, are presented in Table 3.

It will be seen from Table 3 that the measures of the aggregated built-in elasticities are from 15 to 20 percent higher than the elasticities for income levels alone. Even then, however, the elasticity coefficients hardly exceed unity. We may now look at the combined results of Tables 2 and 3. Defining the taxes with an elasticity of greater than one as <u>elastic</u> and those with an elasticity of smaller than one as <u>inelastic</u>, it may be remarked that all taxes are highly elastic according to the buoyancy measures (columns 2 and 6 of Table 3 and the values followed by symbol "b" in column 2 of Table 2). All taxes, except T^{im}, are inelastic according to the estimates of automatic responsiveness when the income levels alone are considered (columns 3 and 7 of Table 3 and the values followed by symbol "e" in column 2 of Table 2); nearly all taxes are inelastic (though they are not too far below the proportionality level) even when the composite effects of changes in the level and the composition of income are measured. The magnitudes of the differences between the computed buoyancies and the computed built-in elasticities are indicated by their ratios given in columns 5 and 9 of Table 3. Roughly speaking, the buoyancies are from 50 to 100 percent higher than the built-in elasticities for a majority of the federal and state taxes. In the case of the local taxes (only the overall local tax category is

TABLE 3

Certain Elasticity Measures from the Covariance-Regression Model Compared

	Elasticities							
	Southern States				Northern States			
		Built-in				Built-in		
Tax (1)	Buoy-ancy (2)	y Alone (3)	Y and Y^{ind}/Y Combined (4)	2/4 (5)	Buoy-ancy (6)	y Alone (7)	Y and Y^{ind}/Y Combined (8)	6/8 (9)
T^{pi}	1.563	.736	.884	1.77	1.275	.583	.582	2.19
T^{c}	1.458	.635	.787	1.85	1.295	.569	.589	2.20
T^{exc}	1.471	.677	.816	1.80	1.610	.475	.578	2.78
T^{stp}	1.399	.688	.802	1.74	1.043	.771	.818	1.27
T^{sle}	1.152	.661	.781	1.48	1.002	.847	.875	1.14
T^{mun}	.971	-.086	-.115	..	.977	-.274	.273	..

Source: Table 2.

analyzed in this section), the automatic responsiveness is negative, though, as may be checked from Table 1, the coefficients are not significantly different from zero. In other words, the local taxes seem to be unresponsive to incomes. The buoyancy elasticity value of nearly one of T^{mun} is apparently due to tax-rate changes, population growth, and similar factors.

It is interesting to note the ranking of various taxes according to the calculated built-in elasticities vis-à-vis the corresponding ranking according to the computed buoyancies. In determining these ranks, we took averages of the alternative regression coefficient from the main model and the covariance-regression model. According to the built-in elasticity the taxes are ranked as follows (the elasticity values are noted in parentheses): T^{im} (1.22), T^{pi} (0.83), T^{exc} (0.77), T^{stp} (0.69), T^{sle} (0.66), and T^{c} (0.65). According to the buoyancy elasticity the ranks come out in the following order: T^{ss} (2.78), T^{fue} (1.99), T^{exc} (1.54), T^{pi} (1.42), T^{c} (1.38), T^{sle} (1.26), T^{stp} (1.22), and T^{mun} (0.91).

In the list of taxes noted here, social security taxes and the fuel tax are the most buoyant taxes; and import duties are the most elastic of the taxes. The buoyancy elasticities of all federal and state taxes are greater than unity, while the built-in elasticities of all taxes other than T^{im} are less than unity. Personal income tax is second in ranking in built-in elasticity with an elasticity of 0.83; it ranks fourth in buoyancy terms. The automatic responses of the sales tax, the corporation tax and stamp duties are almost identical; so also are their buoyancies.

An important inference that may be safely generalized from these results is that the direct taxes of Brazil (leaving aside social security contributions) are not elastic, and the indirect taxes are not more inelastic than the direct taxes. This is a paradox not uncommon of many developing countries: namely, the direct taxes, which are supposed to be the most elastic and progressive of the tax structure, are, in fact, inelastic. The indirect taxes, which are traditionally regarded as regressive, turn out to be less inelastic (and as a possible consequence of this presumably also more progressive) than the direct taxes. Even the aggregate tax (T) equation, which is partly a buoyancy measure, turns out to be just about proportional to income changes.

The villains of the piece are evidently personal and corporate income taxes. Thus, in spite of Brazil's having experienced one of the highest rates of growth in the world during the reference period, and in spite of continual increases of tax rates over the years (see Appendix A), the yield from personal income tax has stayed put at roughly 2.2 percent of GDP (see Chapter 2 and Table 2, column 3). A hypothesis that may be formulated from this result is that the major cause of the inelasticity of Brazil's personal income tax has been an increasing rate of tax evasion. This hypothesis is subjected to an empirical verification in Chapter 8. A possible cause of the inelasticity of the corporation tax (in addition to tax avoidance) seems to be the ever-expanding tax incentive schemes in Brazil. In view of the foregoing results, the observed increases in tax revenue shares during the reference period seem to be explainable more in terms of the Brazilian government's discretionary measures than automatic responsiveness.

NOTES

1. See, for example, the studies of Sahota (1961) for India, Yaqub (1967) for Pakistan, Toukan (1968) for Jordan, Ghai (1968) for Uganda, and Lewis (1965) for Pakistan.

2. See, for example, the studies of Musgrave (1959, pp. 502-14), Brown and Cruizenga (1959), Cohen (1960), Pechman (1956), Prest (1961), and Smith (1963). For similar studies that are parts of bigger economic systems, see Ando and Brown (1968) and Ando and Goldfeld (1968) for the United States, May (1965) for Canada, and Balopoulos (1969) for the United Kingdom.

3. For an exposition of this method, see Suits (1957). For some of the shortcomings of this method, see Griliches (1965).

4. For a discussion of the merits and demerits of the dummy method, see Suits (1957), Schipper (1964, especially pp. 9-107), and Griliches (1965).

CHAPTER

6

IMPACT MULTIPLIERS AND BUILT-IN FLEXIBILITIES

IMPACT MULTIPLIERS

The structure of the model analyzed thus far presents the quantitative and qualitative mechanism of the economy according to economic theory, but this gives only an implicit picture: the structural parameters provide tests of only the direct effects of the predetermined variables. To obtain measures of both direct and indirect influences of the predetermined variables the initial source of such influences has to be traced through the set of all endogenous variables back to the predetermined variables. This is done by deriving the reduced forms for each of the endogenous variables in terms of the predetermined variables alone. The reduced forms are useful for prediction and several other analytical purposes. The reduced-form coefficients, when the corresponding variables are expressed linearly in their units of measurement, are called impact multipliers (see also the note on p. 41 above). In some cases, the influences continue beyond one year. For policy purposes the impact and the intermediate-period multipliers are more useful than the long-period equilibrium multipliers of the textbook type. Indeed, a policy maker has always to deal with disequilibrium situations and rarely if at all with an equilibrium situation. The present empirical analysis, therefore, is focused on impact multipliers and short-period dynamic multipliers.

Several of the equations of our model are in nonlinear form, but since the multipliers are expressed in terms of marginal changes, the equations have to be linearized. A standard procedure of linearization is to take the total derivatives of the equations (Goldberger, 1959, pp. 14-20; Ando and Goldfeld, 1968, pp. 281-82); the same is employed here.

The linearization of nonlinearities requires the absolute values of the variables involved in nonlinearities. Which year values ought to be chosen depends on the purpose of the study. When prediction is the main objective, the final-year values of the variables in current prices are more appropriate, but when the purpose is to explain the data, the sample-mean values in constant prices are more relevant. Since the regressions of this study have been computed in constant-price variables, the calculation of total derivatives in current prices is a laborious process. The analysis of the present study, moreover, consists of both the computation of the total partial effects of the existing instrument values and also predictions, which in any case are a real test of the model. In addition, the varying rates of inflation over time may distort the picture if current-price variables are employed. In view of these considerations, the final-year values of the variables in constant (1963) prices (except money supply, which is in current prices) were employed for the evaluation of the total partial derivatives. As an illustration, the linearization of the demand-for-money equation is given:

$$\frac{M-M_{-1}}{M_{-1}} = -.11 + .06g - .164\frac{P-P_{-1}}{P_{-1}} + .00038X_{-1} + .1386\frac{IL-IL_{-1}}{IL_{-1}} \qquad (24.1)$$

$$\begin{aligned} dM = &\left[.89 + .06g_{68} - .164\left(\frac{P_{68}}{P_{67}} - 1\right) + .00038X_{67} + .1386\left(\frac{IL_{68}}{IL_{67}} - 1\right)\right] dM_{-1} + \\ &M_{67}\left[.06dg - .164\frac{1}{P_{67}}dp + .164\frac{P_{68}}{(P_{67})^2}dP_{-1} + .0038dX_{-1} + \right. \\ &\left. .1386\frac{1}{IL_{67}}dIL - .1386\frac{IL_{68}}{(IL_{67})^2}dIL_{-1}\right] = 1.146dM_{-1} + 1.715dg - \\ &.886dP + 1.10dP_{-1} + 1.14dX_{-1} + 1.46dIL - 1.84dIL_{-1} \qquad (47.2) \end{aligned}$$

The calculated coefficients are the impact multipliers. It is to be noted that the variables are now expressed in first differences, the time unit being a year.*

To get a broad idea of the closeness to the original equations of the linear approximation of those equations (in particular the monetary sector equations) in which the nonlinearity was particularly pronounced, the residuals obtained from the linearized equations at two intervals (1964 and 1966) were compared with the corresponding residuals in the original equations. The discrepancy between the two sets of residuals was found to be tolerable, except for M and P. This does not mean, however, that errors that may be introduced by linearization will be equally negligible for the earlier part of the sample or for the system

---

*For prices such as P^{im}, a value of unity was assumed for the evaluation year 1963. Thus, $P^{foreign}_{1968} = 1$; $P^{domestic}_{1968} = 1$. The value of e^{im} for that year was 0.44.

as a whole, though in the light of the indicated results the errors for most of the variables other than M and P are expected to be of a small order.

In interpreting the impact multipliers at least three important "caveats" (Goldberger, 1959, p. 30) must be kept in mind. First, since the impact multipliers record only one-year responses, while the presence of lagged influences may extend over several years, their values will usually be much smaller than the textbook "timeless multipliers." Second, these are measures of potential impact; the actual impact must in addition depend upon the exogenous stimulus, the multiplicand, which can be large or small depending upon the nature of the exogenous variable concerned. For example, the multiplicand (proportionate change, say) of money supply can be quite substantial in Brazil, whereas that of income tax cannot be very large. Finally, these multipliers are derived under ceteris paribus conditions. For example, the effects of a change in public investment are considered while money supply, tax rates, other public expenditures, and all other listed predetermined variables are held constant. In view of these caveats, the signs and the magnitudes of some of the multipliers may on their face look implausible. These can be explained only by looking through the causal flows and the algebraic relationships of the reduced-form coefficients as well as through the complexities introduced by the listed conditions. The "simplicity" of the multiplier of the elementary Keynesian model, however, "is only a superficial one"; and though complex, the present type of multipliers "may be expected to give more precise and more realistic measures of effects in the economic process" (Goldberger, 1959, pp. 30-31).

The calculated impact multipliers are presented in Table 4. Due to the weakness of the rate of interest variable noted earlier, the multipliers for r have not been reported. For the purpose of Table 4 all coefficients of the chosen equations were used except the coefficient of 1-z for Tr, which was assumed to be zero, as indeed it is according to the required statistical significance levels, and except the coefficient of Y' in the equation for T^{sle}, which is also statistically nonsignificant and has the wrong sign. In addition to the relevant identities, the following structural equations were used for the computation of the multipliers: 2.5.5, 4.3.4, 6, 8, 11.2, 13.2, 14.2, 15, 21, 22, 24.1, 27.3, 29.4, 30, 32.1, 33, 34, 35, 36, 37, 38, 39, 40, 41, and 43.

Effects on Growth

It may be observed that a change of one percentage point in personal income tax (the 1968 rate of this tax is 8.56 percent of taxable income) is indicated to cause a decline of approximately Cr$4 million in GNP

TABLE 4

Impact Multipliers $(\pi)^2$

Dep. Var.	Y_{-1} (1)	Y'_{-1} (2)	Y^{na}_{-1} (3)	Y^{a}_{-1} (4)	C^{p}_{-1} (5)	I^{pl}_{-1} (6)	I^{pl}_{-2} (7)
1 Y'	.4425	.2332	.0940	-.0977	.4424[b]	1.5118	-.2624
2 C^p	.4424	.1478	.0940	-.0977	.4424[b]	.9064	-.1573
3 I^{pl}	.0598	-.0481	.0127	-.0132	.1687	.8092	-.1404
4 I^{pe}		.0002					
5 I	.0598	-.0480	.0127	-.0132	.1687	.8092	-.1404
6 I^{stk}	-.0596	.1334	-.0127	.1316	-.1683	-.2037	.0353
7 Y^{na}	.0605	-.1180	.0062	-.0133	.1708	.2067	-.0359
8 Y^a	-.0351	.0683	-.0036	.0077	-.0990	-.1198	.0208
9 Y	.0255	-.0496	.0026	-.0056	.0719	.0870	-.0151
10 Q	.1416	-.1129	.0301	-.0313	.3998	.4838	-.0840
11 W	.0336	.0178	.0071	-.0074	.0949	.1149	-.0199
12 Y^d	.6848	.2287	.1454	-.1512	1.1595	1.4031	-.2435
13 y	.0005	.00026	.0001	-.0001	.0014	.0017	-.0003
14 g	.0032	-.0063	.0007	-.00071	.0091	.0110	-.0019
15 Ex							
16 Im	.0285	-.0229	.0061	-.0063	.0805	.3860	-.0660
17 F	-.0285	.0229	-.0061	.0063	-.0805	-.3860	.0660
18 M	.01023	-.01995	.0022	-.00226	.0289	.0350	-.0061
19 P	-.00409	.00797	-.00087	.0009	-.01154	-.01397	.00242
20 BL	.0019	-.0036	.0004	-.0004	.0052	.0063	-.0011
21 P^{im}	.0018	-.0035	.0004	-.0004	.0051	.0061	-.0011
22 T^{pi}	-.2697	.00014	.00006	-.00006	.0007	.0009	-.0002
23 T^c	.0065	-.0051	.0014	-.0014	.0182	.0221	-.0038
24 T^{exc}	.0100	-.0080	.0021	-.0022	.0284	.0343	-.0060
25 T^{im}	.0049	.0026	.0010	-.0011	.0137	.0166	-.0029
26 T^{stp}	.0011	.0027	.0005	-.0016	.0148	.0180	-.0031
27 T^{ss}	.0001	-.0003	-.0460	-.00003	+.0004	.0005	-.0001
28 T^{fue}	.0111	.0058	.0023	-.0024	.0312	.0378	-.0066
29 T^{sle}	-.0089	.0053	-.0169	.0630	-.0250	-.0302	.0052
30 T^{os}	-.0020	-.0011	-.0004	.0005	-.0058	-.0060	.0012
31 T^{mun}	.0046	.0025	.0010	-.0010	.0131	.0159	-.0028
32 Tr	.00001	-.00002	-.0034	-.0000	-.00003	.00004	-.00001
33 T^{disp}	-.2422	.0045	-.0515	.0535	.0899	.1088	-.0189
34 T	-.2422	.4473	-.0548	.0535	.0899	.1088	-.0189
35 DEF	.2422	-.0045	.0514	-.0535	-.0899	-.1088	.0189

I^{pe}_{-1} (8)	I^{stk}_{-1} (9)	$(1-Z)_{-1}$ (10)	$(1-Z)_{-2}$ (11)	Q_{-1} (12)	Q_{-2} (13)	Y^{d}_{-1} (14)	g_{-1} (15)
.0022	.6489	172.1735	-85.9618	-.0378	.0002	.0000[b]	.00009
.0013	.3834	103.2213	-51.5358	-.0378	.0002	.0000[b]	.00009
.0003	.1199	92.1500	-46.0082	-.0051	.00003	-.1090	.00001
.0009							
.0012	.1199	92.1500	-46.0082	.0051	.00003	-.1090	.00001
-.0003	.1457	-23.1979	11.5821	.0051	-.00003	.1088	-.00001
.0003	.0887	49.1274	-11.7553	-.0005	.00003	-.1104	-9.3632
-.0002	-.0514	-28.4568	6.8092	.0030	-.00002	.0639	5.4236
.0001	.0373	20.6706	-4.9461	-.0022	.00001	-.0461	-3.9396
.0007	.2829	55.0976	27.5088	.4939	-.0029	-.2583	-.0012
.0002	.0493	13.0815	-6.5312	-.0029	.00002	-.0613	.00001
.0021	.5934	159.7853	-79.7767	-.0585	.0004	-.7491	.00014
.0000	.0007	.1893	-.0945	-.00004	.00001	-.0009	.0000001
.00002	.0047	1.2569	-.6275	-.00028	.000002	-.0059	.0000007
.0006	.0572	43.9556	-21.9458	-.0024	.00001	-.0520	.00001
-.0006	-.0572	-43.9556	21.9458	.0024	-.00001	.0520	-.00001
.00005	.01501	3.9816	-1.9879	-.0009	.00001	-.0187	.000002
-.00002	-.00599	-1.5905	.79407	.00035	-.000002	.00746	-.0000001
.00001	.0027	.7214	-.3602	-.0002	.000001	-.0033	.0000004
.00001	.0026	.6998	-.3493	-.0002	.000001	-.0033	.0000004
.000001	.0004	.1006	-.0502	-.00002	+.0000001	-.0005	.0000001
.00003	.0129	2.5125	-1.2544	.0225	-.00013	-.0118	-.00005
.00005	.0201	3.9074	-1.9509	-.0009	.000005	-.0183	.000002
.00002	.0071	1.8939	-.9456	-.0004	.000002	-.0089	.000001
.00003	.0070	2.0455	-1.0213	-.0001	.000001	-.0096	.0000002
.0000	.0002	.0006	-.0278	-.00001	.0000	-.0003	.0000
.0001	.0162	4.3043	-2.1490	-.0010	.00001	-.0202	.0000022
-.0001	-.0130	-3.4434	1.7192	.0008	-.00001	.0161	-.000002
-.00001	-.0030	-.7920	.3954	.0002	-.000001	.0037	-.0000004
.00002	.0068	1.8078	-.9026	-.0004	.000002	-.0085	.000001
.0000	.00002	.0041	-.0020	-.0000	.0000	-.00002	-.0000
.0002	.0555	12.3928	-6.1851	.0207	-.0001	-.0581	-.00005
.0002	.0555	12.3928	-6.1871	.0207	-.0001	-.0581	-.00005
-.0002	-.0555	-12.3881	6.1851	-.0207	.0001	.0581	.00005

(continued)

TABLE 4, continued

Dep. Var.	M_{-1} (16)	M_{-2} (17)	M_{-3} (18)	P_{-1} (19)	P_{-2} (20)	P_{-3} (21)	BL_{-1} (22)
1 Y'	.00167	-.00101	-.00044	.00809	.01064	.0044	-.0055
2 C^{p}	.00167	-.00101	-.00044	.00809	.01063	.0044	-.0033
3 I^{pl}	.00023	-.00014	-.00006	.00109	.00144	.00059	-.0030
4 I^{pe}							
5 I	.00023	-.00014	-.00006	.00109	.00144	.00059	-.0030
6 I^{stk}	-.00023	.00014	.00006	-.00109	-.00143	-.00059	.0007
7 Y^{na}	.00007	-.00004	-.00002	.0003	.0005	.0002	-.0008
8 Y^{a}	-.00004	.00003	.00001	-.0002	-.0003	-.0001	.0004
9 Y	.00003	-.00002	-.00001	.0001	.0002	.00008	-.0003
10 Q	.00054	-.00032	-.00014	.00259	.0034	.0014	-.0018
11 W	.00023	-.00014	-.00006	-.00262	.00117	.0062	-.0004
12 Y^{d}	.00260	-.00157	-.00068	.01252	.01646	.00681	-.0051
13 y	.000002	-.000001	-.000001	.000009	.00001	.00001	.000006
14 g	.000012	-.000007	-.000003	.00006	.00008	.00003	-.00004
15 Ex							
16 Im	.0001	-.00007	-.00003	.0005	.0007	.0003	-.0014
17 F	-.0001	.00007	.00003	-.0005	-.0007	-.0003	.0014
18 M	.8215	.0311	.0137	-1.2824	-.1806	-.1398	-.00013
19 P	.0313	-.0189	-.0082	.3434	.1070	.0823	.00005
20 BL	.00001	-.000004	-.000002	.00003	.00005	.00002	.0750
21 P^{im}	-.0138	.0083	.0036	-.1511	-.0471	-.0362	-.00002
22 T^{pi}	-.000003	-.000002	+.000001	.00018	.00044	-.000007	-.000003
23 T^{c}	.00002	-.00001	-.00001	.00012	.00016	.00006	-.00008
24 T^{exc}	.00004	-.00002	-.00001	.0002	.0002	.0001	-.0001
25 T^{im}	.00002	-.00001	-.000005	.00009	.00012	.00005	-.00006
26 T^{stp}	.000003	-.000002	-.000001	.00002	.00002	.000008	-.00007
27 T^{ss}	-.0011	.0007	.0003	-.0055	-.0065	-.0029	-.000002
28 T^{fue}	.00004	-.00003	-.00001	.0002	.0003	.00011	-.0001
29 T^{sle}	-.00003	.00002	.00001	-.0002	-.0002	-.00009	.0001
30 T^{os}	-.00001	.00001	.000002	-.00004	-.00005	-.00002	.00003
31 T^{mun}	.00002	-.00001	-.000005	.00008	.0001	.00005	-.00006
32 Tr	-.00008	.00005	.00002	-.0004	-.0005	-.0002	-.0000
33 T^{disp}	-.00092	.0006	.00024	-.0044	-.0058	-.0024	-.0004
34 T	-.0010	.0006	.0003	-.0048	-.0063	-.0026	-.0004
35 DEF	.0009	-.0006	-.0002	.0044	.0058	.0024	.0004

BL_{-2} (23)	P^{im}_{-1} (24)	T^{pi}_{-1} (25)	T^{exc}_{-1} (26)	T^{ss}_{-1} (27)	T^{sle}_{-1} (28)	IL (29)	IL_{-1} (30)
.0020	-.0183	-.8195	-.8006	-.7606	-.8007	.0008	-.0010
.0012	-.0183	-.8192	-.8004	-.7603	-.8004	.0008	-.0010
.0011	-.0025	-.1107	-.1081	-.1027	-.1081	.0001	-.0001
.0011	-.0025	-.1107	-.1081	-.1027	-.1081	.0001	-.0001
-.0003	.0025	.1104	.1079	.1025	.1079	-.0001	.0001
.0003	-.0008	-.1121	-.1094	-.0315	-.1094	.00003	-.00004
.0002	.0004	.0649	.0634	.0182	.0634	-.00002	.00002
.0001	-.0003	-.0472	-.0461	-.0132	-.0461	.00001	-.00002
.0006	-.0058	-.2622	-.2562	-.2433	-.2562	.0002	-.0003
.0002	-.0026	-.0623	-.0608	-.0578	-.0608	.0001	-.0001
.0019	-.0283	-1.2682	-1.2391	-1.1770	-1.2391	.0012	-.0015
.000002	-.00002	-.0009	-.0009	-.0008	-.0009	.000001	-.000001
.00002	-.0001	-.0060	-.0058	-.0056	-.0058	.000006	-.000007
.0005	-.0012	-.0528	-.0516	-.0490	-.0516	.00005	-.00006
-.0005	.0012	.0528	.0516	.0490	.0516	-.00005	.00006
.00005	.5747	-.0190	-.0185	-.0176	-.0186	.7053	-.8866
-.00002	-.0034	.0076	.0074	.0070	.0074	.0143	-.0181
.00001	-.0001	-.0034	-.0034	-.0032	-.0034	.000003	-.000004
.00001	.1504	-.0033	-.0033	-.0031	-.0033	-.0063	.0080
.000001	+.00003	+.4986	-.0005	-.0004	-.0005	.0000	-.000002
.00003	-.0003	-.0120	-.0117	-.0111	-.0117	.00001	-.00001
.00005	-.0004	-.0186	.4818	-.0173	-.0182	.00002	-.00002
.00002	-.0002	-.0090	-.0088	-.0084	-.0088	.00001	-.00001
.00002	-.00003	-.0021	-.0135	-.0014	-.0135	.000001	-.000002
.0000	.0220	-.0003	-.0003	.4998	-.0003	-.0005	.0006
.00005	-.0005	-.0205	-.0200	-.0190	-.0200	.00002	-.00002
-.00004	.0004	.0164	.0160	.0152	.5160	-.00002	.00002
-.00001	.00008	.0038	.0037	.0035	.0037	-.000004	.000005
.00002	-.0002	-.0086	-.0084	-.0080	-.0084	.00001	-.00001
.0000	.0009	-.00002	-.00002	.0365	-.00002	-.00004	.00005
.0001	.0100	.4487	.4384	.4164	.4386	-.0004	.0005
.00014	.0109	.4487	.4384	.4529	.4384	-.00046	.00058
-.0001	-.0100	-.4487	-.4384	-.4164	-.4384	.0004	-.0005

(continued)

TABLE 4, continued

Dep. Var.	X_{-1} (31)	$[\Sigma I/\Sigma K]_{-1}$ (32)	N (33)	K (34)	GNP^{US} (35)	M_1/M (36)	NRA (37)
1 Y'	.0006	-72.4107	5.6428	-.0545		-1.7653	-2.4989
2 C^p	.0006	-72.3888	5.6411	-.0545		-1.7648	-1.4981
3 I^{pl}	.0001	-9.7782	.7620	-.0074		-.2384	-.3374
4 I^{pe}							
5 I	.0001	-9.7782	.7620	-.0074		-.2383	-.3374
6 I^{stk}	-.0001	9.7563	-.7603	.0073		.2379	.3367
7 Y^{na}	.00003	244.0	8.16	.1640		-.0737	-.3417
8 Y^a	-.00002	-91.7	26.7	-.0940		.0427	.1979
9 Y	.00001	152.3	34.0	.0700		-.0310	-.1437
10 Q	.0002	-23.1723	1.8097	-.0174		-.5649	-.7997
11 W	.0001	-5.5017	.4287	-.0041		-.2464	-.1899
12 Y^d	.0009	-.0112	8.7323	-.0844		-2.7319	-2.3191
13 y	.000001	-.0797	.0062	-.00006		-.0019	-.0027
14 g	.000004	-.5285	.0412	-.0004		-.0129	-.0182
15 Ex					1.2400		
16 Im	-.4660	-4.6642	.3635	-.0035		-.1137	-.1610
17 F	.4660	4.6642	-.3635	.0035	1.2400	.1137	-.8390
18 M	.5621	-1.6745	.1305	-.0013		-.0016	-.0578
19 P	.0112	.6689	-.0521	.0005		-33.0194	.0231
20 BL	.000003	-.3034	.0236	-.0002		-.0074	-.0105
21 P^{im}	-.0049	-.2943	.0229	-.0002		14.5285	-.0102
22 T^{pi}	.0000	-.04233	.0031	-.00003		+.0031	-.0015
23 T^c	.00001	-1.0567	.0823	-.0008		-.0257	-.0365
24 T^{exc}	.00001	-1.6433	.1281	-.0012		-.0401	-.0567
25 T^{im}	.00001	-.7965	.0621	-.0006		-.0194	-.0275
26 T^{stp}	.000001	-2.0596	.3462	-.0018		-.0032	-.0297
27 T^{ss}	-.0004	22.4245	.6734	.0151		1.1557	-.0008
28 T^{fue}	.00002	3.2780	-.7372	.0032		-.0441	-.0625
29 T^{sle}	-.00001	21.6062	-3.6319	.0191		.0353	.0500
30 T^{os}	-.000003	.3331	-.0260	.0003		.0081	.0115
31 T^{mun}	.000006	-.7603	.0592	-.0006		-.0185	-.0262
32 Tr	-.00003	1.6370	.0492	.0028		.0844	-.00006
33 T^{disp}	-.0003	39.6462	-3.0895	.0298		.9665	-.1798
34 T	-.00036	41.2832	-3.0404	.0326		1.0509	-.1799
35 DEF	.00033	-39.6462	3.0895	-.0298		-.9665	.1798

Pop (38)	τ^{pi} (39)	τ^{pi}_{-1} (40)	$\tau^{c'}$ (41)	τ^{exc} (42)	τ^{exc}_{-1} (43)	τ^{sle} (44)	τ^{sle}_{-1} (45)
.0160	-4.1794	2.0979	-4.3761	-64.0537	32.0274	-135.9209	68.7610
.0160	-4.1781	2.0973	-4.3748	-64.0353	32.0177	-135.8794	68.7401
.0022	-.5643	.2833	-.5909	-8.6498	4.3249	-18.3545	9.2854
.0022	-.5643	.2833	-.5910	-8.6498	4.3249	-18.3545	9.2854
-.0022	.5631	-.2827	.5896	8.6305	-4.3152	18.3133	-9.2645
.0022	-.5715	.2869	-.5984	-8.7595	4.3798	-18.5872	9.4031
-.0013	.3311	-.1662	.3466	5.0739	-2.5370	10.7665	-5.4469
.0009	-.2405	.1207	-.2518	-3.6856	1.8428	-7.8206	3.9564
.0051	-1.3374	.6714	-1.4004	-20.4983	10.2492	-43.4962	22.0043
.0010	-.3175	.1593	-.3324	-4.8668	2.4334	-10.3270	5.2243
.0247	-6.4677	3.2465	-6.7721	-99.1259	49.5630	-205.212	107.4607
-.0184	-.0046	.0023	-.0048	-.0705	.0352	-.1495	.0756
.0001	-.0305	.0153	-.0319	-.4676	.2338	-.9922	.5020
.0010	-.2692	.1351	-.2819	-4.1259	2.0630	-8.7551	4.4291
-.0010	.2692	-.1351	.2819	4.1260	-2.0630	8.7551	-4.4291
.0004	-.0967	.0485	-.1012	-1.4812	.7406	-3.1432	1.5901
-.0001	.0386	-.0194	.0404	.5917	-.2959	1.2556	-.6352
.00007	-.0174	.0088	-.0183	-.2684	.1342	-.5695	.2881
.00006	-.0170	.0085	-.0178	-.2603	.1302	-.5524	.2795
-.0097	2.5476	-1.2789	-.0023	-.0374	.0187	-.07847	.04025
.0002	-.0610	.0306	2.6061	-.9347	.4674	-1.9834	1.0034
.0004	-.0949	.0476	-.0993	38.5463	-19.2732	-3.0847	1.5605
.0002	-.0460	.0231	-.0481	-.7046	.3523	-1.4951	.7564
.0004	-.0106	.0053	-.0111	-1.0810	.5405	-2.3108	1.1689
.00001	-.0014	.0007	-.0014	-.0207	.0104	-.0439	.0222
.0004	-.1045	.0524	-.1094	-1.6013	.8007	-3.3980	1.7190
-.0003	.0836	-.0420	.0875	1.2811	-.6405	89.7184	-45.3752
-.0001	.0192	-.0097	.0201	.2947	-.1473	.6252	-.3163
.0002	-.0439	.0220	-.0459	-.6726	.3362	-1.4272	.7220
.000001	-.0001	.00005	-.0001	-.0015	.0008	-.0032	.0016
-.0088	2.2883	-1.1486	2.3960	35.1712	-17.5455	76.5243	-38.7005
-.0088	2.2882	-1.1486	2.3958	35.0697	-17.5348	76.5211	-38.6989
.0088	-2.2883	1.1486	-2.3960	-35.0711	17.5356	-76.5243	38.7005

(continued)

TABLE 4, continued

Dep. Var.	e^{im} (46)	e^{ex} (47)	e^{ex}_{-1} (48)	τ^{inf} (49)	τ^{inf}_{-1} (50)	τ^{def} (51)	$\eta^{a}_{-1}\eta^{c}$ (52)
1 Y'	.0534	-.2661	.1330	-62.0207	30.9862	-.0257	-.1152
2 C^p	.0534	-.2660	.1330	-37.1019	18.5768	-.0154	-.0691
3 I^{pl}	.0072	-.0359	.0180	-33.2751	16.5843	-.0035	-.0617
4 I^{pe}							
5 I	.0072	-.0359	.0180	-33.2751	16.5843	-.0138	-.0617
6 I^{stk}	-.0072	.0358	-.0180	8.3564	-4.1749	.0035	.0155
7 Y^{na}	.0053	-.0364	.0182	6.0816	-3.0281	-.0035	-.0158
8 Y^a	-.0031	.0211	-.0105	-3.5227	1.7540	.0020	.0091
9 Y	.0022	-.0153	.0077	2.2589	-1.2741	-.0015	-.0066
10 Q	.0171	3.4749	-1.7374	-19.8474	9.9160	-.0082	-.0369
11 W	.0054	-.0202	.0101	-4.6994	2.3542	-.0020	-.0088
12 Y^d	.0827	-.4117	.2059	-57.4333	28.7567	-.0239	-.1070
13 y	.00006	-.0003	.0002	-.0682	.0341	-.00003	-.0001
14 g	.0004	-.0019	.0010	-.4528	.2262	-.0002	-.0008
15 Ex		+4.8000					
16 Im	.0034	-.0170	3.7079	-15.8722	7.9107	-.0066	-.0294
17 F	-.0034	4.8171	-3.7086	15.8723	-7.9107	.0066	.0294
18 M	-.6825	-.0062	.0031	.0183	.7166	-.0006	-.0027
19 P	.3999	.0025	-.0012	4.3461	-.2862	.0002	.0011
20 BL	.0002	-.0011	.0006	-.2599	.1298	-.0001	-39.1005
21 P^{im}	.8240	-.0011	.0005	-1.9123	.1259	-.0001	-.0005
22 T^{pi}	-.00002	-.0002	.00008	-.03675	.018099	-.00002	-.00007
23 T^c	.0008	.1586	-.0792	-.9050	.4522	-.0004	-.0017
24 T^{exc}	.0012	-.0060	.0030	-1.4075	.7032	-.0006	-.0026
25 T^{im}	-.0194	-.0029	.0015	-.6822	.3408	-.0003	-.0013
26 T^{stp}	.0006	-.0007	.0003	-.7389	.3681	-.0003	-.0014
27 T^{ss}	-.0140	-.0001	.00004	-.1521	.0100	-.00001	-.00004
28 T^{fue}	.0013	-.0067	.0033	-1.5505	.7747	-.0006	-.0029
29 T^{sle}	-.0011	.0053	-.0027	1.2404	-.6197	.0005	.0023
30 T^{os}	-.0002	.0012	-.0006	.2852	-.1425	.0001	.0005
31 T^{mun}	.0006	-.0028	.0014	-.6512	.3253	-.0003	-.0012
32 Tr	-.0010	-.00001	.0000	-.0111	.0007	-.0000	-.0000
33 T^{disp}	-.0292	.1457	-.0728	-4.5874	2.2295	-.0019	-.0083
34 T	-.0303	.1457	-.0728	-4.5985	2.2302	-.0019	-.0083
35 DEF	.0292	-.1457	.0728	4.5873	-2.2295	.0019	.0083

W^{min} (53)	W^{min}_{-1} (54)	W^g (55)	W^g_{-1} (56)	C^g (57)	C^g_{-1} (58)	I^g (59)	H^g (60)	NTR (61)
.0012	-.0006	-.2749	1.3994	-.2749	1.3994	2.4989	.4429	-1.6143
.0012	-.0006	-1.2748	1.3990	-1.2748	1.3990	1.4981	.4428	-1.6138
.0002	-.0001	-.0371	.1890	-.0371	.1890	.3374	.0598	-.2179
.0002	-.0001	-.0371	.1890	-.0371	.1890	1.3375	.0598	-.2179
-.0002	.0001	.0371	-.1885	.0370	-.1885	-.3367	-.0597	.2175
.00016	-.00008	-.0376	.1914	-.0376	.1914	.3417	.0606	-.0221
-.0001	.00008	.0218	-.1108	.0218	-.1109	-.1927	1.6349	1.2788
.00007	-.00003	-.0158	.0805	-.0158	.0805	.1438	1.6955	-.0929
-.0156	.0078	-.0880	.4478	-.0880	.4478	.7997	.1417	-.5166
.0001	-.0001	-.0209	.1063	-.0209	.1063	.1899	.0337	-.1226
.0019	-.0009	-.2551	1.2987	-.2551	1.2987	2.3191	.6854	-2.4981
.000001	-.000001	-.0003	.0015	-.0003	.0015	.0027	.0005	-.0018
.000009	-.000004	-.0020	.0102	-.0020	.0102	.0182	.0032	-.0118
.00008	-.00004	-.0177	.0901	-.0177	.0901	.6380	.0285	-.1040
-.00008	.00004	.0177	-.0901	.0177	-.0901	-.6380	-.0286	.1039
.00003	-.00001	-.0064	.0324	-.0063	.0324	.0578	.0102	-.0373
-.00001	.00001	.0025	-.0129	.0025	-.0129	-.0231	-.0041	.0149
.000005	-.000003	-.0012	.0059	-.0012	.0059	.0105	.0019	-.0068
.000005	-.000002	-.0011	.0057	-.0011	.0057	.0102	.0018	-.0066
.0000	-.0000	-.0002	.0008	-.0002	.0008	.00147	.00026	-.00094
-.0007	.0004	-.0040	.0204	-.0040	.0204	.0365	.0065	-.0236
.00003	-.00001	-.0062	.0318	-.0062	.0318	.0567	.0101	-.0366
.00001	-.00001	-.0030	.0154	-.0030	.0154	.0275	.0049	-.0178
.000003	-.000002	-.0033	.0166	-.0033	.0166	.0297	.0231	-.0192
.0000	-.0000	-.0001	.0005	-.0001	.0005	.0008	.0001	-.0005
.00003	-.00001	-.0069	.0350	-.0069	.0350	.0625	-.0474	-.0404
-.00002	.00001	.0055	-.0280	.0055	-.0280	-.0500	-.2427	.0323
-.00001	.000003	.0013	-.0064	.0013	-.0064	-.0115	-.0020	.0074
.00001	-.00001	-.0029	.0147	-.0029	.0147	.0262	.0047	-.0169
.0000	-.0000	-.00001	.00003	-.00001	.00003	.00006	.00001	-.00004
-.0007	.0003	-.0198	.1007	-.0198	.1007	.1798	-.2425	.8838
-.0007	.0003	-.0198	.1007	-.0198	.1007	.1799	-.2425	.8838
.0007	-.0003	1.0198	-.1007	1.0198	-.1007	.8202	.2425	-1.8839

[a] The determinant of B matrix is .442386. The blanks indicate zero entry. A value of .0000 means a nonzero entry but of a small order (usually less than .000001).

[b] The effects of Y^d_{-1} and C^p_{-1} are lumped together.

(and an almost identical change in aggregate personal consumption). This tax has a negative effect on all of the 36 variables reported in Table 4 except for its own yield, inventory investment, agricultural income, trade balance, price rise, state taxes, overall government revenues, and government disposable income. Similar effects for other tax rates can be read down the respective columns. The other negative effects on real GNP are from the corporation tax, the sales tax, an increase in the exchange rate for exports, the inflation tax, deficit financing, excess reserves, nontax revenues, and (strangely enough) public consumption expenditure. This rather unexpected negative effect of C^g and W^g on GNP does not look so strange when one goes back to the table of structural coefficients and finds that in $Y' = 1C^p + 1C^g + \ldots$, and $C^p = b_c C^g + \ldots$, $b_c = |-1.11| > 1$. Thus the additive direct effect of C^g on GNP is swamped away by a greater negative effect of C^g on C^p—which is somewhat strange, indeed. The following instruments, on the other hand, indicate positive multiplier effects on GNP: an increase of foreign-exchange rate for imports, a high minimum-wage rate, public investment, and expenditure on public health. Exactly the same directions of influences are indicated for the supply side of GDP except that surprisingly enough the inflation tax (τ^{inf}) has a positive multiplier effect on GDP. The opposite effect of τ^{inf} on production as against the demand side of GNP seems to be due to a positive influence of τ^{inf} on capacity utilization, which in turn has a positive influence on production, while capacity utilization does not directly appear on the demand side of GNP. Inflation's effect on investment is negative. While current investment is only a small proportion of capital stock in the production function, it is an important additive component of the demand for GNP. The recorded supply and demand of GNP, however, are the two sides of the same coin. The net effect of inflation on the supply and demand of national product may be read from the growth variable (g) in row 14, where it comes out negative. According to the impact multiplier of g, a treasury deficit equivalent to 1 percent of GNP causes (in probability) a 0.23 percentage point decline in the rate of growth, holding all other listed predetermined variables constant.

Effects on Resource Allocation

As to allocation of resources between present consumption (C^p) and future consumption (as represented by present investment I^{pl}), the tax measures seem to be nearly neutral. For example, the ratios of the coefficients of C^p to the corresponding coefficients of I^{pl} with respect to tax rates and NTR are almost identical (approximately 7) with identical signs. The same is true for wage-control policy. Relative to the tax rates, neither public expenditures (except the welfare expenditures, such as public health) nor monetary instruments (or those fiscal instruments whose effects are mainly monetary) seem to be neutral as between consumption and investment. Thus public investment, which indicates a complementarity relationship with private investment as well as with consumption, induces about twice as much private investment in relation to private consumption as the tax rates do in the opposite direction. Treasury's deficit finance (τ^{inf}) and excess reserve ratio ($\eta^{a}_{-1} - \eta^{c}$) or simply η have almost identical (negative) influences on private investment and private consumption. In other words, the monetary variables affect investment many times more (adversely) than do the fiscal variables.

Effects on Revenues

All the tax instruments studied here naturally have positive effects on the corresponding tax collections (and generally negative on the yields of other taxes). The exception here is the exchange rate, which has a negative effect on revenues from tariffs. Apparently, there is some scope for improving upon the forms of the import and export equations of this study, since with zero tariff rates, tariff revenues should be zero, whereas the present linear equations indicate a positive intercept. Perhaps the equations should depict some kind of inverted parabolas, so as to yield maximum points, which should enable one to locate optimum tariffs from revenue point of view. The present equations depict reality only at the margin, and in any case our sample does not have any observation with zero tariff rates.

The net effects of the monetary instruments as represented by excess reserves (and also deficit financing) on overall tax revenues are negative. The same conclusion applies to minimum-wage policy.

Effects on Price Stabilization

Considering the values from the sums of the current and lagged effects of a given instrument, personal and corporate income taxes exacerbate rather than mitigate the rate of inflation. This is possibly due to their negative effect on real output, cancelling out any decline in effective demand, and possibly also due to a shifting of a part of the corporation tax. The inflation effect of exchange overvaluation and excise and sales taxes is likewise positive, which seems to work from the side of cost rather than real output. The effects of deficit financing and the inflation tax are naturally to increase prices. The same applies to government consumption expenditures. A noteworthy result is the negative effect on price level of w^{min}. That is, a rise in minimum-wage rate causes a decline in inflation. A possible explanation of this result—a reduction in worker efficiency and work incentives when w^{min} is low rather than high—discussed in Chapter 5 applies here also.

Effects on Foreign Balances

Foreign balances are favorably affected on the whole by all taxes and all monetary variables. They are adversely affected by overvalued exchange rate, high minimum-wage rate, public investment, and public health expenditures. A somewhat revealing result in the context of the Brazilian economy pertains to the minimum-wage policy. A harshly austere wage policy has led to very few desirable effects; apart from causing an adverse income distribution it has aggravated inflation, thereby defeating its own purpose.

Effects on Income Distribution

As to the effects of the policy instrument studied here on functional income distribution, only exchange overvaluation and wage policy seem to work in opposite directions. The higher the rate of minimum wage the better, naturally, the lot of the wage earners. Exchange overvaluation for exports seems to favor the nonwage shares, probably because export income appears in the nonwage category. All other instruments seem to have identical (neutral) effects roughly proportional to the existing income shares.

BUILT-IN FLEXIBILITIES*

Absolute and Differential Built-In Flexibilities

Built-in flexibilities corresponding to impact multipliers—i.e., proportionate changes in GNP induced by various exogenous changes

---

*The term built-in flexibility is traditionally defined as the fraction of changes in income (usually national income) that is prevented because of the existence of automatic tax responsiveness. In measuring built-in flexibility the effects of discretionary changes, such as those in tax rates, must be netted out. Where for one reason or another tax rates cannot be assembled and the tax yields due to discretionary measures cannot be separated from automatic responses, the overall

that are prevented by automatic stabilizers in one year—are presented in Table 5. The absolute built-in flexibilities are given in column 5 and the differential built-in flexibilities in column 6. It may be noted that these flexibilities represent the overall effects of all taxes (and other automatic stabilizers/destabilizers), as given by Relation xviii for z and xiii for z'. For the individual tax flexibilities, see below.

Although built-in flexibilities are conventionally computed for exogenous injections coming from autonomous investment or government expenditure and on that basis they are generalized for all autonomous expenditures, there is no reason why they should not be computed for alternative autonomous expenditures (at any rate, there is no autonomous investment in the present model); accordingly, in Table 6 built-in flexibilities are computed for changes in a number of predetermined variables and policy instruments.

It will be seen that the general coefficient of absolute built-in flexibility (z) is 0.116 and that for differential built-in flexibility (z') is -0.134 (see the first row of Table 6).* In other words, the proportion

---

relationship of gross tax collections with income is still useful. The insight into the tax revenues provided by such a relationship, which I have elsewhere called "buoyancy" (Sahota, 1961), is particularly of interest in the underdeveloped countries. It is a gross measure that may reflect upon a government's "tax effort"—in bringing expanded tax bases under the revenue net, enhancing the rates of existing taxes, plugging in the tax loopholes, gearing up tax administration, increasing tax enforcements, and similar measures. Most existing studies are, actually, measures of buoyancies rather than built-in flexibilities, even though they are claimed or supposed to be the latter. In general, if the effects of changes in tax rates can be separated, the remaining factors whose influences may be picked up in built-in flexibility, such as enhanced administrative efficiency, need not be taken too seriously in estimates of automatic responsiveness. Administrative efficiency, for example, can itself be a function of income changes.

*Some of the alternative ways of expressing the coefficient of built-in flexibility are:

$$z = 1 - \frac{\text{multiplier under actual tax rates}}{\text{multiplier assuming lump-sum tax rates}} \qquad (x')$$

or

$$z = 1 - \frac{\text{Det. assuming lump-sum tax rates}}{\text{Det. under actual tax rates}} \qquad (x'')$$

which are equivalent to the text relation:

$$z = 1 - \frac{\Delta Y}{\Delta Y^b} \qquad (x)$$

where the Det. is the denominator of the inverse matrix of $\mathrm{Adj}\,B/|B|$.

TABLE 5

Impact Measures of Absolute Built-in Flexibilities (z) and Differential Built-in Flexibilities (z')

(1)	π Assuming Lump-sum Tax Rates (2)	π Assuming Proportional Tax Rates (3)	π for Actual Tax Rates (4)	z (5)	z' (6)
Det	.3776	.4843	.4272	.116	-.134
y_{-1}		.3904	.4425		-.133
y'_{-1}	.2414	.2145	.2332	.0340	-.0872
Y^{na}_{-1}		.0829	.0940		-.133
Y^{a}_{-1}		-.0862	-.0977		.133
C^{p}_{-1}	1.4136	1.1022	1.2495	.0959	-.149
I^{pl}_{-1}	1.7105	1.3336	1.5118	.117	-.134
I^{pe}_{-1}	.0025	.0020	.0022		-.100
I^{stk}_{-1}	.7502	.5690	.6489	.133	-.141
$(1-Z)_{-1}$	194.7994	151.8858	172.1735	.118	-.141
Q_{-1}		-.0568	-.0378		.334
Y^{d}_{-1}	-.9132	-.7120	-.8071	-.116	.134
g_{-1}		.00014	.00009		.357
M_{-1}	-.0000	.00148	.00167		-.128
M_{-2}	.0000	-.00089	-.00101		.135
M_{-3}	.0000	-.00039	-.00044		.128
P_{-1}	-.0000	.00714	.00809		-.133
P_{-2}	-.0000	.00939	.01064		-.133
P_{-3}	-.0000	.00388	.0044		-.129
BL_{-1}	-.0063	-.0049	-.0055	-.127	.122
P^{im}_{-1}	.0000	-.0161	-.0183		.137
T^{pi}_{-1}		-.7229	-.8195		.134
T^{exc}_{-1}		-.7063	-.8006		.134
T^{ss}_{-1}		-.6709	-.7606		.134
T^{sle}_{-1}		-.7063	-.8007		.134
IL	-.0000	.0007	.0008		-.134
IL_{-1}	.0000	-.0009	-.0010		-.111
X_{-1}	-.0000	.0005	.0006		-.120
$[\Sigma I/\Sigma K]_{-1}$			-72.4107		
N			5.6428		
K			-.0545		
GNP^{US}		.0000	.0000		.000
r	.0000	-.0121	-.0137		.132
M_1/M	.0000	-1.5573	-1.7653		.132
NRA	-2.8273	-2.2044	-2.4989	-.117	.134
Pop		-.0130	.0160		-.231
τ^{pi}		-3.6869	-4.1794		.134
$\tau^{c'}$		-3.8604	-4.3761		.134
τ^{exc}		-56.5070	-64.0547		.134
τ^{sle}		-119.9047	-135.9205		.134
e^{im}		.0471	.0534		-.134
e^{ex}		-.3999	-.2661		-.350
τ^{inf}	-70.3992	-54.7127	-62.0207	-.144	.128
τ^{def}	-.0291	-.0227	-.0257	-.117	.132
$\eta^{a}_{-1}-\eta^{c}$	-.0130	-.1017	-.1152	-.115	.133
w^{min}		.0018	.0012		.333
W^{g}	-.3110	-.2425	-.2749	-.115	.134
C^{g}	-.3110	-.2425	-.2749	-.115	.134
I^{g}	2.8273	2.2044	2.4989	.1162	-.1355
H		-.9916	.4429		
NTR	-1.8264	-1.4243	-1.6143	-.120	.134
$p^{foreign}$		.0083	.0094		-.133

Source: Table 4.

TABLE 6

Dynamic Multipliers

τ	year (1)	Y' (2)	I^{pi} (3)	Y (4)	Q (5)	W (6)	g (7)	F (8)	P (9)	T (10)	DEF (11)	Y' cumu (12)
1. τ^{pi}	1	-4.1800	-.5643	-.2405	-1.3374	-.3175	-.0305	.2692	.0386	2.2882	-2.2882	-4.1800
	2	-1.5539	.1336	2.5700	.7803	-.1189	.0559	-.1351	-.0670	-1.1486	.5577	-5.7339
	3	4.4465	.8049	.1332	1.1003	.3379	.0448	.3838	-.0800	-3.2004	2.5141	-1.2874
	4	4.8925	.7929	-.1751	.5328	.3724	.0001	.3789	-.0278	.6318	1.3371	3.6051
	5	1.7252	.3239	-.2095	.0530	.1315	-.0266	.1547	.0241	2.6121	-.4456	5.3303
	6	-1.3429	-.1227	-.0813	-.1100	-.1020	-.0236	-.0584	.0383	2.0306	-1.2667	3.9874
	C1	5.3303	1.4910	2.4627	1.1290	.4054	.0703	.7823	-.1362	-2.0608	2.1207	..
	C2	3.9874	1.3683	1.9968	1.0191	.3034	.0201	.7239	-.0738	3.2127	.4084	..
2. $\tau^{c'}$	1	-4.3761	-.5909	-.2518	-1.4004	-.3324	-.0319	.2819	.0404	2.3958	-2.3960	-4.3761
	2	-5.4584	-.0205	5.2300	.9182	-.4162	.1005	-.0108	-.1193	-7.7050	-.0419	-9.8345
	3	7.2835	1.4062	.3678	2.0026	.5534	.0968	.6704	-.1642	-7.1460	4.7284	-2.5510
	4	9.7649	1.5901	-.2781	1.0850	.7431	.0130	.7588	-.0732	.1584	3.0667	7.2139
	5	4.3856	.7770	-.4143	.1580	.3342	-.0462	.3711	.0333	4.7026	-.3786	11.5959
	6	-1.8451	-.1282	-.2008	-.2232	-.1400	-.0456	-.0609	.0731	4.2596	-2.3177	9.7544
	C1	11.5995	3.1619	5.3460	2.7634	.8821	.1784	.2711	-.0789	-5.3092	5.3572	..
	C2	9.7544	3.0337	4.4528	2.5402	.7421	.0866	2.0105	-.2099	-3.3346	2.6609	..
3. τ^{exc}	1	-64.0537	-8.6497	-3.6856	-20.4983	-4.8668	-.4676	4.1260	.5917	35.0697	-35.0711	-64.0537
	2	-23.7822	2.0380	39.1571	11.8941	-1.8171	.8537	-3.1669	-1.0222	-47.6184	8.4510	-87.8359
	3	8.6986	4.3054	14.4405	15.2618	.7351	.2500	2.0526	.0577	-16.5930	5.9955	-79.1369
	4	28.1270	5.7124	.0787	11.1731	2.1493	.1347	2.7254	-.1177	-6.3739	10.2168	-51.0103
	5	20.0559	3.9263	-1.1502	5.2902	1.5283	-.0791	1.8745	.0645	10.5402	1.9131	-30.9544
	6	1.2710	.9506	-.8800	1.8089	.0979	-.1516	.4545	.2151	14.5128	-5.6322	-29.6834
	C1	-29.6824	8.2830	49.9907	24.9298	-2.1733	.7708	.9591	-.4305	-35.5156	-8.4947	..
	C2	-29.6834	8.2830	47.9605	24.9298	-2.1733	.5401	8.0661	-.2109	-10.4626	-14.1269	..

TABLE 6, continued

τ	year (1)	Y' (2)	I^{pi} (3)	Y (4)	Q (5)	W (6)	g (7)	F (8)	P (9)	T (10)	DEF (11)	Y' cumu (12)
4. τ^{sle}	1	-135.9209	-18.3545	-7.8206	-43.4962	-10.3270	-.9922	8.7571	1.2556	76.5211	-76.5243	-135.9209
	2	-50.2532	4.4687	22.3281	25.7735	-3.8400	1.8409	-6.7443	-2.2017	-101.6157	18.6000	-186.1741
	3	118.8848	22.7602	2.7635	27.3941	9.0385	1.2707	10.8542	-2.3714	-89.7671	67.5403	-67.2893
	4	135.8438	22.6564	-4.6761	13.8406	10.3403	.0405	10.8131	-.8699	15.2968	37.3474	68.5545
	5	50.7855	9.7535	-5.7985	1.3824	3.8718	-.7175	4.6593	.6092	71.5209	-11.3684	119.3400
	6	-34.6324	-2.8202	-2.3700	-3.0287	-2.6310	-.6607	-1.3424	1.0473	57.2970	-34.8096	84.7076
	C1	119.3400	41.2843	17.2709	24.8944	9.0836	2.1599	2.0128	-4.1874	-114.8617	46.9634	..
	C2	84.7076	38.4641	4.4263	21.8657	6.4526	.7817	26.9970	-2.5309	29.2530	.7854	..
5. e^{im}	1	.0534	.0072	.0022	.0171	.0054	.0004	-.0034	.3999	-.0303	.0292	.0534
	2	.0616	-.0005	-.0161	-.0128	.0023	-.0012	-.0002	.0933	.0968	-.0020	.1150
	3	-.0731	-.0151	-.0033	-.0186	-.0060	-.0010	-.0072	.0331	.0750	-.0476	.0419
	4	-.1017	-.0171	.0028	-.0105	-.0078	-.0002	-.0081	.0115	-.0004	-.0319	-.0598
	5	-.0473	-.0086	.0043	-.0015	-.0036	.0005	-.0041	.0034	-.0485	.0035	-.1071
	6	.0178	.0010	.0022	.0023	.0013	.0005	.0005	.0005	-.0450	.0241	-.0893
	C1	-.1071	-.0341	-.0172	-.0263	-.0097	-.0008	..	..	.1415	-.0523	..
	C2	-.0893	-.0331	-.0079	-.0240	-.0084	-.0010	-.0235	.5417	.0476	-.0247	..
6. τ^{inf}	1	-62.0207	-33.2751	2.2589	-19.8474	-4.6994	-.4528	15.8723	4.3461	-4.5985	4.5873	-62.0207
	2	-5.4203	-3.3243	2.2916	5.1898	-.4196	.4585	-17.4106	-.5014	-28.4307	6.1967	-67.4410
	3	17.8014	.8276	-.4449	2.6366	1.3542	-.1738	.3945	-.3924	-15.9720	13.5724	-49.6396
	4	18.1884	1.5406	-.7589	.0460	1.3849	-.0094	.7357	-.1232	1.0522	6.8305	-31.4512
	5	5.3683	.2093	-.8000	-1.4218	.4096	-.1064	.1007	.0924	8.9617	-.9077	-26.0829
	6	-6.0747	-1.1212	-.2691	-1.5548	-.4616	-.0874	-.5346	.1227	6.6087	-4.2317	-32.1576
	C1	-26.0829	-34.0218	..	-11.9750	-1.7703	.0057	-1.5383	3.3291	..	..	
	C2	-32.1576	-35.1430	2.2776	-14.9516	-2.2319	-.3713	-.8420	3.5442	-32.3786	26.0475	

7. τ^{def}	1	-.0257	-.0035	-.0015	-.0082	-.0020	-.0002	.0066	.0002	-.0019	.0019	-.0257
	2	-.0026	-.0004	.0047	.0049	-.0002	.0004	-.0002	-.0005	-.0213	-.0015	-.0283
	3	.0181	.0024	-.0004	.0037	.0014	.0002	.0012	.0004	-.0118	.0106	-.0102
	4	.0172	.0023	-.0008	.0012	.0013	-.0000	.0011	-.0001	.0032	.0048	.0070
	5	.0045	.0007	-.0008	-.0004	.0003	-.0001	.0003	.0001	.0096	-.0020	.0115
	6	-.0061	-.0008	-.0002	-.0007	-.0005	-.0001	-.0003	.0001	.0066	-.0045	.0054
	C1	-.0102	.0015	.0032	.0016	.0010	.0004	.0064	-.0003	..	.0004	..
	C2	.0054	.0007	.0010	.0005	.0005	.0002	.0087	.0002	-.0156	.0093	..
8. $\eta^{a}_{-i}-\eta^{c}$	1	-.1152	-.0617	-.0066	-.0369	-.0088	-.0008	.0294	.0011	-.0083	.0083	-.1152
	2	.0638	.0411	.0238	.0477	.0041	.0023	.0184	-.0026	-.0897	-.0122	-.0514
	3	.1592	.0413	-.0039	.0419	.0121	.0007	.0197	-.0017	-.0120	.0402	.1078
	4	.1060	.0266	-.0065	.0224	.0081	-.0005	.0127	.0000	.0721	-.0017	.2138
	5	.0021	.0080	-.0046	.0092	.0002	-.0008	.0038	.0011	.0847	-.0378	.2159
	6	-.0592	-.0034	-.0003	.0053	-.0045	-.0005	-.0016	.0010	.0411	-.0401	.1567
	C1	.2159	.0553	.0172	..	.0157	.0022	..	-.0032	..	-.0039	..
	C2	.1567	.0519	.0019	.0896	.0112	.0004	.0824	-.0009	.0879	-.0433	..
9. w^{min}	1	.0012	.0002	.0001	-.0156	.0001	.00001	-.0001	-.00001	-.0007	.0007	.0012
	2	-.0004	-.0001	-.0001	.0037	-.0001	-.0000	.0000	-.0000	.0006	-.0001	.0008
	3	-.0009	-.0001	.0000	.0018	-.0001	-.0000	-.0001	.0000	.0002	-.0004	-.0001
	4	-.0006	-.0001	.0000	.0009	-.0001	.0000	-.0000	.0000	-.0002	-.0001	-.0007
	5	-.0001	-.0000	.0000	.0005	-.0000	.0000	-.0000	-.0000	-.0004	.0001	-.0008
	6	.0003	.0000	.0000	.0003	.0000	.0000	.0000	-.0000	-.0002	.0001	-.0005
	C1	-.0008	.0001	.0000	..	-.0002	.00001	-.0001	-.00001	.0001	.0001	..
	C2	-.0005	.0001	.0000	-.0084	-.0002	.00001	-.0002	-.00001	-.0007	.0003	..

(continued)

TABLE 6, continued

τ	year (1)	Y' (2)	I^{pi} (3)	Y (4)	Q (5)	W (6)	g (7)	F (8)	P (9)	T (10)	DEF (11)	Y' cumu (12)
10. C^g	1	-.2749	-.0371	-.0158	-.0880	-.0209	-.0020	.0177	.0025	-.0198	1.0198	-.2749
	2	.5547	.0752	.2883	.4903	.0419	.0265	-.1445	-.0300	-.1076	-.1316	.2798
	3	.0440	.0061	-.0116	.0259	.0034	-.0001	.0029	-.0009	.0149	.0097	.3238
	4	.0011	.0003	-.0022	.0087	.0001	-.0003	.0002	.0001	.0305	-.0108	.3249
	5	-.0254	-.0033	-.0002	.0032	-.0019	-.0002	-.0016	.0003	.0149	-.0144	.2995
	6	-.0233	-.0031	.0010	.0024	-.0018	.0000	-.0015	.0001	-.0050	-.0063	.2762
	C1	.3249	.0445	.2725	..	.0245	.0245	-.1268	-.0284	-.0671	.8882	..
	C2	.2762	.0381	.2595	.4425	.0208	.0239	-.1268	-.0280	-.0721	.8664	..
11. I^g	1	2.4989	.3374	.1438	.7997	.1899	.0182	-.6380	-.0231	.1799	.8202	2.4989
	2	.2563	.0340	-.4431	-.4764	.0200	-.0382	.0165	.0421	2.0682	.1447	2.7522
	3	-1.7863	-.2411	.0315	-.3669	-.1359	-.0151	-.1150	.0336	1.1572	-1.0437	.9689
	4	-1.7012	-.2335	.0741	-.1218	-.1295	.0019	-.1114	.0091	-.3215	-.4694	-.7323
	5	-.4487	-.0645	.0740	.0380	-.0342	.0103	-.0317	-.0099	-.9511	.1978	-1.1810
	6	.6060	.0773	.0223	.0726	.0460	.0081	.0368	-.0136	-.6517	.4531	-.5750
	C1	-1.1810	-.1677	-.2993	-.1654	-.0897	-.0351	-.6215	.0617	..	-.5484	..
	C2	.5750	-.0904	-.0974	-.0548	-.0437	-.0148	-.8427	.0382	1.4810	.1025	..
12. H^g	1	.4429	.0598	1.6955	.1417	.0337	.0032	-.0286	-.0041	-.2425	.2425	.4429
	2	1.1675	.1579	-.0134	.2744	.0888	.0013	.0752	-.0029	-.3845	.7766	1.6104
	3	.8137	.1136	-.0320	.0888	.0620	-.0034	.0542	.0053	.3770	.1410	2.4241
	4	.0350	.0087	-.0357	-.0074	.0027	-.0062	.0042	.0098	.5550	-.1947	3.2378
	5	-.4489	-.0582	-.0033	-.0230	-.0341	-.0035	-.0277	.0078	.2741	-.2585	2.7889
	6	-.4177	-.0558	.0182	.0035	-.0318	.0005	-.0266	.0020	-.0839	-.1148	2.3712
	C1	..	..	1.6111	.4745	..	-.0086	.1050	..	.5791	.5921	..
	C2	1.5925	.2260	1.6293	.4780	.1213	-.0081	.0507	.0179	.4952	.5921	..

13. NTR	1	-1.6143	-.2179	-.0929	-.5166	-.1226	-.0118	.1039	.0149	.8838	-1.8839	-1.6141
	2	.4386	.0596	.3200	.5008	.0330	.0291	.0283	-.0328	-1.1643	-.2652	-1.1755
	3	1.2684	.1725	-.0688	.3018	.0965	.0057	.0823	-.0187	-.3864	.5807	.0929
	4	.4831	.1156	-.0547	.0895	.0633	-.0041	.0552	-.0013	.4358	.1258	.5760
	5	-.0161	.0016	-.0372	-.0128	-.0011	-.0067	.0008	.0082	.5950	-.2270	.5599
	6	-.5048	-.0659	-.0013	-.0246	-.0384	-.0036	-.0314	.0073	.2686	-.2757	.0551
	C1	.5758	.1314	.2271	.3755	.0702	.0230	..	-.0379	-.6669	-1.4426	..
	C2	.0549	.0655	.0651	.3381	.0307	.0086	.2391	-.0224	.6325	-1.9453	..

Note: The relation for calculating the dynamic multipliers is explained in the text. C_1 means cumulated over one cycle. C_2 means cumulated over the entire six-year period. The coefficients indicate the change in the policy instrument on the left per unit (Cr$1 million) increase in the target variable at the top.

of changes in GNP induced by most external stimuli that are prevented by the automatic stabilizers of the Brazilian economy in one year (i.e., z) is about 12 percent. The corresponding proportion for z', which determines the changes in the relative shares of the automatic stabilizers in GNP is approximately -13 percent. The negative value for z' means that the automatic stabilizers are not strong enough to be able to maintain their relative shares in GNP. Their automatic stabilizing flexibility is roughly about half of what it would be if all revenues were collected by a proportionate income tax, while it is about 12 percent higher than what it would be under lump-sum taxes.* Since comparable estimates are not available for other developing countries, there is no way of telling how Brazil's tax flexibilities fare with them. The corresponding z coefficient derived from a small econometric model for the United States is 33.9 percent (Chalmers and Fischel, 1967, p. 438). The explanation for the rising tax-revenue shares in GNP in Brazil in the reference period must lie in their high buoyancy, rather than in automatic responsiveness alone.

Several other results are worth noting. The built-in flexibility coefficients, though identical for most stimuli, are not identical for all stimuli, as may be seen down the columns of Table 5. For example, the z coefficient of C^{p}_{-1} is about 10 percent; that of I^{pl}_{-1} is about 12 percent; and that for GNP_{-1} is 3.4 percent. Not all these differences can be attributed to random errors and estimation disturbances. Part of the explanation lies in differential interaction effects and lagged influences. Similar results were obtained by Chalmers and Fischel for the United States (1967, p. 438).

Another result worth noting is that the absolute built-in flexibility of public consumption expenditure is negative. This result apparently conflicts with the conventional theory of effective demand and the

---

*It might be mentioned that the estimated flexibilities assume that government expenditures are predetermined and are not significantly influenced by current tax collections. In our model, also, we have assumed that government expenditures (G) are exogenous. If government expenditures were determined by current tax collection, the indicated flexibilities would not really be the implied stabilizers. Indeed, according to the (precarious) balanced-budget theorem, the mere existence of the government budget is inflationary. (For an early exposition of this theorem, see Haavelmo [1945]. For a subsequent long discussion, controversy, and elaboration of this theorem, see a brief summary list in Musgrave [1959, p. 430].) The reality seems to be that government expenditures are planned at the beginning of the (fiscal) year and are thus predetermined for our model, but presumably they are not completely invariant to current tax revenues. By and large, the flexibility formulas can be considered to have their approximate validity in practice.

textbook illustration. An obvious explanation for such an unusual result was noted earlier while discussing the negative coefficient of C^g in the Y' equation: namely, the estimated coefficients indicate that an increase of one unit of government consumption expenditure causes a 1.11-unit reduction in personal consumption.

Finally, it should be noted that the built-in flexibilities of increases in tax rates and other leakages, as expected, are negative. For instance, a decrease in net remittances abroad (NRA) is found to prevent an increase in GNP in presence of the existing tax structure by 13 percent relative to what it would be if all taxes were of lump sum nature.*

Corrected and Uncorrected Built-In Flexibilities

The measures of built-in flexibility of Table 5 are for the overall leakages. The corrected and uncorrected built-in flexibilities (z only) of individual taxes are given below. The uncorrected flexibilities, $\bar{z}_j$,

Tax	Uncorrected $\bar{z}$	Corrected z*
Personal income tax	.0069	.0047
Corporation tax	.0128	.0086
Excise taxes	.0186	.0125
Tariffs	.0129	.0088
Stamp tax	.0116	.0077
Social-security contributions	.0721	.0490
Fuel tax	.0272	.0183
Sales tax	.0046	.0031
Other state taxes	-.0043	-.0029
Municipal taxes	.0095	.0062
Total	.1799	.116

of the first numerical column were computed according to Formula xvi. Since in a linearized model responses are additive, the $\bar{z}_j$ were summed over j and are given in the last row—i.e., $\Sigma_j \bar{z}_j = 0.1799$. A correction factor was obtained from $z/\Sigma_j \bar{z}_j$. The corrected z_j^* were then calculated

*One should not be put off by some of the prima facie implausible signs of some of the lagged endogenous variables. These appear for those equations in which the Cochrane-Orcutt transformations were used, so that the coefficients of the linearized forms of the current and the lagged values of the variables in expressions such as $\alpha(x-\rho x_{-1})$ always take opposite signs.

in accordance with Formula xix. It will be noted that the uncorrected built-in flexibilities are about 50 percent higher than the corrected ones. Other things being equal, the automatic responsiveness of the personal income tax has prevented increases in equilibrium as well as intermediate-period money incomes by about 0.5 percent according to the corrected measure and by about 0.7 percent according to the uncorrected measure. The relative differences between the corrected and uncorrected values of built-in flexibility for Brazil are not much different from those between the built-in flexibility of personal income tax in the United States—about 26 percent in Smith (1963, p. 710) and Chalmers and Fischel (1967, p. 439) from small macro models, and about 33 percent in Musgrave (1959, p. 511) from a simple income determination model. The macro models cited above, however, are quite small as compared to the present one. If the tax structures of the two countries were otherwise similar, and if a model with all the tax equations were estimated (as in this study), the corrected built-in flexibility of the U.S. personal income tax would probably turn out to be no greater than 17 percent. If the basic point being made here is correct then since the built-in flexibilities of the United States and many other countries have been calculated largely for single taxes (mainly for personal income tax), independently of other taxes, they have apparently been grossly overestimated.

Yet, compared with U.S. taxes, the built-in responsiveness of Brazil's personal income tax is insignificant, and that of overall taxes, too, is rather small. In addition, it must also be noted that the calculated levels of flexibilities of Brazilian taxes are due not merely to the existing marginal tax rates but also to rather high values of the combined marginal propensities to spend, for the higher the marginal propensity to consume and invest, the higher the flexibility of a given tax. Of course, the higher the marginal tax yield, the higher the coefficient of flexibility. But the higher the marginal tax yield, the lower the rate at which the coefficient of flexibility increases; the upper limit of the coefficient of flexibility, which is unity, and which means complete flexibility, is approached asymptotically. This is an added reason why it would not be correct to add the flexibilities of individual taxes to arrive at the total flexibility. Such aggregation will always overestimate the true flexibility.

Social security taxes account for about 40 percent of the changes in money income that are prevented automatically in Brazil. If one looks only at the traditional taxes (personal and corporation income taxes, excise and sales taxes, and so on), one will find that the changes in money income prevented by indirect taxes are clearly much higher than those prevented by direct taxes. Excise taxes alone account for a built-in flexibility equivalent to all income taxes (other than social security taxes). The differential flexibilities, in particular, confirm our previous finding that the indirect taxes are somewhat more flexible than the direct taxes of Brazil. In respect to government, federal taxes account for the bulk of built-in stabilization, while state and local taxes contribute proportionately much less.

This casual result in this context should not, however, be taken as evidence for the "perversity hypothesis." For a test of this hypothesis a study of the whole gamut of state-local fiscal practices is required, and such is not an undertaking here.*

---

*In the United States several studies have concluded that state (and local) government taxing and borrowing (as well as spending) behave in a manner that intensifies the cyclical variations in the economy. For one of the most authoritative works on this point, see the study by Hansen and Perloff (1944). Also see the study by Maxwell (1958)

CHAPTER

7

DYNAMIC MULTIPLIERS

Neither the Samuelsonian ultimate indefinite-period multiplier analysis (Samuelson, 1939) nor the one-period impact multiplier analysis of Chapter 6 throws much light on the path that is followed by an endogenous variable in response to an exogenous stimulus. It is this path, the disequilibrium conditions of the medium run (say, half a decade), rather than the immediate impact or the ultimate equilibrium state, with which the policy maker (such as the finance minister or the president of the Central Bank) must deal. Such an intermediate-period analysis of the path of change is provided by dynamic multipliers.

To explain the calculation of dynamic multipliers briefly, let us rewrite Equation 3 of Chapter 1 as xxii—i.e.:

$$B\dot{y} = A_1\dot{z} + A_2\dot{y}_{-1} + D\dot{\tau} \qquad \text{(xxii)}$$

$$\dot{y} = \Pi\dot{z} + \Pi_2\dot{y}_{-1} + \Pi_3\dot{\tau} \qquad \text{(xxiii)}$$

where $\Pi_1 = B^{-1}A_1$, etc., and where x has been partitioned into z and y_{-1}, with appropriately-partitioned coefficient matrices, and where, in keeping with the computed multipliers of the present study, the variables are understood to be expressed in first differences. It is clear that the presence of the y_{-1} vector on the right-hand side introduces dynamic elements into the model. The character of these dynamic effects will depend upon the coefficients Π_2 on the one side, and the patterns of changes in the policy instruments, τ_j, on the other side. The latter may be a one-shot change in a particular τ when, for example, over the years:

$t =$	0	1	2	3, ...
$\dot{\tau} =$	1	-1	0	0, ...

or a once-and-for-all sustained change when, for example:

$$t = 0 \quad 1 \quad 2 \quad 3, \ldots$$

$$\dot{\tau} = 1 \quad 0 \quad 0 \quad 0, \ldots$$

or a repeated change in τ, for example:

$$t = 0 \quad 1 \quad 2 \quad 3, \ldots$$

$$\dot{\tau} = 1 \quad 1 \quad 1 \quad 1, \ldots$$

or any other scheme. For calculating the dynamic multipliers of this chapter, a once-and-for-all sustained change of one unit in the given τ_j is specified. The initial conditions are assumed to conform to a static equilibrium: for example, the tax instrument τ^{pi} which is expressed as percentage tax rate on taxable income is specified to change by one percentage point in period t=0. During this period, therefore, the only nonzero change is in τ^{pi}; all other exogenous changes are set to zero. That is, in period t=0 Equation xxiii becomes simpy:

$$\dot{y} = \Pi_3 \tau^{pi} \qquad \text{(xxiv)}$$

In subsequent periods, $\dot{\tau}^{pi}$ equals zero, while the vector of $\dot{y}_{-1}$ acquires nonzero values, so that:

$$\begin{aligned} \dot{y}_{+1} &= \Pi_2 \dot{y}, && \text{in period } t=1 \\ \dot{y}_{+2} &= \Pi_2 \dot{y}_{+1}, && \text{in period } t=2 \end{aligned} \qquad \text{(xxv)}$$

and so on. For some of the equations—e.g., the monetary-sector equations—there are more than one lag, resulting in higher-order difference equations of the form:

$$\dot{y}_{+3} = \Pi_{21} \dot{y}_{+2} + \Pi_{22} \dot{y}_{+1} + \Pi_{23} \dot{y} + \ldots \qquad \text{(xxvi)}$$

Furthermore, the present model contains several equations that are estimated by COT. The multipliers derived from such equations may contain terms $\dot{\tau}_{-1}$, the absolute values of whose coefficients ($R\Pi_3$) are necessarily smaller than the corresponding (Π_3) coefficients of $\dot{\tau}$ and have the opposite signs. The subsequent-period multiplier equations in these cases are of the form:

$$\begin{aligned} \dot{y}_{+1} &= \Pi_2 \dot{y} - R\Pi_3 \dot{\tau}, && \text{in period } t=1 \\ \dot{y}_{+2} &= \Pi_2 \dot{y}_{+1}, && \text{in period } t=2 \end{aligned} \qquad \text{(xxvii)}$$

and so on.

The calculations of xxv-xxvii for one to six years for a set of ten selected endogenous variables in response to changes in 13 diffe policy instruments studied here are presented in Table 6. It may be recalled that the income categories—and so most of the endogenous variables and those instrument variables that consist of public expenditures—are expressed in millions of 1963 cruzeiros, while tax and monetary instruments are in percentages of one form or another. In the case of the latter instruments (and those endogenous variables that are expressed in millions of cruzeiros), the coefficients denote millions-of-1963-cruzeiro changes in the variables at the top per one-percentage-point change in the instrument on the left. The indicated percentage-point is from the final-year (1968) values of the variables concerned, for which reference can be made to Appendix A, Table A. 1. The coefficients in Period 1 are the impact multipliers of Table 4. These are the effects that appear in the current (or zero) period. The effects that appear in the second year are given under Year 2, and so on. The effects of one complete cycle are cumulated in the seventh row and the effects of the first six years are cumulated in the last row of each instrument.

As stated earlier, the primary purpose of the model is to appraise the influences of different policy instruments on four main objectives: growth and resource allocation, income redistribution, stabilization, and resource mobilization through revenues. The endogenous variables studied here are in a way the intermediate objectives derived from the stated final objectives (which, too, in a way are the derived objectives from the ultimate objective of social welfare). In Table 6, ten such derived objective variables have been chosen, which may be visualized to be related to the final four objectives somewhat as follows:

Any instrument τ that positively influences

- Y, increases growth (of course the equation for growth [g] itself is also included through an identity relation);
- I, allocates resources to future consumption;
- Q, increases I but may or may not worsen income distribution in the long run;
- W, promotes desirable income redistribution;
- F, promotes stabilization of foreign balances and exchange rate;
- DEF, is stabilizing of budgetary balance;
- Y', increases growth and may also increase inflationary pressures;
- P, is destabilizing of price level;
- T, increases revenues.

Table 6 requires a careful analysis. Explanations for the magnitudes and signs of particular coefficients of this table have to be found by tracing them back to all the structural coefficients involved in the calculation of the respective reduced-form coefficients. Only the salient features and important results will be noted here. For the rest of the story, the reader may examine Table 6 in detail according to his particular interest.

It will be seen that the system is oscillatory and, by and large, stable. Different equations of the system reach their highest displacements in the very first, the second, or the third year. After that they start converging toward their long-run equilibrium values. There are relatively few equations that fail to display convergence to equilibrium during the six-year period of Table 6. With these few exceptions the system is, by and large, damped and tends towards equilibrium. We shall come back to this feature in Chapter 8, where the dynamic properties of the system are studied.

DYNAMIC EFFECTS ON GROWTH

Most tax instruments affect the growth variables (I^{pi}, Y, and g) negatively during the early period but positively after a lapse of time. For instance, the impact effect on GDP and g of income taxes (τ^{pi} and $\tau^{c'}$) is negative. The six-year cumulative effect, on the other hand, is positive. A change of one percentage point in the rate of personal income tax (from its 1968 value of 8.56 percent) causes GDP to grow by a little less than Cr$2 million (1963) in the course of a six-year period. An increase of Cr$2 million in GDP (Y) when the (1968) level of Y is Cr$11,673 million and of T^{pi} is Cr$93 million—so that a one-percentage-point increase in τ^{pi}, other things being equal, means an increase of nearly Cr$11 million in T^{pi}—is of a small order. It should be seen, however, that the ultimate change is not negative. Thus, Brazilian income taxation is, at best, growth-promoting; at worst it is growth-neutral. The influence of the corporate tax is similar: a one-percentage-point change in corporate tax $\tau^{c'}$ (from its 1968 value of 22.5 percent) leads to a change in GDP of about Cr$4.5 million in six years. The indirect taxes (τ^{exc} and τ^{sle}) have similar influences, the influence of τ^{exc} being much higher than any other tax. The net effects of the indicated indirect taxes on the rate of growth (g) is also positive, the negative impact of the first year and some of the later years being swamped by the subsequent years' positive effects. The dynamic effects of borrowed deficit financing (τ^{def}) and excess bank-balance effects are not much different from those of income taxes. The effects of e^{im} are the reverse of income taxes. Surprisingly enough, the impact effects of treasury's deficit finance (τ^{inf}) on GDP (Y) are positive. The reasons for this were explained earlier. The explanation is entirely in terms of capacity utilization: real output (GDP) is positively related to capacity utilization; capacity utilization is positively related to inflation tax; hence, real output is directly related to the inflation tax.

The relationship of income and inflation must be put in proper perspective. Actually, inflation is like narcotics. When through long inflationary periods the economy becomes "addicted" to inflation, inflation must continue to keep the economy stimulated. Once the inflation doses are stopped, the economy becomes depressed, excess capacity emerges, and the rate of growth of real output falls (which explains the calculated relationship). It must at the same time be noted that, somewhat in the tradition of the Harrod-Domar growth models, the rate of growth in our model is related through an identity to the demand side of income (GNP) and not to the real-supply side (GDP). The demand side of income is positively related to private investment, which in turn is negatively related to the inflation tax. Therefore, GNP is negatively related to inflation. Hence, the rate of growth (g) is negatively related to the inflation tax.

Another puzzling result is the net negative influence (over six years) of public investment on growth. On the other hand, the effects of public consumption outlays are positive. If these rather unusual results with regard to I^g and C^g are not due to statistical errors, they deserve careful attention. What is conventionally categorized as consumption outlays might, in fact, be highly productive expenditure on education, health, and technical information, while what is put under physical public investment may be supplanting relatively more productive private investment or be less productive than alternative public outlays as such. This result seems to provide empirical basis for Aaron's recommendation for Brazil, to the effect that the economic classification of the budget as between physical investment and current consumption "is bound to do more harm [to growth] than good" and "should be abandoned" (Aaron, 1968, p. 16). The effect of public health expenditures (H^g) on growth is clearly positive. Public investment in human beings, indeed, turns out to be highly productive in Brazil.

In analyzing these results it must be noted that the present (dynamic) multiplier effects are not a mere empirical counterpart to the textbook type of consumption multipliers. Unlike the textbook case, in the present multipliers most variables are allowed to have feedback effects on the variable under study, and most prices are allowed to float. In the present model what is held constant is the autonomous changes in variables other than the particular one that may be under reference. All the endogenous variables are allowed to react; all endogenous prices are allowed to float; and so on. Thus, when a change in a tax instrument is assumed, all public expenditures are held constant (having been determined autonomously) only in the first year. From the second year, public expenditures are influenced by the past year's changes in taxes (see the public-expenditure equations in the text and Appendix B). Hence, from the second year onward public expenditures are not held constant. In the dynamic system, they are determined within the system, as was explained in the relevant theoretical discussion above.

DYNAMIC EFFECTS ON RESOURCE ALLOCATION BETWEEN PRESENT AND FUTURE CONSUMPTION

The influence on private investment (I^{pl}), which represents allocation to future consumption vis-à-vis present consumption, is negative in the immediate period but positive in subsequent periods, giving a net positive effect by the sixth year from the personal income tax, the corporation tax ($\tau^{c'}$), the excise tax, deficit financing through borrowing (τ^{def}), excess reserves, and public consumption (apparently because C^g reduces private consumption more than itself). The instrument whose negative impact effect exceeds the dynamic positive effects at least up to the sixth year is τ^{inf}. The effect of e^{im} is the reverse of that of income taxes: it is positive at first and becomes negative in later years. Disturbingly enough, moreover, it indicates an explosive behavior, at least insofar as the six-year outcome is concerned. The effects of I^g are similar to those of e^{im}, except that the changes in I^g are damped. To take a typical instance from Table 6, a one-percentage-point rise in the excise-tax rate for 1968 (amounting to approximately Cr$50 million) might be expected to lead to the following changes: a lowering of the demand for GNP by Cr$64 million in the impact year and by about Cr$24 million in the second; a rise of Cr$9 million, Cr$28 million, Cr$20 million, and Cr$1 million in the next four years, respectively, resulting in a net reduction of demand of Cr$30 million in six years; a fall in private investment of Cr$8.6 million in the year of change but a net rise by the sixth year of Cr$8.3 million; a lowering of wage incomes by the sixth year by Cr$2.2 million and a raising of nonwage income by Cr$25 million; a lowering of the rate of growth of GNP by 0.47 percentage points in the first year, reversing into a positive change of 0.85 percentage points in the second, +0.25 percentage points in the third, -0.08 percentage points in the fifth, and -0.15 percentage points in the sixth year, to a cumulative net increase of 0.54 percentage points in six years; and so on.

DYNAMIC EFFECTS ON REDISTRIBUTION OF INCOME

Most of the policy variables studied here come out neutral as between income shares. There are some, however, that have non-neutral effects; there are others whose influences change over the cycle. Wage rate (w^{min}) is redistributive in favor of wage earners. According to these estimates, a one-cruzeiro rise in the minimum wage rate might be expected to raise wage income by Cr$100 (in 1963 prices) and lower nonwage income by Cr$15,600 in the initial year, and to lower nonwage income by Cr$8,400 and the wage bill by Cr$200 in six years.

As another case, a million-cruzeiro increase in public health expenditure might be expected to raise nonwage income by Cr$142, 000 and the wage bill by Cr$34, 000 in the year of change; both categories by approximately twice as much in the second year; nonwage income by approximately half as much and wage income by approximately twice as much in the third year; to about four times as much (a combined total of about Cr$0. 6 million) cumulatively over six years, approaching almost exactly a million-cruzeiro rise in the combined wage and nonwage incomes in the long run.

However, when the effects on relative shares, the redistributional effects proper, are considered, even identical changes of direction in Q and W must be analyzed in relative terms. Thus, while a rise in the corporate tax rate begins with proportionate impingement upon income shares in the first year and ends with proportionate cumulative effects in the sixth year, a similar rise in the excise tax rate begins with proportionate effects but ends with a pro-nonwage bias.

DYNAMIC EFFECTS ON REVENUE YIELD AND RESOURCE MOBILIZATION

The effects of all the tax instruments on revenues are naturally positive. Interestingly enough, only the impact effects of the corporation tax and the excise tax on revenues are positive; their subsequent effects become negative, presumably due to their negative effects on growth of the demand for and the supply of income. The impact and dynamic effects on tax revenues of nontax instruments—including w^{min}, τ^{inf}, τ^{def}, and public consumption outlays—are negative. A reduction of one percentage point in the rates of the personal income tax (amounting in 1968 to about Cr$10 million in 1963 prices, according to legal rates) and the sales tax (Cr$75 million) might be expected to reduce overall tax revenue in 1963 prices by Cr$2. 28 million and Cr$76. 55 million, respectively, in the first year and Cr$3. 2 million and Cr$29. 3 million, respectively, in six years. It is easy to visualize the error there would be in the changes in revenue yields if they were (as they traditionally are) projected in proportion to changes in their tax rates. Over longer periods, the projections could go wrong by greater margins. It so happens that in the first year the changes in the yields from excise taxes and sales tax are nearly proportional to changes in their tax rates. The change in the yield from the personal income tax, however, is only about one-fifth of that predicted by legal tax rates. There may be various causes of a miserably low rate of revenue response from the personal income tax, including tax evasion, adverse effects on output and employment, and so on. Among these a high level of evasion of this tax probably remains the foremost, as will be seen in Chapter 9.

DYNAMIC EFFECTS ON STABILIZATION

Inflation (P)

Since the oscillatory behavior of the system and its convergence or divergence to equilibrium values are the topics of inquiry in another section, these will not be taken up here. The relevant questions here are whether and by how much a change in an instrument is inflationary or deflationary. Contrary to common belief, positive changes in tax rates are inflationary in the impact year, presumably through their adverse impact effects on real output. This is true of the changes in the rates of the excise tax, the sales tax, and the income tax. Only in the long run do these effects become deflationary. Expansionary monetary policy as represented by the inflation tax is inflationary (as expected). On the other side, interestingly enough, an increase in the minimum wage rate has a negative dynamic effect on inflation. Its explanation is the same as noted earlier in the case of regression coefficients and impact multipliers.

Foreign Balances (F)

The dynamic effects on foreign balances of all taxes except personal income tax are positive. The negative effects are indicated by four instruments: namely, e^{im}, w^{min}, H^g, and I^g. Public investment (like private investment) seems to be import-intensive, while its contribution to exports is relatively small. An increase of a million-cruzeiro expenditure on physical investment in the public sector might be expected to run down foreign balances by Cr$0.64 million in the first year and by Cr$0.83 million by the sixth year. A one-percentage-point increase in personal income tax, on the other hand, might be expected to augment foreign balances by Cr$0.27 million in the first year and Cr$0.72 million by the sixth year. Other tax instruments present a similar story.

Budget Balance (DEF)

It may be noted that the variable DEF is defined as budgetary deficit. Positive values stand for deficit, negative values for surplus. It may be seen that the signs of the values in the DEF column are invariably the reverse of those of T. If a policy of "balanced budget" is to be followed, then any combination of taxes and expenditures over various periods may be read from Table 6. For example, a million-cruzeiro expenditure on public investment and public health apiece can be met by an increase of half of one percentage point in personal income tax, if the budget balance is to be assured in the first year. Over six years, however, the same combination might result in a deficit of about a million cruzeiros. For other such comparisons and combinations, one can refer to Table 6.

The dynamic multipliers calculated in Table 6 throw a good deal of light on policy alternatives. However, all the calculations of this table represent only one experiment: a sustained once-and-for-all change of one unit in an instrument. Such an experiment reveals important dynamic properties of the system. Yet it is incomplete in at least two important respects. First, it is a partial analysis in that the effects of a perturbation in an instrument are measured by solving each equation individually. The results of such partial solutions will differ from those of general solutions when the equations involved have lagged endogenous variables in them, as is the case in the present model. For a general-equilibrium solution the difference equations must be solved simultaneously. This analysis is carried out in Chapter 8. Second, when one's purpose is to achieve a certain outcome in terms of GNP and other objective variables, a single experiment will not tell how close one is to an optimal solution. Numerous simulated experiments are usually carried out by researchers looking for optimal solutions. Simulation is a trial-and-error method for the purpose. One can never know whether one has found <u>the</u> optimal solution. A relatively more economical method for the purpose is that of the optimal-control theory, which will be discussed in Chapter 12.

CHAPTER

8

DYNAMIC PROPERTIES OF THE MODEL

It must be apparent by now that dynamic multipliers themselves provide a good deal of information for the time patterns of responses of the model. Additional insight into the dynamic properties of the model can be gained from the solutions of the difference equations suggested by the impact multipliers of Table 4, which under certain approximations summarize dynamic responses.

The dynamic paths followed by different endogenous variables when various instruments are increased by one unit from their 1968 levels are sketched in Figures 1 and 2. The dynamic paths of the same endogenous variables, when injections come from other instruments, tell a similar story, though differences do appear here and there. It will be seen that the system appears somewhat unstable during the first one or two years following the exogenous injections of policy instruments and becomes generally damped thereafter. The oscillations are much more pronounced on the demand side of income (Y')—the Keynesian multiplier case—than on the supply side of real income (Y). The dichotomy seems to underline the doubts about the applicability of the Keynesian multiplier in the developing countries that were voiced by V. K. R. V. Rao in the early 1950s (Rao, 1952). The oscillations are not uniform or regular in all cases. More than any other equation, the price equation seems to follow a regular cyclical path that tends gradually to be damped. In the case of inflation tax, price fluctuations stretch and become extremely wide over long periods. The absolute influence of changes in tax rates and public expenditures on price variations is mild. Price changes seem to be determined mainly by monetary variables.

The solutions of difference equations in simultaneous systems are explained in Goldberger (1959, Chapter 6) and Baumol (1959, Chapters 15 and 16). More recently, Chow (1970, 1972a, 1972b, and 1973) has extended the methodology and its uses for optimal-control solutions.

Figure 1: Dynamic Paths Followed by Endogenous Variables Y', Y, and T in Response to a Unit Increase in Instrument

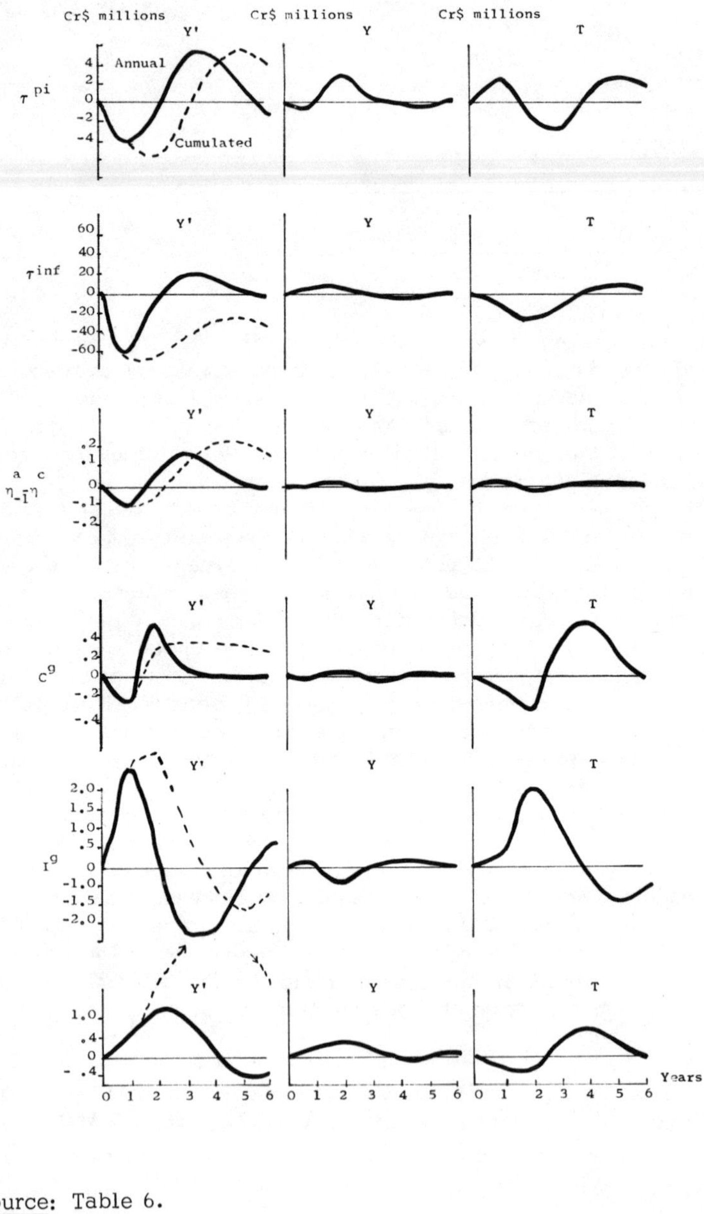

Source: Table 6.

Figure 2: Dynamic Paths Followed by Endogenous Variables W, F, and P in Response to a Unit Increase in Instrument

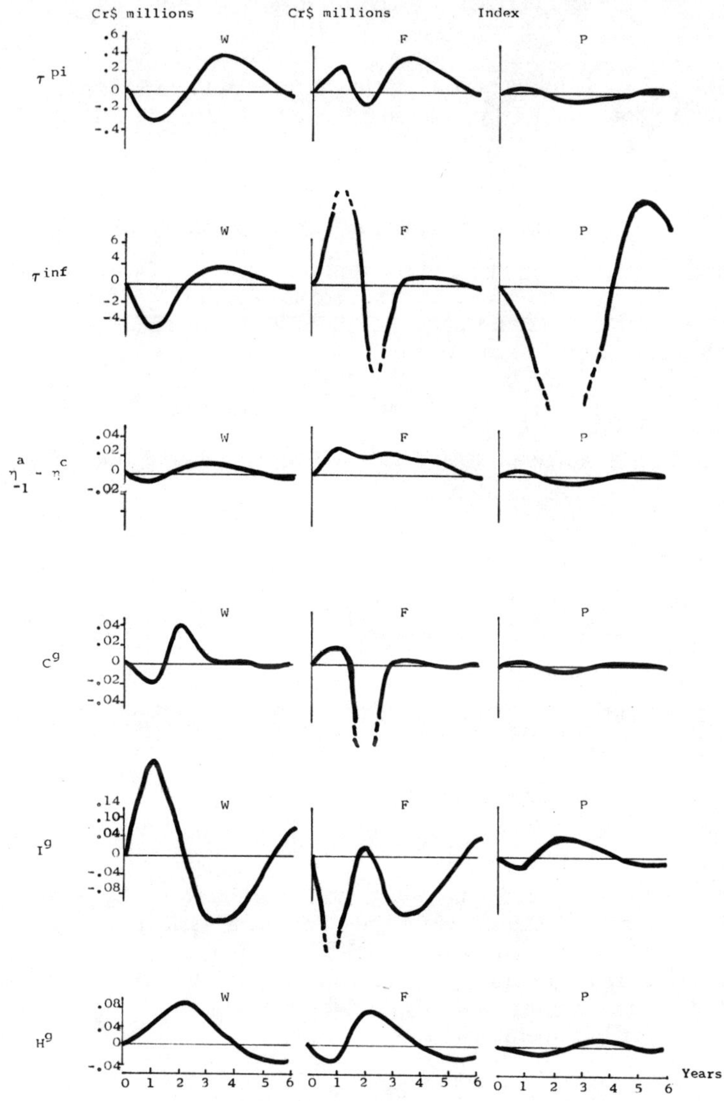

Source: Table 6.

In this chapter we largely follow Goldberger's method, briefly explained below. The control-theory solutions will be taken up in Chapter 12.

DERIVATION OF DYNAMIC RELATIONS

Using our earlier notation, dating the variables and letting $\dot{y}$ have any lags i, Equation xxii of Chapter 7 can be written as follows:

$$B_i \dot{y}_{t-i} = A_1 \dot{z}_t + D\dot{\tau}_t \qquad i = 0, 1, 2, \ldots, \text{ at most } I \qquad \text{(xxii)}$$

where $B_i = (b_{g,h+i})$ is a rectangular matrix of G structural-difference equations for $\dot{y}_g$ endogenous variables, $(g = 1, \ldots, G; h = 1, \ldots, H = G)$. $B = (b_{gh})$, for $i = 0$, is a square matrix of $G = H$ endogenous variables. For the present purpose, the $\dot{z}$ vector of strictly endogenous variables can be ignored. Hence:

$$B_i \dot{y}_{t-i} = D\dot{\tau}_t \qquad i = 0, 1, \ldots, \text{ at most } I \qquad \text{(xxviii)}$$

where I is the largest lag of the system. In its expanded form, Equation xxviii has the following elements:

$$\begin{aligned} & b_{g10}\dot{y}_{1t} + b_{g11}\dot{y}_{1,t-1} + b_{g12}\dot{y}_{1,t-2} + \ldots + b_{g1I}\dot{y}_{1,t-I} + \\ & b_{g20}\dot{y}_{2t} + b_{g21}\dot{y}_{2,t-1} + b_{g22}\dot{y}_{2,t-2} + \ldots + b_{g2I}\dot{y}_{2,t-I} + \\ & \vdots \qquad\quad \vdots \qquad\qquad \vdots \qquad\qquad \vdots \qquad \vdots \\ & b_{gH0}\dot{y}_{Ht} + b_{gH1}\dot{y}_{H,t-1} + b_{gH2}\dot{y}_{H,t-2} + \ldots + b_{gHI}\dot{y}_{H,t-I} \\ & \qquad = d_{g10}\dot{\tau}_{1t} + d_{g20}\dot{\tau}_{2t} + \ldots + d_{gK0}\dot{\tau}_{Kt} \end{aligned}$$

$$g = 1, \ldots, G \qquad \text{(xxviii')}$$

In the model of this study the endogenous variables y_G have lags in general from zero to three, while the instrument variables $\dot{\tau}_k$ also have lags from zero to one. It is important to note that all y variables, including the lagged values of them, are now treated as endogenous. This is because they are autoregressive.

Using the lag operator E* and defining:

$$Ey_t = y_{t-1}$$

---

*For the definition and algebra of lag operators, the reader is referred to Allen (1956, Appendix A).

and $$(b_0 + b_1E + b_2E^2)y_t = b_0y_t + b_1y_{t-1} + b_2y_{t-2}, \text{ etc.}$$

the system (xx) may be rewritten as:

$$B_iE^i\dot{y}_t = D\dot{\tau}_t \quad i = 0,\ 1,\ 2,\ \ldots \tag{xxix}$$

where the coefficient B_0 is understood to be normalized to unity. Note that the $\dot{y}$ variables are autoregressive. The $\dot{\tau}$ variables are not autoregressive even if they may have lagged effects on $\dot{y}$. Hence, the lag operator applies only to $\dot{y}$ and not to $\dot{\tau}$. Letting:

$$\bar{B} = B_iE^i = B_0 + B_1E + B_2E^2 + \ldots B_iE^i \tag{xxx}$$

we have:

$$y_t = \bar{B}^{-1}D\dot{\tau}_t \tag{xxxi}$$

or for convenience (as will soon be evident):

$$|\bar{B}|\,y_t = (Ad_j\bar{B})D\dot{\tau}_t \tag{xxxii}$$

where $|\bar{B}|$ is the determinant and $(Ad_j\bar{B})$ the adjoint of $\bar{B}$.

The determinant $|\bar{B}|$ is a polynomial in E of degree at most GI. The complete solution of the system of difference equations is found in three distinct steps, each giving one important component of it: (a) a general solution of the set of homogeneous difference equations reflecting the endogenous (inherent) response characteristics, (b) a particular solution corresponding to the time path of changes in the vector of instrument variables $\dot{\tau}$, and (c) the full, complete solution resulting from the introduction of the "initial conditions" and from the superimposition of the general solution on the particular solution. The particular component of the complete solution is to be interpreted as an equilibrium path, a constant trend, or an equilibrium (ultimate) multiplier of the Keynesian type that would emerge were there no inherent autoregressiveness (zero inherent responsiveness) in the system.

The calculated polynomial $|\bar{B}|$ in E gives a characteristic equation of the form:

$$\bar{b}_0\lambda^{GI} + \bar{b}_1\lambda^{GI-1} + \ldots + \bar{b}_{GI-1}\lambda + \bar{b}_{GI} = 0 \tag{xxxiii}$$

giving a maximum of GI characteristic roots. In vector form:

$$\lambda \equiv (\lambda_1,\ \lambda_2,\ \ldots,\ \lambda_{gi}) \tag{xxxiv}$$

It is important to note that these roots are common to all equations of the system. The general solution, then, is:

$$\hat{\dot{y}}_t = C\lambda^t \qquad \text{(xxxv)}$$

where $C = C_{gi}$ are arbitrary constants to be determined (later) from the initial values of y_{t-j}, $j = 1,2,\ldots,J$ (to be specified below). The eigenvalues λ_{gi} define the dynamic path—the oscillation (if any), stability, damping, and periodicity characteristics of the endogenous response mechanism of the system. If the roots are real and all less than unity, the system is stable (damped); otherwise it is explosive. In the short run all roots might have influence on the endogenous variable concerned; in the long run only the dominant roots matter. (Also see below.) Positive real roots give monotonic paths; negative real roots give saw-toothed paths. If the roots are complex (which case must appear in pairs only—e. g., $\lambda_j = \alpha + \beta i$ and $\lambda_g = \alpha - \beta i$), the system is oscillatory. The oscillations increase and become explosive when the damping factor $a = (\alpha^2 + \beta^2)^{\frac{1}{2}} > 1$; they are damped for $a < 1$. A convenient functional form in which such an oscillatory system can be captured is the sinusoidal function* of the type:

$$\hat{\dot{y}}_t - \bar{\dot{y}}(t) = Ce^{at}\operatorname{Cos}(\theta t + b) \qquad \text{(xxxvi)}$$

From this function, the period in which one complete oscillation occurs is $2\pi\ \theta$, where $\theta = \text{Arctan}\ \beta/\alpha$ and π is a radian measure = 180 degrees of an Argand diagram; the amplitude (damping factor) $a = (\alpha^2 + \beta^2)_{\frac{1}{2}}$; the phase at which the peak occurs is specified at time t by $\hat{\dot{y}} = -b/\theta$ plus or minus a multiple of 2π; the amplitude of oscillations is given at $\pm C$.

The next step is to obtain the particular component of the solution. For this purpose, it is essential to assume the time path of changes in the instrument variable $\dot{\tau}$ concerned. Here a few solutions will be presented under the simple assumptions of a sustained unit change in τ. In Chapter 12 additional solutions will be attempted in which changes in τ will be controlled to achieve specified objectives or to minimize the deviations from the desired values of the given objectives in the specified time period.

Under the indicated assumption a particular solution is obtained simply by omitting the operator E^i in xxix and summing the respective b coefficients for each $\dot{y}$ over i, precisely in a way that is used to find the equilibrium consumption multipliers in macro models. That is:

$$\bar{\dot{y}}_t = B^{-1} D\dot{\tau}_t \qquad \text{(xxxvii)}$$

and if τ changes at a constant annual rate of one unit, simply:

$$\bar{\dot{y}} = B^{-1} D = A \qquad \text{(xxxviii)}$$

---

*For the derivation and characteristics of sinusoidal functions, see Allen (1956, Chapter 4).

It may be noted at this point that the inherent dynamic properties given by A carry over from first differences ($\dot{y}$) to absolute values (Y), as shown by Goldberger (1959, pp. 113-14).

Finally, the general solution is obtained by combining the two components (xxx) and (xxxiii), thus:

$$\hat{\dot{y}}_t = C\lambda^t + \bar{\dot{y}}_t \qquad \text{(xxxix)}$$

where the initial conditions are reflected in C, the endogenous response mechanism in λ (which will now be defined as a matrix Λ), and the autonomous forces in A. The final step, therefore, is to derive the solution by introducing the "initial conditions," the values of the endogenous ($\dot{y}$) variables in an appropriately-determined number of previous years. The "initial conditions" are contained in the observed data matrix (see Appendix A).

$$Y = (y_{gt})$$

$g = 1, \ldots, G,$ where G is the total number of endogenous variables (equations)
$t = 0, 1, \ldots, N-1,$ where N is the number of calculated characteristic roots (xl)

Then defining:

$$\Lambda = (\lambda_n^t)$$

$n = 1, \ldots, N \leq GI,$ where I is the largest lag
$t = 0, \ldots, N-1$ (xli)

$$C = (C_{gt})$$

$g = 1, \ldots, G$
$t = 0, \ldots, N-1$ (xlii)

$$\bar{Y} = (\bar{y}_{gt})$$

$g = 1, \ldots, G$
$t = 0, \ldots, N-1$ (xliii)

Note that the matrix of characteristic roots is raised to the power t, while Y and $\bar{Y}$ variables are simply recorded at time t. Solving xxxix for C:

$$\check{C} = (Y-\hat{Y})\Lambda^{-1} \qquad \text{(xliv)}$$

or

$$\check{C} = (Y-A)\Lambda^{-1} \qquad \text{(xlv)}$$

and our basic complete solution is

$$\hat{\dot{y}}_t - \bar{\dot{y}}_t = \check{C}\lambda^t \qquad \text{(xlvi)}$$

It is important to note that the λ's are now weighted by the calculated constants matrix $\check{C}$. In the short run, thus, a dominant root may be dominated by a dominant $c_n\lambda_n^t$, though in the long run a $c_j\lambda_j^t$ with dominant λ_j^t will always exceed other values. The initial conditions, thus, may play a critical role in the dynamic behavior of a system. Thus, as pointed out by Goldberger (1959, p. 119), the conventional procedure of testing the dynamic properties of an equation or systems by an examination of eigenvalues alone may not be adequate.

APPLICATION

In our model the monetary and the real sectors have important feedbacks and are not dichotomous. Nevertheless, we shall present analyses of three alternative subsets of equations: (a) the monetary sector, consisting of $\dot{M}$ and $\dot{P}$ equations, (b) the real (sub)sector, consisting of four equations (objectives), namely, growth ($\dot{Y}$), distribution ($\dot{S}$), balance in foreign exchange ($\dot{F}$), and government revenues ($\dot{T}$), and (c) the combined real and monetary sectors, consisting of the equations for $\dot{P}$, $\dot{g}$, $\dot{S}$, and $\dot{T}$. The first solution will be explained in detail. For the latter two solutions only the final results will be presented. The coefficients used here are the "impact multipliers" of Table 4. The solutions are obtained for eight different instruments, evaluating (changes in) one at a time. The following solution uses treasury's deficit τ^{inf}, shortened here as simply τ, as the exogenous influence. Thus from Table 4 we have:

$$\dot{M} - .81\dot{M}_{-1} - .0382\dot{M}_{-2} - .0167\dot{M}_{-3} - 1.408\dot{P}_{-1} + .2205\dot{P}_{-2} + .17\dot{P}_{-3} = 183.4\dot{\tau}$$

$$\dot{P} - .038M_{-1} + .023\dot{M}_{-2} - .01\dot{M}_{-3} - .417\dot{P}_{-1} - .13\dot{P}_{-2} - .1\dot{P}_{-3} = 4.58\dot{\tau}$$

Using the lag operator E, we can write the homogeneous form of the above equation as:

$$\begin{bmatrix} [1 - .81E - .0382E^2 - .167E^3] & [-1.41E + .2205E^2 + .17E^3] \\ [-.038E \quad .023E^2 - .01E^3] & [1 - .417E - .13E^2 - .1E^2] \end{bmatrix} \begin{bmatrix} \dot{M} \\ \dot{P} \end{bmatrix} = \begin{bmatrix} 0 \\ 0 \end{bmatrix}$$

The determinant solution of this system gives a polynomial in E of sixth degree (GI = 2 × 3 = 6) as follows:

$$1 - 1.22E + .45142E^2 + .04531E^3 + .12952E^4 + .00427E^5 + .00337E^6 = 0$$

from which follows the characteristic equation:

$$\lambda^6 = 1.22\lambda^5 + .45142\lambda^4 + .04531\lambda^3 + .12952\lambda^2 + .00427\,\lambda + .00337 = 0$$

The characteristic roots, which were computed by the Vanderbilt University Computer Time-Sharing Users' Library Program Rooter 300-003, are given below:

$$\begin{aligned} \lambda_1 &= .646 \\ \lambda_2, \lambda_3 &= .151 \pm .295i \\ \lambda_4 &= .982 \\ \lambda_5, \lambda_6 &= -.081 \pm .373i \end{aligned}$$

The general solution, therefore, is

$$\hat{\dot{y}}_t = C\lambda^t$$

where C is a matrix of arbitrary constants to be determined from the initial values of $\dot{y}$.

The absolute values of the real roots are less than unity; hence, the system is stable. However, since λ_4 is close to unity, the system, once displaced, will not rapidly converge to the equilibrium values. The system has monotonic components corresponding to the two positive real roots (λ_1, λ_4) and an oscillatory component given by the two complex conjugates. The sinusoidal function from the dominant pair of complex eigenvalues is:

$$\hat{\dot{y}}_t - \bar{\dot{y}}(t) = Ce^{.3314t}\,\mathrm{Cos}\,(88t + b)$$

giving a period of complete regular oscillation of four years and a damping factor of 0.3314^t. The constants C and b, which give the phase and the peak of oscillations, will be determined in the complete general solution below.

Assuming $\dot{\tau} = 1$—i.e., that government incurs constant annual deficit financing (collects an inflation tax) of one percent of GNP—then a particular solution of this system would be:

$$\begin{bmatrix} [1\ -.81\ \ -.0382\ -.0167]\ [\ -1.408 + .2205 + .17] \\ [\ -.038 + .023\ \ -.01\ \ ]\ [1\ -.417\ \ -.13\ \ -.1] \end{bmatrix} \begin{bmatrix} \dot{M} \\ \dot{P} \end{bmatrix} = \begin{bmatrix} 183.4 \\ 4.58 \end{bmatrix} [\dot{\tau}]$$

or

$$\begin{bmatrix} \bar{\dot{M}} \\ \bar{\dot{P}} \end{bmatrix} = \begin{bmatrix} .1351 & -1.0175 \\ -.025 & .353 \end{bmatrix}^{-1} \begin{bmatrix} 183.4 \\ 4.58 \end{bmatrix} [1] = \begin{bmatrix} 3100.8 \\ 23.4 \end{bmatrix}$$

Any multiple or fraction of $\dot{\tau}$, instead of $\dot{\tau} = 1$, can be easily incorporated in this final result by simple (scalar) multiplication. Thus, $\dot{\tau} = .1$ would imply $\bar{\dot{M}} = 310.08$ and $\bar{\dot{P}} = 2.34$. These changes will, however, also affect the values of C matrix (see below). At $\dot{\tau} = 1$ we observe that the equilibrium values of the changes in monetary stock and price level are Cr$3, 101 million and 23.4 percent, respectively. The moving equilibrium path of both variables is upward sloping.

To get the complete solution, initial conditions must be specified. We specify two alternative sets of initial conditions: a) a "static equilibrium" position in which $\dot{M}_{-1} = \dot{M}_{-2} = \ldots = \dot{P}_{-1} = \dot{P}_{-2} = \ldots = 0$; b) the actual $\dot{M}$ and $\dot{P}$ during the last six years of the sample period—i.e., for 1966, 1967, and 1968 (see data in Appendix D). Under a) we get:

$$\begin{bmatrix} \dot{M}_t \\ \dot{P}_t \end{bmatrix}_{2\times6} = [C_t]_{2\times6}\,[\Lambda^t]_{6\times6}\,[A_t]_{2\times6}$$

$$\begin{bmatrix} 0 & 0 & 0 & 0 & 0 & 0 \\ 0 & 0 & 0 & 0 & 0 & 0 \end{bmatrix} = [C]_{2\times6} \begin{bmatrix} 1 \\ 1 \\ 1 \\ 1 \\ 1 \\ 1 \end{bmatrix} \begin{bmatrix} .646 \\ .151+.295i \\ .151-.295i \\ .982 \\ -.081+.373i \\ -.081-.373i \end{bmatrix} \begin{bmatrix} .646 \\ .151+.295i \\ .151-.295i \\ .982 \\ -.081+.373i \\ -.081-.373i \end{bmatrix}^2 \cdots \begin{bmatrix} .646 \\ .151+.295i \\ .151-.295i \\ .982 \\ -.081+.373i \\ -.081-.373i \end{bmatrix}^5 +$$

$$\begin{bmatrix} 3100.8 & 3100.8 & 3100.8 & 3100.8 & 3100.8 & 3100.8 \\ 23.4 & 23.4 & 23.4 & 23.4 & 23.4 & 23.4 \end{bmatrix}$$

It may be verified that a change in $\dot{\tau}$ will result in a change in the constants C through their influence on A.

The matrix Λ with complex entries was inverted by the Vanderbilt University Computer Program Complex-Inverter GSS. The resulting constants are:

$$C = \begin{bmatrix} -791.5 & 132.5 \pm 427.5i & 3497.2 & 64.99 \pm 139.1i \\ -5.97 & 1.001 \pm 3.23i & 26.4 & .49 \pm 1.05i \end{bmatrix}$$

The full complete solution therefore is:

$$\begin{bmatrix} \dot{M}_t \\ \dot{P}_t \end{bmatrix} = \begin{bmatrix} -791.5 & 132.5 \pm 427.5i & 3497.2 & 64.99 \pm 139.1i \\ -5.97 & 1.001 \pm 3.23i & 26.4 & .49 \pm 1.05i \end{bmatrix} \begin{bmatrix} .646^t \\ (.151 \pm .295i)^t \\ .982^t \\ (-.081 \pm .373i)^t \end{bmatrix} + \begin{bmatrix} 3100.8 \\ 23.4 \end{bmatrix}$$

or:

$$\dot{M}_t - 3100.8 = 791.5(.646)^t - (132.6 \pm 427.5)(.151 \pm .295i)^t - (3497.2)(.982)^t - (65.0 \pm 139.1i)(-.081 \pm .373i)^t$$

$$\dot{P}_t - 23.4 = 5.97(.646)^t - (1.001 \pm 3.23i)^t - (26.4)(.982)^t - (.49 \pm 1.05i)(-.081 \pm .373i)^t$$

The values of $\dot{M}_t$ and $\dot{P}_t$ calculated for 1, 2, ..., 10, 20, 30, 40, and 50 years are presented in Table 7. It will be seen that the numerical values of the simulated annual changes are not out of line with the order of magnitude of the observed annual money and price changes (whenever $\dot{T}^{inf}$ has been close to one percent of GNP). Of course, $\dot{T}^{inf}$ has not been close to 1 percent all the time, nor has it been the only exogenous influence or the only instrument at work.

The solution with actual initial values of $\dot{M}$ and $\dot{P}$ for the years 1966, 1967, and 1968 is given below. For this solution the zero matrix for $\hat{\dot{y}}$ on the left-hand side of the expanded complete solution above (or Equation xxxix) is replaced by an actual $\hat{\dot{y}}_t$ matrix, thus:

$$\begin{bmatrix} 15 & 44 & 43 & 38 & 28 & 24 \\ 15 & 44 & 43 & 38 & 28 & 24 \end{bmatrix}$$

The new solution is:

$$\dot{P}_t - 23.4 = 40.7(.646)^t - (59.5 \pm 40.6i)(.151 \pm .295i)^t - 2.7(.982)^t + (36.6 \pm 3.61i)(-.081 \pm .373i)^t$$

$$\dot{M}_t - 3100.8 = 826.4(.646)^t - (191.2 \pm 383.7i)(.151 \pm .295i)^t - 3474.2(.982)^t - (28.0 \pm 134.5i)(-.081 \pm .373i)^t$$

It is evident that due to the λ's being weighted in this solution the values of $\dot{M}$ and $\dot{P}$ might differ significantly in the early years from those given by the general solution of the characteristic equation in

TABLE 7

Simulated Values of M and P from Full, Complete Solution of Monetary Sector, When Initial Solution Assumed One of Static Equilibrium

	Annual Change		Cumulated Change	
Year	M	P	M	P
1	116.84	1.3934	116.836	1.3934
2	159.56	33.29	276.400	31.60
3	57.68	45.06	334.08	79.74
4	15.66	30.61	349.74	120.35
5	20.19	23.42	369.93	143.77
6	44.88	23.13	414.81	166.90
7	80.38	22.70	495.19	189.60
8	122.00	21.93	619.19	211.53
9	167.16	21.50	784.35	233.03
10	214.48	21.27	998.83	254.30
20	685.34	21.12	1684.17	275.42
30	1086.48	21.43	2770.65	296.85
40	1421.08	21.69	4191.73	318.54
50	1700.11	21.91	5891.84	340.45

step one above. It is for this reason that the conventional practice of testing the dynamic properties of a model by characteristic roots alone might not be quite adequate, as it might fail to reveal the complete picture of a dynamic path.

From the real (sub)sector, we selected the following four basic equations, where $\dot{Y}$ is the derivative of GDP and represents the output expansion objective; $\dot{S} = \dot{W} - \dot{Q}$ is the relative improvement of the wage bill over nonwage income and symbolizes the income-redistribution objective;* $\dot{F}$ stands for the rate of change of exports minus imports and describes the foreign-balance objective; $\dot{T}$ stands for government-revenue

*The average ratio of W^m/Q^m during the four years 1950-53 was approximately 0.45; the average of the same ratio during 1966-68 was approximately 0.32. A minimal national objective would perhaps revolve around raising the low 1966-68 ratio steadily to the level of the early 1950s. Some such policy to recoup the lost relative-income position of manufacturing labor was being evolved at the ministry of finance during 1967-69 while the present author was an economic advisor to the Brazilian government. That situation forms the basis of fixing a desired value of the redistribution objective in the social welfare function of Chapter 12 below.

objective. To keep the calculations simple, only the relatively more important variables have been included.

$$(1)\quad 1\dot{Y} + .0030\dot{Y}_{-1} + 0.00\dot{S} - .0022\dot{S}_{-1} + .00001\dot{S}_{-2} + 0.00\dot{F} + 0.00\dot{T} + .0472T_{-1}$$

$$= -.2405\dot{\tau}^{pi} + .1207\dot{\tau}^{pi}_{-1} - 3.6856\dot{\tau}^{exc} + 1.8428\dot{\tau}^{exc}_{-1} + 2.2589\dot{\tau}^{inf}$$

$$-1.2741\dot{\tau}^{inf}_{-1} - .0066\dot{\eta} + .1438\dot{I}^{g} + 1.6955\dot{H}^{g} + .0022\dot{e}^{im}$$

$$+ .00007\dot{w}^{min} - .00003\dot{w}^{min}_{-1}$$

$$(2)\quad 0.00\dot{Y} + .0009\dot{Y}_{-1} + 1\dot{S} + .4968\dot{S}_{-1} - .00292S_{-2} + 0.00\dot{F} + 0.00\dot{T} + .26\dot{T}_{-1}$$

$$= -1.02\dot{\tau}^{pi} + .5112\dot{\tau}^{pi}_{-1} - 15.6275\dot{\tau}^{exc} + 7.8158\dot{\tau}^{exc}_{-1} - 15.1480\dot{\tau}^{inf}$$

$$+ 7.574\dot{\tau}^{inf} - .0281\dot{\eta} + .61\dot{I}^{g} + .108\dot{H}^{g} + .0117\dot{e}^{im} - .0157\dot{w}^{min}$$

$$+ .0078\dot{w}^{min}_{-1}$$

$$(3)\quad 0.00\dot{Y} - .0002\dot{Y}_{-1} + 0.00\dot{S} + .0024\dot{S}_{-1} - .00001\dot{S}_{-2} + 1\dot{F} + 0.00\dot{T} - .0516\dot{T}_{-1}$$

$$= .2692\dot{\tau}^{pi} - .1351\dot{\tau}^{pi}_{-1} + 4.126\dot{\tau}^{exc} - 2.063\dot{\tau}^{exc} + 15.8723\dot{\tau}^{inf}$$

$$-7.9107\dot{\tau}^{inf}_{-1} + .0294\dot{\eta} - .638\dot{I}^{g} - .0286\dot{H}^{g} - .0034\dot{e}^{im}$$

$$- .00008\dot{w}^{min} + .00004\dot{w}^{min}_{-1}$$

$$(4)\quad 0.00\dot{Y} + .0013\dot{Y}_{-1} + 0.00\dot{S} + .0207\dot{S}_{-1} - .0007\dot{S}_{-2} + 0.00\dot{F} + 1\dot{T} - .4348\dot{T}_{-1}$$

$$= 2.2882\dot{\tau}^{pi} - 1.1486\dot{\tau}^{pi}_{-1} + 35.0697\dot{\tau}^{exc} - 17.5348\dot{\tau}^{exc}_{-1} - 4.5987\dot{\tau}^{inf}$$

$$+ 2.2302\dot{\tau}^{inf}_{-1} - .0083\dot{\eta} + 18\dot{I}^{g} - .2425\dot{H}^{g} - .0303\dot{e}^{im}$$

$$- .0007\dot{w}^{min} + .0003\dot{w}^{min}_{-1}$$

This model has the following characteristic equation:

$$\lambda^4 - .9322\lambda^3 + .226\lambda^2 - .000707\lambda + .000005 = 0$$

which gives the following characteristic roots:

$$\lambda_1, \lambda_2 = .001542 \pm .004276i$$

$$\lambda_3, \lambda_4 = .464558 \pm .085437i$$

Using the dominant (complex) characteristic roots, we can specify a sinusoidal function for $\dot{Y}$ (and for $\dot{S}$, F, and T, for that matter) as:

$$\dot{Y}_t - \bar{\dot{Y}}(t) = Ce^{.473t} \operatorname{Cos}(10.4t + b)$$

where $\bar{\dot{Y}}(t)$ is the constant trend value of $\dot{Y}$ to be determined by a specified exogenous influence and C and b are constants to be calculated from the initial conditions. The oscillations are damped as indicated by the exponential value $0.473 < 1$.

The oscillations of the present real sector as divorced from the monetary sector are rather weak: a regular oscillation takes 35 years to complete one cycle. Reciprocally stated, the frequency of the oscillation, $\theta/2\pi$, is only 0.038 per year. On the other hand, the frequency of the oscillation in the monetary sector, as was seen in the preceding section, is 0.25. The main source of instability in Brazil is evidently money.

A set of particular solutions under the assumption of a sustained annual unit change in the respective instrument, taken piecewise, one instrument at a time, for this purpose, is given below:

Objective	Instrument
$\bar{\dot{Y}}$	1. $- .4150(\dot{\tau}^{pi}) + .2088(\dot{\tau}^{pi}_{-1})$
	2. $- 6.3754(\dot{\tau}^{exc}) + 3.1877(\dot{\tau}^{exc}_{-1})$
	3. $2.1867(\dot{\tau}^{inf}) - 1.3265(\dot{\tau}^{inf}_{-1})$
	4. $- .0208(\dot{\eta})$
	5. $.1388(\dot{I}^{g})$
	6. $1.7093(\dot{H}^{g})$
	7. $.0045(\dot{e}^{im})$
	8. $.00024(\dot{w}^{min}) -.00011(\dot{w}^{min}_{-1})$
$\bar{\dot{S}}$	1. $- 4.0345(\dot{\tau}^{pi}) + 2.0254(\dot{\tau}^{pi}_{-1})$
	2. $-61.8239(\dot{\tau}^{exc}) + 30.9159(\dot{\tau}^{exc}_{-1})$
	3. $-31.3978(\dot{\tau}^{inf}) + 14.7048(\dot{\tau}^{inf}_{-1})$
	4. $- .2162(\dot{\eta})$
	5. $+ 1.1836(\dot{I}^{g})$
	6. $+ .4254(\dot{H}^{g})$

Objective		Instrument
	7.	$+ \ .0499(\dot{e}^{im})$
	8.	$- \ .0298(\dot{w}^{min}) + .0149(\dot{w}^{min}_{-1})$
$\dot{\bar{F}}$	1.	$.4624(\dot{\tau}^{pi}) - .2321(\dot{\tau}^{pi}_{-1})$
	2.	$7.0865(\dot{\tau}^{exc}) - 3.5432(\dot{\tau}^{exc}_{-1})$
	3.	$15.9447(\dot{\tau}^{inf}) - 7.8494(\dot{\tau}^{inf}_{-1})$
	4.	$.0450(\dot{\eta})$
	5.	$- \ .6329(\dot{I}^{g})$
	6.	$- \ .0489(\dot{H}^{g})$
	7.	$- \ .0060(\dot{e}^{im})$
	8.	$- \ .0003(\dot{w}^{min}) + .00013(\dot{w}^{min}_{-1})$
$\dot{\bar{T}}$	1.	$3.9317(\dot{\tau}^{pi}) - 1.9736(\dot{\tau}^{pi}_{-1})$
	2.	$60.2590(\dot{\tau}^{exc}) - 30.1297(\dot{\tau}^{exc}_{-1})$
	3.	$2.8480(\dot{\tau}^{inf}) - 1.424(\dot{\tau}^{inf}_{-1})$
	4.	$.3187(\dot{\eta})$
	5.	$.0418(\dot{I}^{g})$
	6.	$- \ .4206(\dot{H}^{g})$
	7.	$- \ .0522(\dot{e}^{im})$
	8.	$- \ .0023(\dot{w}^{min}) + .0011(\dot{w}^{min}_{-1})$

If the change in a τ is altered by a fraction/multiple of the assumed sustained change of $\dot{\tau} = 1$, the above constants will be lowered/raised proportionately. For example, if $\dot{I}^g = 1$, the trend value of $\dot{\bar{Y}} = 0.1388$ (in terms of the respective measurement units of this study); if $\dot{I}^g = 2$, $\dot{\bar{Y}} = 0.2776$; if $\dot{I}^g = 0.1$, $\dot{\bar{Y}} = 0.01388$, and so on.

The final complete solution (say, with $\dot{I}^g = 1$), therefore, is

$$\begin{bmatrix} \dot{Y}_t \\ \dot{S} \\ \dot{F} \\ \dot{T} \end{bmatrix}_t = [C]_{4\times 4} \begin{bmatrix} .0015 \pm .0043i \\ .4646 \pm .0854i \end{bmatrix}^t \begin{bmatrix} .1388 \\ -1.1836 \\ -.6329 \\ .0418 \end{bmatrix}$$

where the constants C are to be derived from the initial conditions as in the case of the foregoing monetary submodel, but will not be calculated here. Since the roots are substantially below unity, the divergences from equilibrium will not persist for long periods.

Finally, we present the solution for the combined monetary and real sectors. To keep the calculations and analysis manageable, only four equations (objectives) are selected. These include the equation for $\dot{P}$ from the monetary subsector discussed above, the $\dot{S}$ and $\dot{T}$ equations from the immediately preceding real subsector, and a fourth equation for growth (this time), $\dot{g}$. The characteristic roots for this submodel are:

$$\lambda_1 = .7795$$

$$\lambda_2, \lambda_3 = .003232 \pm .017793i$$

$$\lambda_4 = .426585$$

$$\lambda_5 = .000026$$

$$\lambda_6, \lambda_7 = -.218582 \pm .431369i$$

All roots are less than unity; hence, the system is stable.

The entire system has a monotonic component corresponding to the three positive real roots (and no saw-toothed component) and an oscillatory component corresponding to the two pairs of conjugate complex roots. The sinusoidal function based on the dominant complex roots is:

$$\hat{y}_t - \bar{y}(t) = Ce^{.48t}\,\mathrm{Cos}\,(90t + b)$$

which gives a period of complete oscillation of four years, or an annual frequency of the oscillation of one-quarter. The damping factor (0.48) is almost exactly the same as was found for the pure real sector above.

Particular solutions under the assumption of sustained annual unit changes in the respective instruments, considering each instrument piecewise, are noted below:

$$\dot{g} = -.0330\dot{\tau}^{pi} + .01657\dot{\tau}^{pi}_{-1};\ -.81317\dot{\tau}^{exc} + .4066\dot{\tau}^{exc}_{-1};$$

$$-.3956\dot{\tau}^{inf} + .1999\dot{\tau}^{inf}_{-1};\ -.0007\dot{\eta};\ .0161\dot{I}^{g};$$

$$.0856\dot{H}^{g};\ .0010\dot{e}^{im};\ .00003\dot{w}^{min} - .000011\dot{w}^{min}_{-1}$$

$$\dot{P} = .0257\dot{\tau}^{pi} - .0125\dot{\tau}^{pi}_{-1};\ 2.8138\dot{\tau}^{exc} - 1.4070\dot{\tau}^{exc}_{-1};$$
$$1.164\dot{\tau}^{inf} - .6883\dot{\tau}^{inf}_{-1};\ .0027\dot{\eta};\ -.0556\dot{I}^{g};$$
$$-.0195\dot{H}^{g};\ 1.0875\dot{e}^{im};\ -.00008\dot{w}^{min} + .00004\dot{w}^{min}_{-1}$$

$$\dot{S} = -1.8119\dot{\tau}^{pi} + .9106\dot{\tau}^{pi}_{-1};\ -27.7419\dot{\tau}^{exc} + 13.8711\dot{\tau}^{exc}_{-1};$$
$$-13.2449\dot{\tau}^{inf} + 6.7\dot{\tau}^{inf}_{-1};\ -.0249\dot{\eta};\ .2405\dot{I}^{g};$$
$$.1917\dot{H}^{g};\ .0357\dot{e}^{im};\ -.0153\dot{w}^{min} + .0076\dot{w}^{min}_{-1}$$

$$\dot{\bar{T}} = 3.9838\dot{\tau}^{pi} - 2.002\dot{\tau}^{pi}_{-1};\ 60.99\dot{\tau}^{exc} - 30.50\dot{\tau}^{exc}_{-1};$$
$$-8.887\dot{\tau}^{inf} + 4.2\dot{\tau}^{inf}_{-1};\ -.0156\dot{\eta};\ .339\dot{I}^{g};$$
$$-.4218\dot{H}^{g};\ -.0787\dot{e}^{im};\ -.0018\dot{w}^{min} + .0008\dot{w}^{min}_{-1}$$

It is interesting to note that in spite of wide fluctuations indicated by the general solution, the ultimate effects of different instruments are not out of line with the impact influences. For example, in the case of the redistribution variable, the one-year and the long-period equilibrium multipliers, respectively, are -1.02 and -1.81 with respect to $\dot{\tau}^{pi}$, -15.63 and -27.74 with respect to $\dot{\tau}^{exc}$, 0.1081 and 0.1917 with respect to $\dot{H}^{g}$, and so on. Most values of the long-period equilibrium multiplier are approximately twice the values of the one-year impact multipliers. The short-period effects of different control variables, thus, are crucially important in the Brazilian economy. The finance minister can expect to realize roughly 50 percent of the ultimate effects of his policy changes in about one year. As was seen earlier from dynamic multipliers (Table 6), however, the intermediate multipliers often overshoot both the impact multipliers and the equilibrium (ultimate) multipliers. It seems, therefore, as if the achievable targets can be pursued confidently with appropriately-determined instrument levels. The major concern might then dwell upon reducing unnecessary fluctuations due to changes in those instruments that cause variations in the supply of and the demand for money. This is a control theory problem to which we shall return in Chapter 12.

The complete solution of this segmented model under the assumption that an initial position is one of "static equilibrium" and, for simplicity, considering only the dominant characteristic roots, is given below for a sustained annual change of Cr$1 million in public-health expenditure.

$$
\begin{bmatrix} \dot{g}_t \\ \dot{P}_t \\ \dot{S}_t \\ T_t \end{bmatrix} = \begin{bmatrix} -.1210 & .0177 + .0196i & .0177 - .0196i \\ .2766 & -.0405 + .0436i & -.0405 - .0436i \\ -.0276 & .0040 + .0043i & .0040 - .0043i \\ .5962 & -.0872 + .0940i & -.0872 - .0940i \end{bmatrix} \begin{bmatrix} (.7795)^t \\ (-.2186 + .4314i)^t \\ (-.2186 + .4314i)^t \end{bmatrix}
$$

$$
+ \begin{bmatrix} .0856\dot{H}^g \\ -.0195\dot{H}^g \\ .1917\dot{H}^g \\ -.4218\dot{H}^g \end{bmatrix}, \ldots, \text{ and other endogenous changes piecewise (see above)}
$$

Since the differences among various weights, as given by the matrix of constants C, are not inordinately large, the inclusion in the calculations of the remaining four characteristic roots is not likely to make perceptible differences. From this complete solution, the values of $\dot{g}$, $\dot{S}$, $\dot{P}$, and $\dot{T}$ in any year can be calculated, as was done for the monetary subsector in Table 8.

Sufficient computations and analysis have been presented to indicate that the present model is a reasonably accurate portrayal of the Brazilian economic system and that it is stable and has other desirable dynamic properties. Possible weaknesses remaining in the data, in the specification of the theoretical system, and in the estimation of the econometric model, seem to have been swamped by the dominant responses and influences that have been captured by the model and have been made possible in part by the wide variations (high noise) that have occurred during the sample period. Including the impact and dynamic multipliers, the complete solutions of the estimated difference equations, and the tests of dynamic properties, we now have the necessary information for an analysis of tradeoffs among alternative policy instruments and competing objectives. That analysis will be taken up in Chapter 12.

Before coming to that analysis we will probe somewhat deeper into three aspects of our model. First, in Chapter 9 we test the predictive power of the model; second, in Chapter 10 we analyze the causes of the relative ineffectiveness of the income-tax instrument that was revealed in Chapter 6 above; third, in Chapter 11 we measure the effects of public-finance instruments on size distribution of income (which measurement could not be made within the econometric model).

CHAPTER

9

PREDICTIONS

In selecting our regression equations in Chapter 4 we relied primarily on the goodness of fit, significance levels of the individual coefficients, consistency with a priori knowledge, and, in some cases, plausibility of the computed estimators. Nevertheless, as stated earlier, a real test of an equation is not how best it describes the data but how well it predicts outside the sample period. Indeed, an important use of empirical research is to determine equations that can predict as well as possible the future course of economic variables.

In this study we make two types of predictions: (a) ex post predictions for the sample period for which the values of all predetermined and exogenous variables are available (these calculations are made from the reduced-form coefficients and will be presented in Chapter 12, where they are needed for a relevant analysis); (b) ex ante predictions for the years following the sample period—i. e., 1969 and 1970—for which even some of the exogenous variables have to be forecast before the endogenous variables can be predicted. The latter set of predictions is analyzed in the present chapter.

With a view to testing whether the predictions made from the elaborate models of this study deserve reasonable confidence, several tests are provided. One of the tests is the so-called naive-model test. We set up a naive model based on an exponential growth trend, called "naive model III" (Christ, 1966, pp. 571-75). The model is described in Note b of Table 8. The idea is to compare the prediction errors of the model under reference with the prediction errors of a very simple model based on some average or trend characteristics. If a sophisticated model cannot predict better than a simple naive model then either the equation in question is inadequately specified as an explanation of the combined sample and forecast periods or a structural

TABLE 8

Predictive Behavior of the Model[a]

		Econometric Model				Naive Model[b]	
Variable (1)	Year (2)	Observed (3)	Predicted (4)	Absolute Error (5)	Relative Error (6)	Predicted (7)	Relative Error[c] (8)
Y	1969	12,914	12,384	-530	-.041	12,282	-.05
	1970	14,144	13,657	-487	-.03	14,287	.01
C^p	1969	11,884	11,850	-34	-.003	11,010	-.07
	1970	..	12,329	..	..	14,130	..
I^{pl}	1969	1,805	1,960	155	.09	1,928	.07
	1970	..	2,325	..	..	..	..
P	1969	122.3	128.0[d]	5.70	.05	129.2[d]	.06
	1970	119.8	121.8[d]	2.00	.02	117.7[d]	-.02
$T^{pi} + T^c$	1969	447	468	21	.05	392	-.12
	1970	575	515	-60	-.10	583	.01
T^{exc}	1969	790[e]	916[e]	126	.15	1,079	.36
	1970	1,012	1,086	74	.07	809	-.20
T^{im}	1969	139	158	19	.14	62	-.55
	1970	171	175	4	.02	268	.56
T^{fue}	1969	279	245	-34	-.16	291	.04
	1970	278	302	24	.09	322	.16

Note: The two dots in the body of the table indicate that the data for the relevant cell could not be assembled.

[a] All predictions are from the structural equations: (11.2) for Y; (2.5.5) for C^p, omitting all those coefficients that have a t value of one or less; (4.33) for I^{pl}; (29.4) for P; (32.2) and (33.2) for $T^{pi} + T^c$; (34) for T^{exc}; (35) for T^{im}; and (38) for T^{fue}. Predictions could be made only for those variables for which the relevant data for the postsample period were available at the time when the study was completed.

[b] The naive model used here is set up on the assumption of an exponential growth trend. That is, the current period's value of T is assumed to be equal to the value of the respective T one period earlier times the ratio of the last period's to the preceding period's value, plus a random normal disturbance with mean zero and variance increasing exponentially over time. (For suggestions for different naive models, see Christ [1966, pp. 571-75].) As an illustration: $T = (T_{-1}/T_{-2})T_{-1} + w$; its forecast is $\hat{T} = (T_{-1})^2/T_{-2}$; and the relative forecast error $\hat{w}/T = (T-\hat{T})/T = (T_{-1})^2/(T_{-2}-T) - 1$.

[c] Calculated according to the formula of Relation 24 in note b.

[d] The prediction for $\dot{P}$ was first made in terms of percentage change. It was then expressed as an index — i.e., $100 + \Delta P$.

[e] The observed values of T^{exc} for the three years 1968, 1969, and 1970 are Cr$771, Cr$790, and Cr$1,012, respectively. The low figure for 1969 is apparently due to the effects of tax holidays (referred to earlier in the text) and similar discretionary measures. The prediction was adjusted accordingly.

TABLE 8A

Predictive Tests for Ex Ante Predictions for 1969 and 1970 and Ex Post Predictions for Sample Period 1950-68[a]

	Econometric Model			Naive Model		
Variable (1)	Algebraic Mean Relative Error[b] (1969-70) (2)	Root-Mean-Square Relative Error[c] (1969-70) (3)	Relative Sampling Period, Standard Error of Estimate[d] (1950-68) (4)	Algebraic Mean Relative Error (1969-70) (5)	Root-Mean-Square Relative Error (1969-70) (6)	Relative Sample Period, Standard Error (1950-68) (7)
Y	.035	.035	.005 (.032)[e]	-.020	.036	.044
C^p	.003	.003	.040	-.070	.07	.085
I^{pl}	.09	.09	.17	.07	.07	.27
P	.035	.038	.08	.02	.045	.68
$T^{pi} + T^c$	.075	.08	.18	.055	.085	.19
T^{exc}	.11	.11	.17	.08	.288	.12
T^{im}	.08	.10	.07	.05	.556	.15
T^{fue}	.12	.13	.09	.10	.117	.14

[a] For C^p and I^{pl} the relative errors are only for one year: 1969. The values of columns 2, 3, and 5 were calculated from the relative errors of Table 8. The standard errors of estimate of column 4 are directly from the estimated econometric model. Their relative measures were caluclated by dividing the same by the respective sample means.

[b] Denoting the relative errors of column 6 of Table 8 by v_i (i = 1969, 1970), column 2 of this table is simply: $\frac{1}{2}(v_{69} + v_{70})$.

[c] That is: $[\frac{1}{2}(v_{69}^2 + v_{70}^2)]^{\frac{1}{2}}$.

[d] That is: $\left[\sum_{t=3}^{n}\left(\frac{Y_{t-1}^2}{Y_t \cdot Y_{t-2}} - 1\right)^2 \Big/ (n-2)\right]^{\frac{1}{2}}$

[e] As calculated from the reduced form.

change has occurred between the sample and prediction period.* It might be noted that naive model III employed here requires two lagged values. This is the main reason why ex post predictions for a presample year (1949) were not made.

Certain other predictive tests are made by computing various forecast-error statistics, including the absolute and relative errors of forecasts for individual predictions, and certain summary statistics, such as the algebraic mean errors and the root-mean-square errors for sets of predictions. The summary statistics are presented in Table 8A. The former statistics along with predicted and observed values are presented in Table 8.

Ordinarily predictions should be made in current-year prices. To be consistent with the main estimates of this study, however, we used the same measures of the variables that were employed for the econometric model. As a rule predictions should appropriately be made from the reduced-form estimates, since only the reduced-form tests can test them.[1] However, more than half of the 61 variables appearing in the reduced-form equations are not available for the forecast years. For some of the structural equations, on the other hand, almost all variables are available. Moreover, structural equations can be tested individually for their own validity in the pre- and postsample periods by making predictions from them. In the present chapter we report predictions from the structural equations. Forecasts for some of the endogenous variables from the reduced form, as mentioned earlier, will appear in the final chapter. We tried to make the latter forecast for GDP alone for the year 1969. The resulting errors from this forecast are noted in parentheses in the relevant columns of Tables 8 and 8A.

It will be seen from Table 8 that the predictions are not too inaccurate. Most tax-revenue predictions are off the mark by 7 to 8 percent. The predictions for the national-account variables are superior to those of tax collections. About half the forecast errors are larger (as expected) and the other half smaller (against expectations) than the

---

*It might, incidentally, be noted that the forecast variances will, on the average, be larger than the computed variances in the sample period. This will be so because, apart from the errors due to incorrect model or structural change between the sample period and the prediction period, forecast errors arise from three sources: (a) the errors in forecasting the exogenous variables, (b) the errors in the estimates of the parameters (both the intercepts and the slope parameters), and (c) the random disturbances during the forecasting period. The standard formulas for the forecast variances, which are available from textbooks (see, for example, Christ [1966, pp. 550-56]), are laborious to calculate for multiple regressions, so we will use only the standard errors of estimate (which are only one of the components of forecast errors) as computed from the sample-period data.

respective standard errors of estimate. The root-mean-square error test suggests that the econometric model predicts considerably better than the naive model that has been employed as a benchmark. The test based on the standard error of estimate (Table 8, column 4) and the corresponding standard error of the naive model also support the above result, except in the cases of fuel-tax and import-tax revenues. These are the two taxes that have undergone substantial discretionary changes in the late 1960s; we were not able to incorporate these autonomous changes into an econometric prediction. It is to be noted that while the predictions based on the econometric model deviate generally to the same direction of the regression line for the two contiguous forecast years, the deviations of the predictions of the naive model reverse their signs more frequently, even for contiguous years. It is mainly due to this difference of the two forecasts that according to the algebraic mean-relative error test the econometric predictions do not always come out superior to the predictions made from the naive model. Two implications of this result might be mentioned. First, it is possible that some systematic influences have either been left out while estimating the respective equations or have arisen during the forecast period; if these systematic components can be identified, the predictive power of the equations might be improved. Second, it is also probable that stochastic predictions (not attempted here) might improve upon the deterministic predictions made in this study.[2]

NOTES

1. The appropriateness of various predictive tests in different situations is discussed by Christ (1966, Ch. x).

2. Econometric calculation of stochastic predictions is of relatively very recent interest. An empirical study of this nature has recently been published (Chow, 1972b). A computer program to handle stochastic predictions from econometric models is currently being developed by Cliff Huang and Stanley Black at Vanderbilt University.

CHAPTER

10

AN ANALYSIS OF THE RELATIVE INEFFICIENCY OF THE PERSONAL-INCOME-TAX INSTRUMENT

The regression analysis of the preceding chapters reveals that while Brazil's personal income tax is quite buoyant, even by the standards of advanced countries, the built-in elasticity of this (traditionally most progressive) tax is, indeed, very low, even less than one. The present chapter is devoted to an analysis of the causes of the calculated low built-in elasticity of this tax.

One can test the built-in elasticity of a tax against two norms: first, it may be tested against unity to see whether the tax in question, according to the terminology defined above, is elastic or inelastic; second, one can determine, under certain assumptions, the norm of elasticity that is supposed to be legally built into the prescribed tax schedule and the slab structure by the legislating body and then compare the actual (calculated) elasticity against the legal one. In the case of personal income tax, for example, one can put the whole tax schedule in the computer and, under certain ceteris paribus conditions determine the changes in tax collections that may be contingent upon a given increase in incomes. The ceteris paribus conditions in the present case would include an unchanged pattern of income distribution among different income slabs, unchanged family size, unchanged degree of compliance, and unchanged administrative efficiency. In this study, the first norm will be referred to as "the proportionality norm" and the second as "the legal norm" or "the socially desired norm."

Ordinarily, this kind of analysis should be carried out in respect to each of the taxes studied here. In this study, the analysis is confined for the following main reasons to personal income tax. First, it is the major progressive tax all over the world (and supposedly also in Brazil), the built-in flexibility of which has been the subject of extensive inquiries in many countries; the causes of the lack of positive differential built-in flexibility of this tax are of particular concern. Second, the analysis in respect to one tax can serve as a prototype study for other taxes, which can be taken up by other researchers.

Finally, the tax-evasion estimates of one tax can provide an index of evasion for other taxes also.

Broadly speaking, the computed inelasticity of the basically progressive personal income tax of Brazil can be due to continual changes in one or all of the following three factors:

1. a redistribution of income against high-income taxpayers and in favor of low-income taxpayers or nontaxpayers, or shifts of income from high-income regions to low-income regions;
2. an increase of exemptions and deductions and similar tax allowances relative to the increase of incomes (while the existing levels of administrative efficiency may be low, a deterioration of them is unlikely and can be ignored);
3. an increase in tax avoidance and evasion.

The low buoyancy can be due, in addition to all of the above-mentioned causes, to the continuation of a suboptimal slab structure and/or rate schedule and, of course, to a lowering of tax rates.

THE SLAB STRUCTURE AND RATE SCHEDULE

An analysis of the slab structure and the rate schedule of Brazil's personal income tax will be taken up first.

For the purpose of this study, an average family—broadly, consisting of four members, say, husband, wife, and two children—rather than a single taxpayer, has been chosen as the typical taxpaying unit, for comparison among different countries. This taxpaying unit will be referred to as the "average family."

The lower limit of income at which personal income tax liability currently begins seems to leave little that is wanting in this aspect of the Brazilian tax structure. As shown in Table 9, the income of an average family has to rise nearly six times before tax liability begins. This is almost exactly equal to the rise in the United States in the 1930s and in Israel in the 1950s. In the Asian countries of the British Commonwealth (e.g., India, Pakistan, and Ceylon), even though, due to the British legacy, the income tax is a fairly traditional one, still incomes have to rise above their respective per capita incomes by about twice as much as the corresponding multiple in Brazil before an average family becomes liable to tax.

As to changes in the minimum exemption limit over time, it appears that in 1965 the exemption limit was at an all-time low of Cr$235 in 1963 prices. The changes in actual minimum incomes at which tax liability started in different years, with the corresponding effective tax rates, can be seen in Table 10.

Temporal equity requires, perhaps, that the minimum-exemption limit be raised somewhat. To the extent, however, that the minimum-exemption limit has gone down over the years, the computed elasticity

TABLE 9

Disposable Personal Income of Married Taxpayer With Two Children at Which Personal-Income Tax Liability Begins, in Various Countries

Country (1)	Year (2)	Income at Which Tax Liability Begins: In 1960 U.S. $ (3)	Income at Which Tax Liability Begins: As a Multiple of National Per Capita Income (4)
Australia	1954-55	866	.9
Brazil*	1961	1,564	9.7
	1964	1,056	6.4
	1968	1,056	5.9
Canada	1954-55	2,512	1.8
Ceylon	1954-55	1,330	11.4
Egypt	1951	..	3.6
France			
From labor	1950	2,080	2.6
From other incomes	1950	320	0.4
	1961	1,380	1.0
India	1954-55	931	14.3
	1957-58	665	12.0
Israel	1952-53	3,317	6.6
Italy	1961	240	4.2
Japan			
From labor	1953-54	466	2.3
From other incomes	1953-54	385	1.9
Mexico	1950	..	1.5
New Zealand	1953-54	1,593	1.5
Norway	1952-53	887	1.1
Pakistan	1954-55	1,342	18.0
Sweden	1951-53	1,513	1.5
United Kingdom	1954-55	1,522	1.8
	1961	2,739	1.3
United States*	1913	11,976	11.7
	1917	5,543	4.1
	1929	7,427	6.1
	1939	7,021	5.9
	1944	3,367	1.7
	1955	2.534	1.3
	1961	2,400	1.1
	1966	4,800	1.5
West Germany	1954	711	1.3
	1961	4,000	1.2

*In Brazil, there are two categories of income tax levied on persons: (a) the "schedular" income tax, ranging from 1 percent on labor income to 10 percent on interest income, levied at flat rates, and (b) the "complementary" income tax levied at a progressive rate. The tax rates noted in this table and the following tables are only the complementary tax rates.

Sources: UN (1957, Table 1); UN Yearbook; Sahota (1961, Table 11); Eckstein (1964, Text Tables 6 and 7 and Appendix Tables 3 and 5); and Goode (1964, Table 19).

TABLE 10

Minimum Income at Which Personal-Income Tax Liability Begins in Brazil and Tax Rate on Initial Income Bracket, 1951-69

Year (1)	Per Capita Income in 1963 Cruzeiros (2)	Income in 1963 Cruzeiros at Which Tax Liability Begins (3)	Tax Rate on the Lowest Slab (percent) (4)
1951	102	470	3
1952	108	490	3
1953	108	447	3
1954	117	350	3
1955	121	472	3
1956	123	472	3
1957	128	406	3
1958	132	361	3
1959	136	429	3
1960	145	329	3
1961	153	663	2
1962	158	698	1
1963	155	505	3
1964	155	536	3
1965	153	475	3
1966	158	371	3
1967	159	414	3
1968	170	394	3
1969	177	436	3

of the income tax should be higher rather than lower in relation to the designed elasticity.

The graduation of the rate structure can be seen in Table 11 and Figure 3, which present a comparative picture of the rate structures of several countries. One gets the impression that the Brazilian tax rates rise rather sharply at all ranges of income in comparison with the other countries included. Thus, the starting effective tax rates in Brazil, when an average family's income rises to ten times the national per capita income, are higher than those in countries such as Argentina and Mexico. They became consistently higher in 1968 in comparison with the rates in 1961 (Table 11). The progression is the steepest between 10 and 20 multiples of per capita income. Thus, when taxable incomes rise to 20 times the national per capita income, the Brazilian tax rates become comparable to those of Canada and France. They are

TABLE 11

Personal-Income Taxes as Percent of Earned Income at Selected Income Levels for Married Taxpayer With Three Children, in Various Countries

Country	Income as a Multiple of National Per Capita Income			
	10	20	50	100
Argentina				
From labor	1.4	5.1	11.9	16.4
From commercial income	2.4	7.0	14.9	17.3
Australia	25.4	42.0	57.0	62.3
Brazil				
1961	..	17.5	39.1	51.5
1968	5.5	22.0	44.0	55.0
Burma	..	..	6.3	10.7
Canada	16.1	27.6	41.4	50.9
Ceylon	..	0.3	10.0	18.0
Egypt				
From labor	3.0	3.9	9.6	13.3
From commercial income	16.0	16.0	18.2	21.6
France				
From labor	9.7	17.5	31.1	44.5
From other incomes	23.6	29.8	40.4	51.4
India				
1950	..	2.3	5.7	11.6
1964-65	1.2	4.6	13.6	18.4
Japan				
From labor	26.7	38.5	48.3	51.7
From other incomes	29.2	39.8	48.9	52.0
Mexico				
From labor	1.5	2.3	4.2	7.7
From commercial income	4.1	5.7	9.2	13.2
Philippines	..	1.2	9.1	17.8
United Kingdom: 1970	35.0	54.2	70.9	76.7
United States: 1960	8.5	22.9	33.3	32.9

Note: Figures are for 1950 unless otherwise stated.

Sources: Sahota (1961, Table 12). For Brazil: computed from sources given in Appendix A and from Anuário Estatístico do Brasil (annual), which gives data on the number of wives and the number of children and other dependents and the total exemptions in cruzeiros for each category; it also gives the average tax rates paid in various income brackets.

Figure 3: Personal-Income Taxes as Percent of Earned Income at Chosen Multiples of Income for an "Average Family," 1950 (Brazil, 1970)

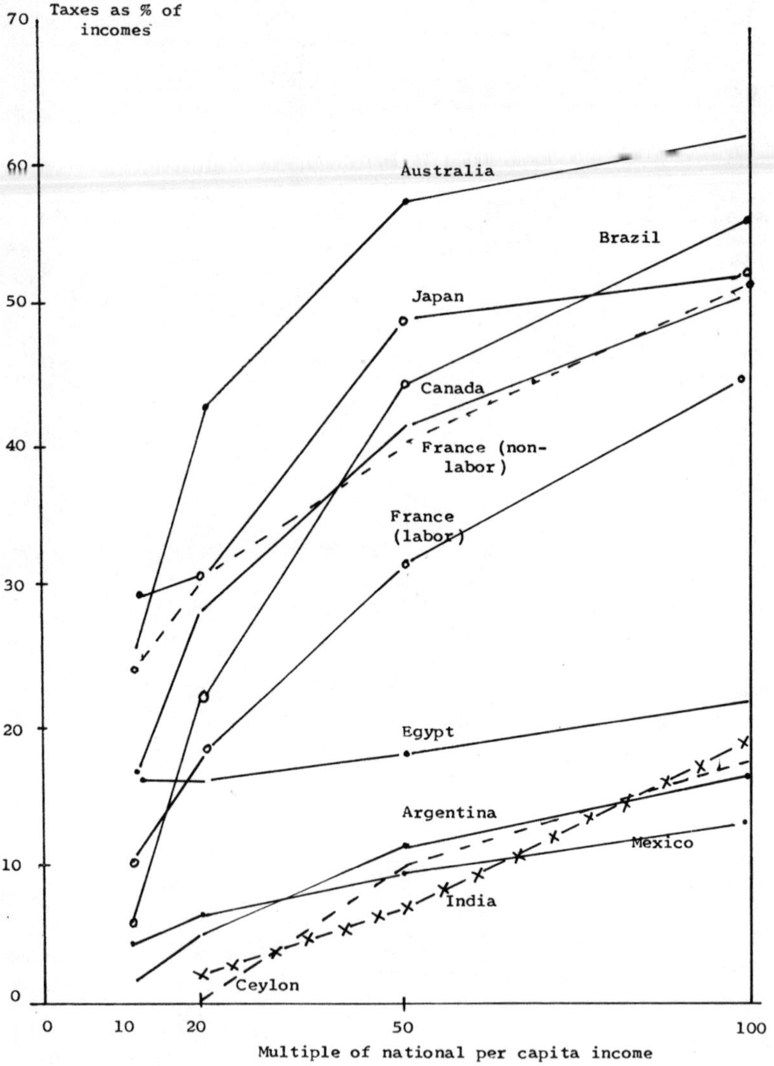

Source: Table 11.

far below those of the latter countries at a multiple of 10. At a multiple of 50 only Japan and Australia (among the countries of Table 11) lie above Brazil. An overall view of the rate line for Brazil indicates that the middle- and lower-middle classes of Brazil seem to be overtaxed relative to the other income classes of Brazil, when compared with the corresponding relatives for other countries. The tax graduation for families with incomes up to 50 times the national per capita income seems to be particularly steep.

A comparative view of the interbracket progressivity of Brazil's income tax is given in Table 12. Using Pigou's relation (see the note to Table 12), progressivity coefficients for the corresponding multiples of per capita income are computed for Brazil, India, the United States, and the United Kingdom. The results are also sketched in Figure 4. Several interesting points can be noted from Table 12. While in the advanced countries, the United Kingdom and the United States, the effective tax liability starts at less than half their per capita incomes, in Brazil it starts only when her per capita income reaches about .5 times. In India, on the other side, the effective tax liability does not start until about 15 times her per capita income. It is interesting to note that the high marginal tax rates at upper income brackets in India and the United Kingdom (over 90 percent at the uppermost bracket) do not have as close a relationship with their interbracket progressivity as have the relatively low U. S. and Brazilian marginal rates at upper income levels (roughly about 60 percent). The calculated measure of progressivity (the change in marginal tax rates) declines rather sharply up to about 35 multiples of per capita income in all four countries. It declines but slowly thereafter, except in India, where it starts rising again. The changes in the magnitudes of progressivity are much less pronounced in India and the United States, as compared with those in Brazil and the United Kingdom (observe the scale for these countries in Figure 4), perhaps because of the spread of tax rates over a wider range of incomes in India and the United States. Considering the fact that the effective rates calculated here are the result of all sorts of exemptions and deductions, as well as concessions in the form of lower rates of capital gains tax, etc., the high marginal tax rates at upper-income levels are a poor index of interbracket income-tax progressivity. In fact, the present results indicate that while the British income tax system has the highest marginal tax rates among the countries studied here, the rate of change of marginal rates or progressivity over the rising income brackets is no higher than those in other countries. It is clearly much lower than in India. The progressivity line for Brazil conforms to the patterns seen for the United States and the United Kingdom, except that the peak is shifted to the right (to the middle-income ranges). This shift occurs in two steps instead of one, as is the case in the two developed countries under reference. Let us put equity aside for a moment and look at the expected buoyancy. Since tax compliance among middle-income classes, a majority of whom

TABLE 12

Interbracket Progressivity

Multiple of Per Capita Income (1)	Brazil, 1961 Income (Cr$) (2)	Effective Tax Rate (percent) (3)	Marginal Tax Rate (percent) (4)	Progressivity Coefficient (5)	Progressivity Index (6)
1	56.3	0	0	..	..
2	113	0	0	..	..
5	282	2.7	2	.00506	22
10	563	9.2	17	.03385	142
15	845	12.4	26	.02386	100
20	1,126	17.3	29	.01684	71
25	1,410	20.2	32	.01168	49
30	1,689	24.7	35	.00935	39
35	1,971	26.2	35	.00562	24
40	2,252	27.0	38	.00559	23
45	2,534	28.6	40	.00543	23
50	2,815	30.2	40	.00426	18
75	4,230	36.0	45	.00421	18
100	5,630	42.2	50	.00284	12
200	11,260	42.2	50	.00092	4

India, 1964-65 Income (Rs) (7)	Progressivity Coefficient (8)	Progressivity Index (9)	United States, 1960 Income (U.S.$) (10)	Progressivity Coefficient (11)	Progressivity Index (12)
422	.00	..	2,400	.003822	458
844	.00	..	4,800	.002950	354
2,110	.00	..	12,000	.001750	210
4,220	.00	..	24,000	.001130	136
6,330	.001135	100	36,000	.000833	100
8,440	.000845	74	48,000	.000855	103
10,550	.000624	55	60,000	.000663	79
12,660	.000622	55	72,000	.000518	62
14,770	.000576	51	84,000	.000418	50
16,880	.000508	45	96,000	.000342	41
18,990	.000543	48	108,000	.000356	43
21,100	.000521	46	120,000	.000311	37
31,600	.000792	70	180,000	.000232	28
42,200	.000627	55	240,000	.000169	20
84,400	.000416	37	480,000	.000100	12

United Kingdom, 1969		
Income (£) (13)	Progressivity Coefficient (14)	Progressivity Index (15)
730	.03869	946
1,460	.02201	538
3,650	.00395	97
7,300	.00366	89
10,950	.00409	100
14,600	.00311	76
18,500	.00181	44
21,900	.00150	37
25,550	.00109	27
29,200	.00082	20
32,850	.00067	16
36,500	.00048	12
54,750	.00031	8
73,000	.00016	4
146,000	.00003	0.7

Note: For Brazil the schedular part of personal-income tax is ignored. The progressivity relation used is:

$$\frac{m_j - e_{j-1}}{\frac{1}{2}(Y_j + Y_{j-1})}$$

where m is the marginal tax rate, e stands for the effective tax rate obtained by dividing actual tax collections by actual taxable incomes of the respective brackets, and Y is taxable income. The subscript j stands for the level of income at which progressivity coefficients are being computed. This is the progressivity relation suggested by Pigou (1928, Part II, Chapter ii). The dated version of it distinguishes it from a similar relation employed by Slitor (1948). In addition, since the discrete jumps in income levels are too large to be approximated to differentiation (in terms of which Pigou derived his relation), the denominator is the mean of Y_j and Y_{j-1}, rather than the Y_j used by Slitor.

Sources: Brazil: Appendix A; United Kingdom: Prest (1970, Table 16) and Inland Revenue Statistics; United States: Goode (1964, Tables A-10 and A-11); India: Statistical Abstract of India.

Figure 4: Interbracket Progressivity Lines: Brazil, India, United Kingdom, and United States

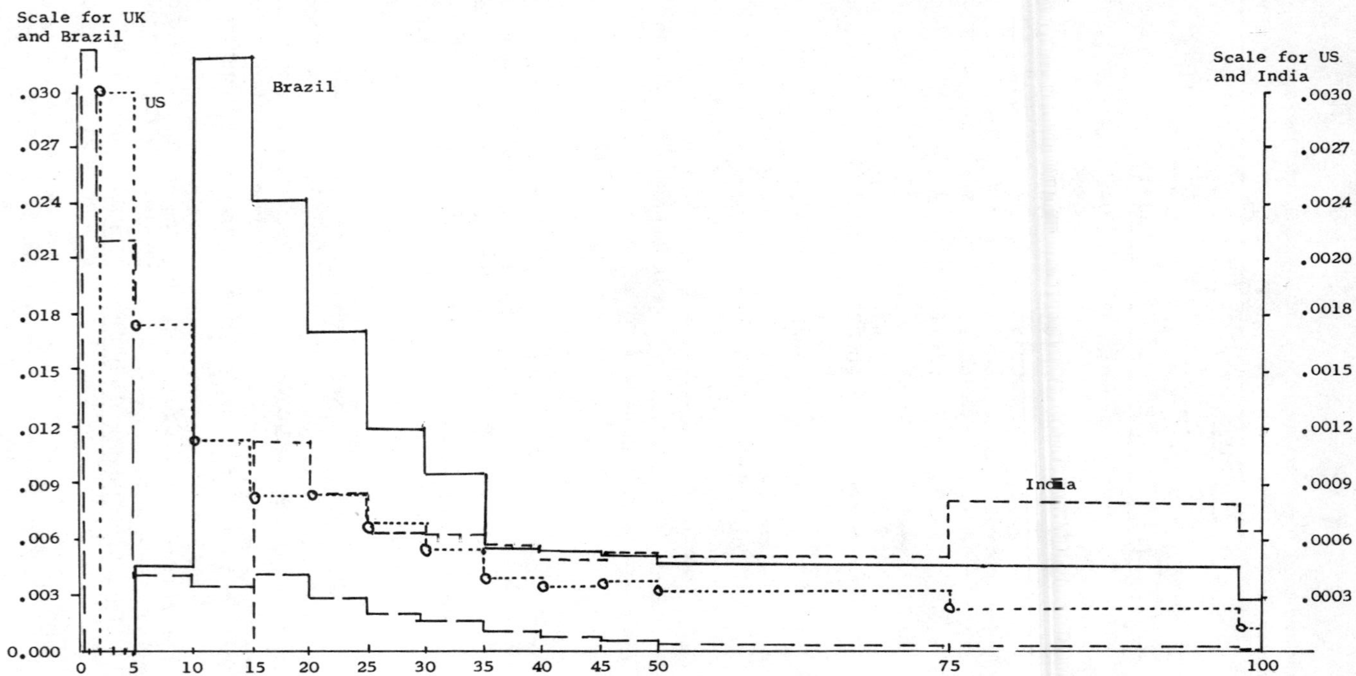

Source: Table 12.

TABLE 13

Highest and Lowest Personal-Income-Tax Rates in Various Countries

		Minimum		Maximum	
Country (1)	Year (2)	Income Above (U.S.$) (3)	Rate (percent) (4)	Income Above (U.S.$) (5)	Rate (percent) (6)
Argentina	1964	522	4.0	65,200	45.0
Australia	1953-54	238	0.4	38,047	66.7
	1967	462	0.4	35,555	66.7
Belgium	1966	3,200	27.5	80,000	60.0
Brazil	1964	570	3.0	19,000	65.0
	1968	422	3.3	12,800	55.0
Cambodia	1968	..	13.0	..	62.0
Canada	1954-55	1,092	15.0	436,860	80.0
Ceylon	1953-54	1,330	10.0	26,604	85.0
	1968	..	7.5	..	65.0
Chile	1966	..	10.0	45,400	60.0
China (Taiwan)	1968	175	6.0	50,000	60.0
Colombia	1963	278	0.5	222,222	51.0
El Salvador		2,800	2.85	100,000	60.0
France	1961	460	5.0	45,640	65.0
India	1957-58	665	3.0	66,500	78.0
	1970	640	5.0	33,333	75.0
Indonesia	1968	500	15.0	13,000	50.0
Italy	1961	500	2.0	800,000	50.0
Japan	1953-54	590	15.0	146,670	65.0
	1968	456	10.0	165,700	75.0
Korea	1968	300	7.7	3,500	55.0
Malaysia	1968	660	6.0	16,660	50.0
New Zealand	1953-54	296	12.8	11,252	60.0
Pakistan	1953-54	1,118	1.2	31,956	78.1
	1968	842	2.0	21,050	70.0
Philippines	1968	450	3.0	125,000	70.0
Singapore	1968	660	6.0	16,660	55.0
Spain	1964	1,666	2.5	100,000	44.0
Sweden	1951-52	205	10.0	41,055	70.0
Switzerland	1965	1,550	0.228	28,480	23.5
Thailand	1968	200	10.0	20,000	50.0
United Kingdom	1954-55	296	12.5	57,642	95.0
United States	1954-55	2,111	20.0	316,754	92.0
	1965	3,500	14.0	1,003,500	70.0
Venezuela	1968	2,667	4.5	1,777,778	45.0
West Germany	1953-54	1,118	1.2	31,956	78.1
	1961	428	20.0	52,900	53.0

Sources: G.S. Sahota (1961, Table 13) and Japan Tax Association (1970) for some of the Asian countries; Appendix for Brazil; Arthur Anderson and Co. (1968) and Oldman (1964) for some of the Latin American countries.

are salaried people, is likely to be high relative to other income groups, the numerical value of buoyancy should, other things being equal, be higher than the legally-designed elasticity measure. On the other hand, if heavy taxation causes the earning capacity and the incentives to work and invest of these people to be more adversely affected than would be the case with the taxpayers as a whole if the same revenues were to be collected by alternative graduation, the buoyancy will be reduced. The effects of these changes on the pretax incomes of the taxpayers under reference are not known for certain—they could even be favorable if the work-leisure and saving-income schedules of these people are not very elastic. In other words, the direction of effects of the increased graduation of tax rates on the economic incentives of the people in the low and middle ranges of incomes is not always certain. On the other hand, the cut in their posttax incomes and, pari passu, an increase in tax collections due to rapidly rising tax rates are determinate measures. Accordingly, one can postulate that the effects of the stated steepness in tax graduation on the buoyancy of income tax will, on the whole, be positive.

On the upper end of the ladder, the maximum tax rates in Brazil do not appear to be excessive. Among the 30 countries studied in Table 13, they are higher only than those prevailing in Argentina, Colombia, Indonesia, Malaysia, Spain, Switzerland, Thailand, Venezuela, Italy, and West Germany. They are substantially lower than the rates prevalent in the British Commonwealth countries. As can be seen from Table 13, most countries seem to be settling around an upper limit of 60 to 65 percent (except for the British Commonwealth countries). Since high marginal tax rates at upper-income levels are commonly believed to be one of the major causes of tax avoidance, the relative absence of this cause in Brazil should provide less ground for tax erosion in this country. Certain other aspects of the slab structure and rate schedule will come under investigation in the section that follows.

TAX AVOIDANCE AND EVASION

A Look into the Changes in Income Distribution

As noted at the beginning of this chapter, the elasticity of income tax can be influenced by changes in income distribution, exemptions and deductions, and tax avoidance and evasion. A comparison of personal incomes for the census years 1960 and 1970 clearly indicates that income distribution in Brazil worsened during the decade (the 1950 census of population collected no information on income distribution).*

---

*For the Gini indexes prepared from the 1960 and 1970 census reports, see Chapter 11. A detailed analysis of size-income distribution appears in Chapter 11, too.

Therefore, the elasticity of revenues from personal income tax should have gone up—but it has not. A common though (insofar as most developing countries are concerned) often misleading method of gauging changes in income distribution consists of an analysis of income tax returns. It is misleading because of the distorting effects of evasion and because the tax covers a small proportion of the population of these countries. In Brazil the tax-return data are an even weaker index of income distribution, as only a relatively small and declining category of nonwithheld income is published by income classes while little is known about the income classes of those for whom taxes are deducted at source. It is believed that taxpayers are increasingly shifting from the nonwithheld to the withheld tax category, or have been at least during recent years. Nonetheless, it may be worth the while to get an overall view of how the changing tax rates appear to have affected at least the direction of income distribution.

The changing marginal rates on various income classes are given in Table 14 for the years 1951, 1961, 1965, and 1970. The values for the first three years are also sketched in Figure 5. The curve for 1970 was not sketched as it tends to clutter up with the 1961 and 1965 curves in several income brackets and may spoil a clear view of the others. Broadly speaking, the 1970 curve lies a little above the 1961 and 1965 curves up to about Cr$2, 200 (in 1964 prices) and below them thereafter. Accordingly, this tax became less progressive in 1970 than it had been in the early and mid-1960s. Apart from that, the 1970 schedule seems closer to the 1961 schedule than to the 1965 schedule. It is transparent from Figure 5 that, in absolute terms, the middle- and upper-income classes were subjected to marginal tax rates substantially higher in the 1960s than in the early 1950s, when compared with the taxpayers at the lower rungs of the ladder. In proportionate terms, however, though tax rates have been raised at all levels, they have gone up relatively more for the lower- and middle-income taxpayers. For instance, in 1965, as compared to 1950, while the marginal tax rates were only about 30 to 45 percent higher on taxpayers with incomes of Cr$31, 000 or above, and approximately double on those with incomes between Cr$12, 000 and Cr$31, 000 (in 1964 prices), they ranged from a little less than three to more than three times as high on the lower-, lower-middle, and middle-income ranges with incomes (in 1964 prices) of Cr$1, 320 to Cr$9, 600. This bias was rectified somewhat between 1961 and 1965 but was reestablished between 1965 and 1970, as can be seen from column 6 of Table 14. From the equity point of view the taxpayers in the middle ranges of income were definitely worse off in 1965 than in 1951, as compared with the taxpayers in the starting and the end ranges of income. From the considerations of compliance and the cost of collection, however, this change appears to be preferable in that the degree of compliance and the cost of collection are minimal in these ranges of income. From the point of view of economic effects, it is difficult to say whether the changes under discussion have served the objectives of collecting revenues with a minimum of adverse effects

TABLE 14

Changes in Marginal-Income-Tax Rates on Different Income Brackets, Brazil, 1951, 1961, 1965, and 1970 (in percentages)

1964 Bracket (thousands of 1964 Cruzeiros)	Marginal-Tax Rate						
	1951	1961	1965	1970	3 - 2	4 - 3	5 - 3
(1)	(2)	(3)	(4)	(5)	(6)	(7)	(8)
706 – 829	0	0		0			
830 – 882	0	0	3	3	0	3	3
882 – 1,152	3	0	5	5	-3	5	5
1,152 – 1,323	3	2	5	8	-1	3	6
1,323 – 1,440	3	2	8	8	-1	6	6
1,440 – 1,680	3	3	8	8	0	5	5
1,680 – 1,764	3	6	8	12	3	2	6
1,764 – 1,920	5	6	12	12	1	6	6
1,920 – 2,160	5	10	12	12	5	2	2
2,160 – 2,205	5	14	12	12	9	2	-2
2,205 – 2,400	5	14	16	12	9	2	-2
2,400 – 2,646	5	17	16	16	12	3	-1
2,646 – 3,024	7	17	20	16	10	3	-1
3,024 – 3,360	7	20	20	16	13	0	-4
3,360 – 3,528	7	23	20	20	16	2	-3
3,528 – 3,840	9	23	25	20	14	2	-3
3,840 – 4,410	9	26	25	20	17	4	-6
4,410 – 4,800	9	26	30	25	17	4	-1
4,800 – 5,292	12	29	30	25	17	4	-4
5,292 – 5,760	12	29	35	25	17	4	-4
5,760 – 7,350	15	32	35	30	17	4	-2
7,350 – 7,680	15	32	40	30	17	8	-2
7,680 – 9,600	15	35	40	30	20	5	-5
9,600 – 10,209	18	38	40	35	20	5	-3
10,209 – 12,000	18	38	45	35	20	7	-3
12,000 – 13,230	21	40	45	40	19	7	0
13,230 – 14,400	21	40	51	40	19	11	0
14,400 – 17,640	24	45	51	40	21	11	-5
17,640 – 21,600	27	45	57	45	18	12	0
21,600 – 23,520	30	50	57	45	20	12	-5
23,520 – 31,230	30	50	65	50	20	15	0
31,230 – 62,460	35	50	65	50	15	15	0
62,460 – 93,690	40	50	65	50	10	15	0
93,690 – over	50	50	65	50	0	15	0

Note: Income brackets were converted to 1964 prices for all years. In 1961, in addition to the income-tax rates given, there was a surtax of 3 percent on all taxable incomes.

Sources: LEX (1951, Article 26, p. 432; 1961, and 1965); Government of Brazil (1965); Appendix A.

Figure 5: Changes in Marginal-Income-Tax Rates on Different Income Brackets, Brazil, 1951, 1961, and 1965

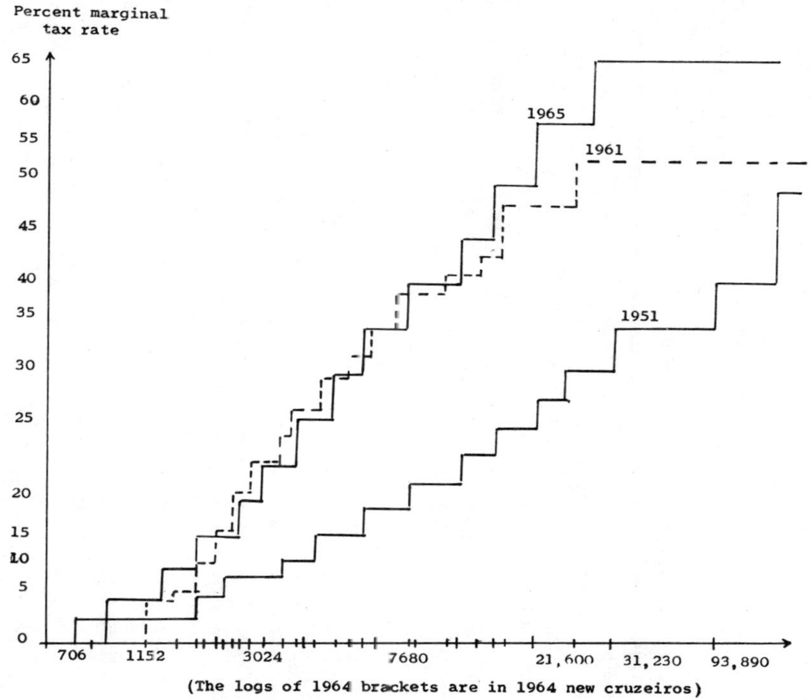

Source: Table 14.

on economic growth and checking inflation more effectively than the alternative changes of, say, equal proportionate increases in tax rates at all levels.* For, while the disincentive effects on economic growth of high marginal tax rates on higher incomes are fairly well recognized, the disincentive effects on lower- and middle-income classes—much of whose investment, in general, goes into education of themselves and their children—are not fully appreciated. In any case, the main concern of this section is to see whether there have been any changes in income distribution that may have influenced tax elasticity via income redistribution. As far as the changes in the tax rates are concerned, the cut in the middle ranges of income has been deeper than that in the upper ranges. Hence, the changes in the marginal income tax rates are likely to have caused a redistribution of incomes in favor of the higher-income people.

Similar conclusions follow from the effective rates of taxation given in Table 15 and sketched in Figure 6. The effective rates of income taxation have risen more in the middle ranges of income relative to at least the three uppermost income brackets (see Table 15, column 4). Once again, however, there is no sanctity about the 1950 tax rates with which the 1960 tax rates are being compared. It could be that the middle-range taxpayers were relatively lightly taxed in 1950. It is difficult to assign absolute values to equity. All one can do is compare the two schedules over time.

The results are very disappointing when we measure the intra-bracket changes in assessed incomes between 1950 and 1960. A glance at Table 16 and Figure 7 indicates that, even though per capita incomes rose by about one-third between 1950 and 1960, the real incomes assessed to income tax declined, by and large, uniformly over all ranges, except for a few. Some of the exceptions to this decline, however, are noteworthy. For instance, the assessed incomes of the two topmost brackets have more than doubled; the same is true for the first two brackets. Note that these changes pertain to the 1950s when there were no significant shifts of the taxpayers from the tax category under discussion to the tax category of "deductions at source." As a matter of fact, a study of the said shifts, to be presented in Table 21, indicates a lower amount of income shifts to "deductions at source" than would account for the fall in assessed incomes in Table 16. It will be seen in Table 21, for instance, that between 1952 and 1960, on balance, no more than Cr$210,000 (in 1949 prices) worth of assessed incomes shifted to the tax category of "deductions at source."

Given no large-scale shifts of taxpayers to other tax-payment categories, a fall of the assessed incomes in the lower and middle ranges of the tax brackets can still be explained perhaps by the

---

*Of course, there is no presumption that the relative tax rates of 1951 were optimal.

TABLE 15

Personal Income Tax Actually Paid as Percentage of Income Assessed, by Income Bracket, Brazil, 1950 and 1960

1959 Income Bracket in 1949 Cruzeiros	Taxes Paid as Percentage of Income Assessed		
	1950	1960	3 - 2
(1)	(2)	(3)	(4)
15– 22	0.0	2.9	2.9
22– 27	1.93	3.8	1.9
27– 30	2.67	3.8	1.1
30– 37	2.67	4.7	2.0
37– 49	2.67	6.3	3.6
49– 54	2.67	8.9	6.2
54– 74	3.98	11.5	7.5
74– 81	3.98	11.5	7.5
81– 98	4.97	13.9	9.0
98– 107	4.97	13.9	9.0
107– 123	4.97	16.1	11.1
123– 134	6.09	16.1	10.0
134– 148	6.09	18.4	12.3
148– 172	7.56	22.1	14.5
172– 179	7.56	29.0	21.4
179– 246	9.97	29.0	19.3
246– 268	9.97	29.0	19.3
268– 358	12.24	29.0	16.8
358– 447	14.22	35.8	21.6
447– 492	16.11	35.8	19.7
492– 536	16.11	35.8	19.7
536– 626	17.89	35.8	17.9
626– 738	20.46	43.5	23.0
738– 894	20.46	43.5	23.0
894–1,788	25.65	43.5	17.9
1,788–2,682	31.12	43.5	12.4
Over 2,682	37.33	43.5	6.2

adverse distribution of incomes indicated in Tables 14 and 15. One fails to rake up any reasons other than outright evasion, however, for the failure of the assessed incomes of the upper-income brackets to rise substantially more than they were observed to rise when changes in tax laws and income distribution were favorable to them.

Figure 6: Effective Tax Rates, Brazil, 1950 and 1960

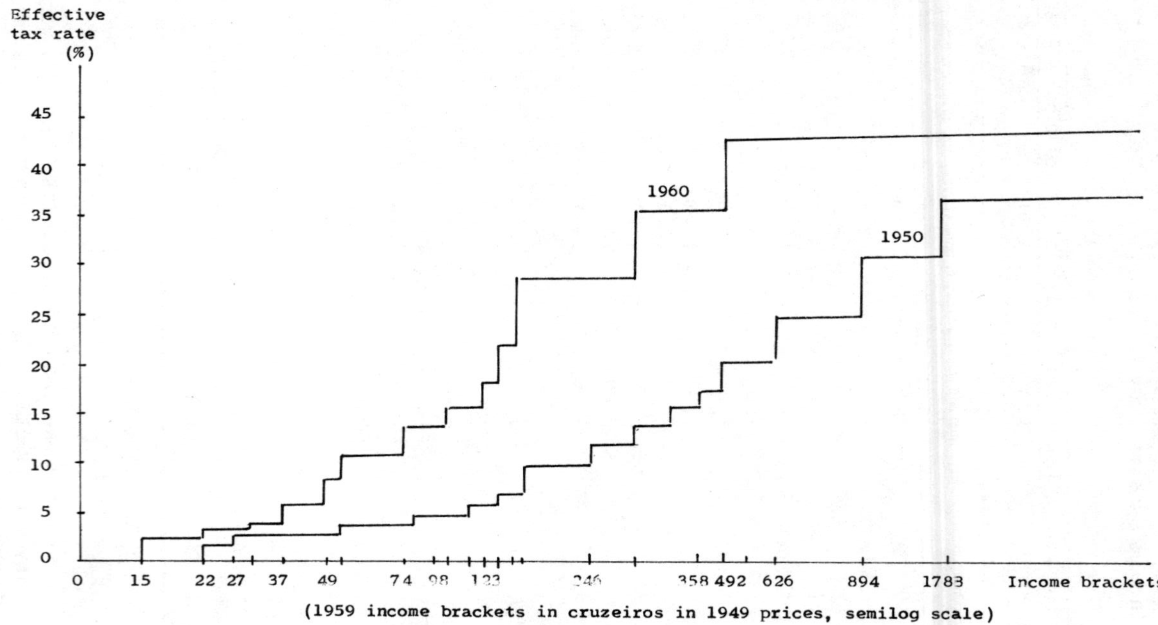

Source: Table 15.

TABLE 16

Shifts in Incomes among Personal-Income Taxpayers, Brazil, 1950 and 1960 (Cr$ thousand)

Income Brackets	Total Income Assessed		
	1950	1960	3 - 2
(1)	(2)	(3)	(4)
15- 22	0	1, 681	+1681
22- 27	1, 085	1, 855	+ 800
27- 30	3, 979	1, 855	-2124
30- 37	3, 979	1, 789	-2190
37- 49	3, 979	2, 248	-1731
49- 54	3, 979	2, 631	-1348
54- 74	2, 257	2, 631	+ 374
74- 81	2, 257	1, 585	- 672
81- 98	1, 537	1, 585	+ 48
98- 107	1, 537	1, 585	+ 48
107- 123	1, 088	1, 068	- 20
123- 134	1, 088	733	- 355
134- 148	1, 299	733	- 566
148- 172	1, 299	554	- 745
172- 179	1, 299	1, 037	- 262
179- 246	1, 599	1, 037	- 562
246- 268	1, 599	1, 074	- 525
268- 358	979	1, 074	+ 95
358- 447	643	1, 074	+ 431
447- 492	439	1, 074	+ 635
492- 536	439	298	- 141
536- 626	323	298	- 25
626- 738	582	298	- 289
738- 894	582	317	- 265
894-1, 788	546	317	- 229
1, 788-2, 682	130	317	+ 187
Over 2, 682	140	317	+ 177

Note: Income brackets pertain to 1949 and are in cruzeiros in 1949 prices. All other figures are in thousands of 1949 cruzeiros. What is called a "cruzeiro" here is the currency unit of 1971. In the late 1960s the Brazilian currency unit was called "new cruzeiro," equivalent to 1, 000 earlier cruzeiros.

Figure 7: Shifts in Incomes among Personal Income Taxpayers, Brazil, 1950 and 1960

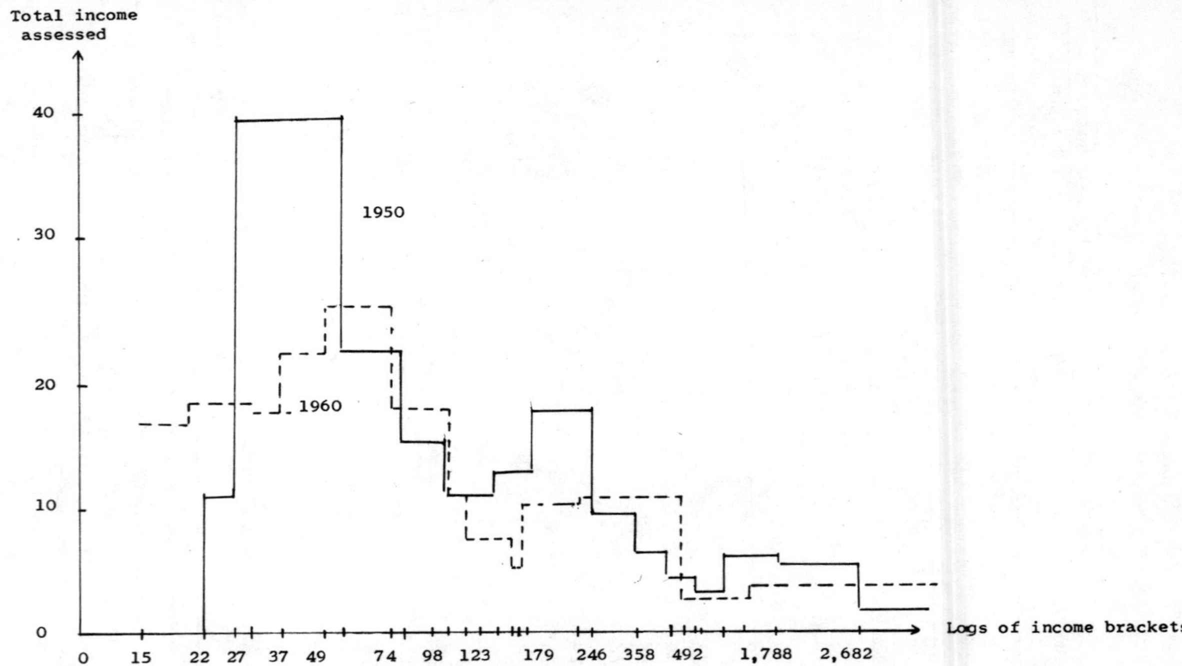

Note: The income brackets pertain to 1949 and are in new cruzeiros in 1949 prices. The incomes assessed are in thousands of new cruzeiros.

Source: Table 16.

A Brief Analysis of Exemptions and Deductions

Between 1950 and 1964 national income more than doubled, and per capita income rose by about 40 percent. The exemptions for wife, children, and other dependents remained, by and large, steady: around Cr$15 (in 1949 prices) for wife and half of that per child or other dependent.* From these two pieces of information—40-percent rise in per capita incomes and almost constant exemption per dependent in constant-price cruzeiros during the reference period 1950-64—one would expect the total exemptions as a percentage of assessed incomes to decline. The observed values of the exemption rate point to the contrary, as can be seen from the following figures pertaining to the annual income-tax payments via nonwithheld tax returns†:

	Exemptions as a Percentage of Gross Personal Incomes
1952	24.5
1953	21.9
1954	25.5
1955	34.5
1956	32.5
1957	30.7
1958	32.3
1959	31.2
1960	27.6
1961	29.6

It will be seen, thus, that the actual exemption rate rose by more than 20 percent during the nine years noted. Even if it can be argued that the rates of marriage and population growth rose during the period under study, it is difficult to believe the same could more than cancel out the effects of rising per capita incomes with constant exemption allowances. Moreover, it cannot be said that in the process of the shifts of taxpayers or part of their incomes from one category of tax-payment to another the taxpayers with a relatively larger number of exemptions have remained in the category under investigation. To the contrary, the analysis of these shifts (as seen in Table 16) indicates that the highest-income taxpayers, for whom exemptions are an insignificant part of their incomes, have not shifted out of this tax category to the same extent that the remaining taxpayers have. At best the

*The exemption for wife, in 1949 prices, came to approximately NCr$14.2 in 1952, NCr$10.3 in 1956, NCr$15.3 in 1957, NCr$15.00 in 1961, and about NCr$11.00 in 1969.

†The source for these figures is IPEA, Diagnóstico (1965).

shifts, if any, seem to have been, by and large, proportionately distributed over all income brackets.

There is the possibility that personal deductions, such as those on medical expenses, philanthropic contributions, local taxes paid, and so on, have increased at a higher rate relative to gross personal incomes. Herein also may lie the major loopholes for tax avoidance. No published data are available in Brazil to throw light on personal deductions.

An Estimate of Tax Evasion

The ultimate explanation of the inelasticity of the Brazilian income tax might be an increasing rate of outright evasion. In this study, an approximate method has been worked out to estimate tax evasion. Briefly, the method consists of computing the elasticity of personal income tax as legally designed by the government and as found in actual performance under constant-income-distribution and certain other plausible assumptions. The effects of changes in income distribution can be superimposed on the estimates above. The difference between the designed (or legal or nominal) elasticity and the computed or actual elasticity is tax avoidance and tax evasion.

Legal tax avoidance is, perhaps, not of great consequence in the underdeveloped countries (though it is so in developed countries such as the United States). Outright tax evasion, such as underreporting of incomes (or the failure to report them at all), appears to be a serious matter. Low tax-paying morality and inefficient tax administration might be partly responsible for it. In any case, a distinction between the two is not of great importance insofar as the government's revenue and economic aims are concerned. Both avoidance and evasion frustrate the efforts of the government to achieve its public-finance objective. As a shorthand expression for the purpose of this chapter, therefore, all discrepancies between the personal income taxes that ought to be paid and those that are actually collected will be termed "evasion."

The detailed computations of tax evasion appear in Tables 16 to 22. The derivation of the final estimates of the evasion of Brazil's personal income tax can be better understood and appreciated if Tables 17 to 22 are followed through in sequence, column by column. The tables are fairly self-explanatory. The key steps underlying these calculations, where not explicit enough from the column headings, are explained in footnotes to various tables, in particular in the footnotes to Table 18.

The data by income brackets are available only for nonwithheld tax payments. As a first approximation, therefore, the interbracket and intrabracket evasions over time were estimated for personal income tax payments from this source only. Subsequently, corrections were made to account for shifts between various tax categories—e. g., between this category and the personal income tax deductions at source.

Table 17 presents an estimation of the nominal built-in elasticity, or the elasticity legally provided for in the progressively-designed slab structure and rate schedule. According to these estimates, using the 1960 structure of income tax, a 1-percent increase in national income should lead to a 2.19-percent rise in tax collections, provided there were no changes in income distribution and no disproportionate shifts of taxpayers of different brackets out of this component of tax payments.

The calculated elasticity coefficient $\alpha = 2.19 > 2$ is apparently too high and seems to exceed the upper bound of the progressivity norm suggested elsewhere by the present author (Sahota, 1974). However, this is the progressivity of legal taxation, which may be unachievable in practice.

Similar estimates were made for actual changes in national income for each of the reference years. The detailed steps for 1956 are presented in Table 18, which also contains the footnotes explaining each step. It will be observed that, for this year, tax evasion in 1949 prices comes to Cr$457,400 (column 13) out of an actual collection of Cr$2,117,000 (column 6), giving a tax-evasion rate of approximately 20 percent.

Similar calculations were made for each of the reference years and are presented in Table 19. The last row of this table shows tax-evasion figures for given years.

Having obtained these evasion figures, corrections were then made for shifts of taxpayers between different income tax categories. This is done in Table 20. The last row of Table 19 appears in column 4 of Table 20. From this column are subtracted the tax-payments corresponding to the shifts of taxpayers out of this tax category. These shifts are computed in Table 21 under the same principle as was used in Tables 17 and 18. That is, the required tax-payments for the actual increase in personal incomes under an unchanged income distribution were computed for the category "deductions at source" pertaining to personal income taxation only. These were subtracted from the actual tax deductions. This was repeated for each year for which the data are available. The balance was considered to be the shift into this category from the category of annual returns. The resulting figures (Table 21, column 6) were subtracted from Table 20, column 4, and appear in Table 20, column 7. This is the net tax evasion in New Cruzeiros in 1949 prices. The percentage of annual tax evasion is given in Table 20, column 8, cumulated cruzeiro evasion in column 9, and cumulated percentage of evasion in column 10. The results are also sketched in Figures 8 and 9.

Similar estimates of tax evasion could not be extended to the more recent years due to lack of published data by income brackets since 1961. To estimate evasion for the more recent years a slightly different method was employed: this is presented in Table 22. In this table the

TABLE 17

Estimation of Nominal Built-In Elasticity of Personal-Income Tax, Brazil, as of 1960

Income Bracket (1)	1960 Assessed Incomes (2)	1960 Income Tax Yield as Percent of Assessed Incomes = Average Tax Rates (3)	Col. 2 × Col. 3 = Tax Yield (4)	Required Increases and Shifts of Assessed Incomes When National Income Increases by 1 Percent and Distribution Remains Unchanged: Col. 4 × $\frac{101}{100}$ = Initial Increase (5)	Shift into (from Lower Slabs) (6)	Shift out (to Upper Slabs) (7)	5 + 6 - 7 = Final Position (8)	Col. 3 × Col. 8 = Required Tax Yield When Incomes Increase at 1 Percent (9)	Col. 9 - Col. 4 = Required Additions in Tax Yield When Incomes Rise by 1 Percent (10)
91– 135	10,256	2.9	297.4	10,358	299*	299	10,358	300	3
136– 180	11,314	3.8	430.0	11,427	299	457	11,269	428	-2
181– 225	10,909	4.7	512.7	11,018	457	538	10,937	514	1
226– 300	13,706	6.3	863.4	13,843	538	553	13,828	871	8
301– 450	16,046	8.9	1428.0	16,206	553	378	16,381	1,458	30
451– 600	9,667	11.5	1111.7	10,067	378	402	10,043	1,155	43
601– 750	6,517	13.9	905.8	6,582	402	325	6,659	926	20
751– 900	4,468	16.1	719.3	4,513	325	608	4,567	735	16
901–1050	3,377	18.4	621.3	3,410	608	251	3,454	636	15
1051–1500	6,325	22.1	1397.8	6,388	251	108	6,402	1,414	16
1501–3000	6,548	29.0	1898.9	6,613	108	309	6,693	1,940	41
3001–4500	1,818	35.8	650.8	1,836	309	549	1,913	684	33
Over 4500	1,931	43.5	840.0	1,950	549	..	2,005	872	32
Total	102,888	11.3	11.677	103,917	..	..	103,917	11,933	256

Notes: Nominal built-in elasticity = (256)100/11677 = 2.19. The absolute cruzeiro figures are in thousands of 1960 cruzeiros. Brackets are in absolute-value cruzeiros. For explanations of various columns, see footnotes to Table 18.

*For the purpose of the lowest income bracket, it was assumed that as many new taxpayers enter this bracket as leave it to enter the next higher bracket.

TABLE 18

Estimation of Income-Tax Evasion, Brazil, 1956

1955 Income Bracket (1)	1955 Assessed Income (2)	1956 Income Bracket (3)	1956 Assessed Income (4)	1956 Tax Rate (5)	4 × 5 (6)	1955 Incomes Distributed among 1956 Brackets (7)	Col. 7 × $\frac{1019}{1000}$ (8)	Required Shifts of 1955 Assessed Incomes in 1956: Shifts into (from Lower Slabs) (9)	Required Shifts of 1955 Assessed Incomes in 1956: Shifts out (to Higher Slabs) (10)	8 + 9 - 10 = Final Position (11)	5 × 11 (12)	12 - 6 = Tax Evasion (13)
21– 26	385	17– 21	468	3.12	14.6	308	314	+6	6	314	9.8	-4.8
26– 39	1,516	21– 31	1,782	3.95	70.4	870	887	+6	13	880	34.8	-35.6
39– 52	1,513	31– 41	1,673	5.05	84.5	697	710	+13	10	713	36.0	-48.5
52– 64	1,287	41– 51	1,419	6.10	86.6	1,714	1,747	+10	33	1,770	108.0	21.4
64– 86	1,630	51– 69	1,733	7.65	132.6	1,387	1,413	+33	26	1,420	108.6	-24.0
86– 129	2,225	69– 103	2,111	9.94	209.8	1,630	1,661	+26	31	1,656	164.6	-45.2
129– 172	1,502	103– 137	1,565	12.42	194.4	2,267	2,310	+31	43	2,298	285.4	91.0
172– 215	1,106	137– 172	1,171	15.15	177.4	1,362	1,388	+43	26	1,371	207.7	30.3
215– 257	841	172– 206	852	16.19	137.9	1,036	1,056	+26	20	1,062	171.9	34.0
257– 300	680	206– 240	676	18.09	122.3	641	654	+20	12	662	119.8	-2.5
300– 429	1,282	240– 343	1,198	20.98	251.3	1,510	1,539	+12	29	1,522	319.3	68.0
429– 858	1,547	343– 686	1,387	24.08	334.0	1,884	1,920	+29	36	1,913	460.7	127.6
858–1,287	471	686–1,029	417	31.55	131.4	800	815	+36	15	836	263.8	132.4
Over 1,287	495	Over 1,029	427	39.99	170.8	682	695	+15		710	284.0	113.2
Total	16,480	..	16,879	..	2,117.0	16,788	17,109	306	300	17,115	2,574.3	457.4

(continued)

NOTES TO TABLE 18

Notes: Tax evasion as a percentage of total collections in 1956 = 100(457.4)/2117 = 20.6 percent. The absolute figures are in thousands of 1949 cruzeiros, except for the figures for income brackets, columns 1 and 3, which are in 1949 cruzeiros. This table is an illustration for the year 1956 of the main calculations done in Table 20 for each year of the reference period. The calculations of each column were as follows:

Cols. 1 and 3. Tax brackets were deflated by the index of the general price level with a view to determining the changes in tax rates on given real incomes resulting from inflation and the yearly modifications of the tax schedule by the finance minister.

Cols. 2 and 4. These include incomes assessed for income-tax payment, net of exemptions and deductions, which are reported through the annual tax returns, and exclude incomes from which taxes are deducted at source.

Col. 5. These are average tax rates—i.e., tax payments in an income bracket as a percentage of the corresponding income assessed.

Col. 6. That is, actual tax payments. Due to rounding errors, however, these totals may not exactly tally with the published aggregates.

Col. 7. When income brackets change from one year to the next, assessed incomes are distributed proportionately among the corresponding and neighboring new brackets. It may be observed that in 1956 a new initial bracket was created (i.e., the income bracket Cr$17-21), whereas in 1955 the initial bracket was Cr$21–26 (in 1949 prices). In view of this it was assumed that new taxpayers proportionate to the Cr$21–26 bracket for 1955 entered the new 1956 initial bracket with an aggregate (or proportionately calculated assessed incomes) of Cr$308,000.

Cols. 8-11. For the purpose of computing these columns it was assumed that incomes of all income brackets as well as income taxpayers increase at the same rate as the rate of growth of national income. This is equivalent to assuming constant income distribution. As a first approximation, it was further assumed that between the two years studied the taxpayers have not shifted between the two major categories of income-tax payments—i.e., "income-taxpayments by tax returns," and "income-tax deductions at source." (This assumption will be dropped subsequently, and the shifts under reference will be taken into account. See Tables 18 and 19.)

Col. 8. This gives an estimate of the assessed incomes in 1956 under the foregoing assumptions, applying the 1956 rate of growth of the national income of 1.9 percent to the assessed incomes of 1955.

The results of an increase of incomes appears in two forms, insofar as the income brackets are concerned: a) the assessed incomes of each bracket expand by the given rate of growth of incomes as accounted for in column 8, and b) some assessees shift up to the next higher bracket or brackets. The latter effect was taken into account by the following method. By assuming that the assessees of each income bracket are uniformly distributed over the whole range of the respective income bracket, it is easy to find out the number of assessees and their incomes that will cross over to the next higher slab when a certain increase takes place in their incomes. For instance, when there is a 1-percent increase in per capita income, those assessees in the income bracket Cr$257–300, whose incomes were Cr$297.07(= 300[100/101]) or higher previously, will move up to the next income bracket of Cr$300–429. Similarly, when family incomes rose by 1.9 percent, all the assessees from the income bracket Cr$257–300 getting an income of Cr$294.4(=300[1000/1019]) or higher crossed the Cr$300 level and entered the next income bracket, in which bracket they started paying an average tax rate of 20.98 percent instead of the 18.09 percent in the lower bracket.

Cols. 9 and 10. These take care of these shifts "out of" and "into" various brackets. For the initial bracket it was assumed that there are as many taxpayers entering this bracket from the previously below-the-tax-limit families as there are those who shift out of this bracket to the next higher one. The final position, comprising the expansion and the shift effects of a rise in incomes, is given in column 11.

Col. 12. This shows the required tax collections, under the given assumptions.

Col. 13. This is obtained by subtracting the actual from the required tax collections; the result is tax evasion. The total of column 13 will give the aggregate evasion in 1956, which comes to Cr$457,400 or 20.6 percent of that year's actual collections. This figure looks rather high. A major correction of this, however, has to be made when account is taken of the shift of taxpayers to the "deductions at source" category. Formally, the tax evasion relation can be stated as follows:

$$\text{Evasion} = Y_j(\zeta_j + g\zeta_{j+1})$$

where ζ_j is the effective tax rate for the jth bracket; g is rate of growth; and Y_j is taxable income of the jth bracket, with adjustments for the lowest and the uppermost brackets, as usual.

TABLE 19

"Unadjusted" Estimation of Evasion of Personal-Income Tax Not Deducted at Source,*
by 1959 Income Brackets, Brazil, 1951-61

Income Bracket 1959	Personal-Income-Tax Evasion											
	1951	1952	1953	1954	1955	1956	1957	1958	1959	1960	1961	Total
19 – 25	4.8	1.2	-1.5	16.2	11.5	-4.8	-17.0	-15.0	-9.5	-10.8	10.0	-33.0
25 – 31	6.0	0.3	-1.5	16.3	-49.0	-35.0	-15.6	-14.1	-6.5	-8.6	10.0	-81.6
31 – 39	8.0	0.4	-0.7	20.2	-10.1	-23.8	-20.3	-11.3	2.4	-12.3	35.0	-12.2
39 – 49	-7.9	0.3	1.7	9.3	0.8	-12.1	-29.9	-12.5	-7.4	-10.1	12.0	-55.6
49 – 62	-0.8	0.3	2.6	6.3	45.9	-3.0	12.0	-8.0	-88.9	-6.0	17.0	-23.0
62 – 82	-3.0	3.0	3.6	4.3	9.9	4.0	90.9	-3.2	45.1	-4.7	65.0	214.6
82 – 103	4.6	-2.9	3.9	16.5	81.9	4.0	103.1	24.5	47.4	2.6	57.0	342.6
103 – 124	8.3	-2.9	4.1	-15.0	62.9	7.2	74.6	29.7	41.6	3.4	52.0	266.3
124 – 144	-5.3	-2.8	2.4	-14.0	57.2	10.4	88.0	-0.4	65.8	13.2	37.0	246.9
144 – 206	-15.0	-9.6	16.1	-2.3	-93.6	23.7	-18.2	-49.0	156.8	-3.0	60.0	61.9
206 – 412	-6.5	-2.1	12.6	-155.2	13.1	82.9	135.9	11.0	185.9	49.0	112.0	444.2
412 – 618	-18.6	-12.6	0.6	-55.7	388.6	82.4	-363.2	33.5	218.3	66.0	100.0	539.4
Over 618	26.8	-21.4	44.7	125.2	-22.6	313.6	490.4	60.0	196.6	116.3	95.0	1393.4
Total	1.4	-48.7	89.6	-27.9	501.2	457.4	530.7	34.3	847.7	195.0	608.0	3193.0

Notes: All figures in thousands of 1949 cruzeiros. Bracket figures in absolute cruzeiros. Evasion figures were first computed for various years in the respective year's income brackets; the same were then distributed proportionately among corresponding and neighboring 1959 income brackets. Thus, evasions of all years of reference period were converted in terms of the 1959-income brackets. The expression "unadjusted" implies that no account as yet is being taken for the shift of taxpayers from "incomes not deducted at source" to those "deducted at source."

*Impôsto de renda, pessoas físicas.

TABLE 20

"Adjusted" Estimation of Evasion of Personal-Income Tax Not Deducted at Source, Brazil, 1953-61

Year (1)	Tax Yield due Under Unchanged Distribution of Income[a] (000 Cr$) (2)	Actual Yield (T) (000 Cr$) (3)	Col. 2 - Col. 3 = Year-to-Year Tax Evasion (000 Cr$) (4)	Col. 4 as Percent of Col. 3 = Percent Year-to-Year Tax Evasion (5)	Computed Shifts of Tax Payments to "Tax Deductions at Source"[b] (000 Cr$) (6)	4 - 6 = Year-to-Year Tax Evasion (000 Cr$) (7)	Col. 7 as Percent of Col. 3 (8)	$\Sigma^{n}_{t=1953}$ (Col. 7)$_t$ = Cumulated Tax Evasion (000 Cr$) (9)	$\frac{100(\text{Col. 9})}{2193.0}$ = Cumulated Percent Tax Evasion Since 1953 (10)
1953	2,282.6	2,193.0	89.6	3.6	30.0	49.6	2.3	49.6	2.3
1954	2,193.8	2,221.7	- 27.3	- 1.3	30.0	- 57.3	- 2.5	- 7.7	- 0.3
1955	2,708.6	2,207.4	501.2	22.7	80.0	481.2	21.8	473.5	21.6
1956	2,574.3	2,117.0	457.4	21.6	- 50.0	407.4	19.2	880.9	40.2
1957	2,569.0	2,038.3	530.7	26.0	290.0	240.7	11.8	1,121.8	51.2
1958	2,392.9	2,358.6	34.3	1.5	480.0	- 445.7	-18.8	676.6	30.8
1959	2,889.7	2,042.0	847.7	41.5	-560.0	1,407.7	68.9	2,083.7	95.0
1960	2,110.0	1,915.0	195.0	10.2	- 90.0	285.0	14.9	2,369.8	108.1
1961	1,686.0	1,078.0	608.0	56.4	70.0	538.0	50.0	2,906.6	132.6

Notes: Cruzeiro figures are in thousands of 1949 cruzeiros. Data by income classes no longer published after 1961; hence, estimates based on the method of Table 18 could only be made up to 1961. On the lower side of the reference period the calculations could not be done before 1953, as the shift components of column 6 are not available.

[a] From Table 19, last row. Negative signs indicate negative evasion or overpayment of tax.

[b] From Table 21, col. 6.

TABLE 21

Calculation of Shift of Tax Payments from "Personal-Income-Tax Payments Not Withheld at Source" to "Personal-Income-Tax Deductions at Source"

Year (1)	Actual Personal Income Tax (2)	Rate of Growth of GNP (3)	$(\text{Col.}3)s_t^* =$ Required Percent Increase in Tax Yields[a] (4)	$(\text{Col. }2)_{t-1} \times \left(\frac{100+\text{Col. }4}{100}\right)_t$ = Required Tax Yield[b] (5)	2 - 5 = Shift of Tax Payments from Other Categories to This Category[c] (6)
1952	300	5.6	12.32	..	..
1953	360	3.2	7.12	330	30
1954	410	7.7	16.47	380	30
1955	550	6.8	14.89	470	80
1956	580	1.9	4.16	630	- 50
1957	890	6.9	15.11	600	290
1958	1,500	6.6	14.45	1,002	480
1959	1,160	7.3	16.07	1,720	-560
1960	1,250	6.7	14.56	1,340	- 90
1961	1,500	7.3	15.90	1,430	70
1962	1,740	5.4	11.81	1,740	0
1963	1,510	1.6	3.52	1,950	-440
1964	2,760	3.1	6.85	1,560	1,200

Notes: All cruzeiro figures in thousands of 1949 cruzeiros. Personal income taxes include taxes deducted at source from incomes subject to personal-income tax. In Brazilian terminology, these include, impôsto de renda: fonte (exceto juros de dívida pública, juros diversos, e pessoas jurídicas).

[a]The symbol s_t^* stands for the legal or nominal elasticity in year t. Computation of this elasticity for 1956 was illustrated in Table 17. Similar calculations were done for the other years (not reported here).

[b]That is, these would have been the tax collections from the personal-income-tax deductions at source ("fonte"), had these taxpayers' incomes increased at the rate of growth of national income and without any change in income distributions.

[c]Given the above assumptions, and an additional assumption that it is generally not possible to evade tax payments from personal incomes subject to deductions at source, this column shows the shift of incomes from other categories of personal-income-tax payments to the category of personal-income-tax deductions at source.

Figure 8: Current and Cumulated Rate of Personal-Income-Tax Evasion, Brazil, 1952-68

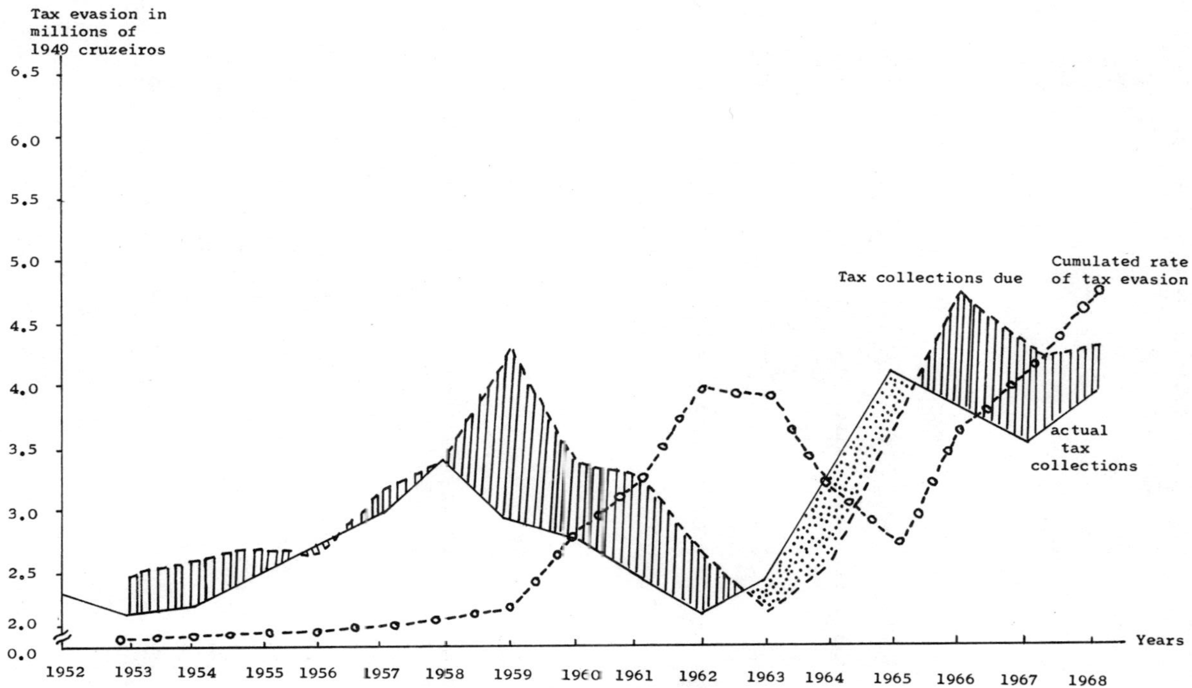

Source: Table 20.

Figure 9: Tax Evasion, by Income Bracket, Personal-Income Tax Collected via Nonwithheld Returns, 1951-61

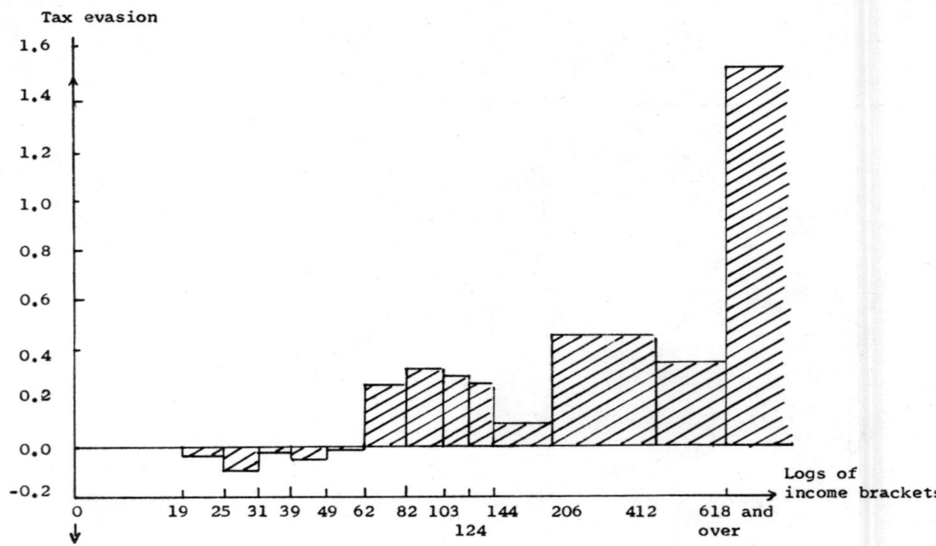

Note: Semilog scale—i.e, income brackets—along the abscissa, are in logs of 1949 new cruzeiros, and the evasion estimates, along the ordinate, are in absolute values in millions of new cruzeiros in 1949 prices. Tax brackets pertain to 1959.

Source: Table 19, last column, adjusted for the corrections of Tables 20 and 21.

nominal tax elasticity of Table 17 was applied to all personal income tax yields, however collected. The estimates were made for the entire period, from 1952 to 1968. These estimates, in a way, provide a check on our earlier measures, shown in Table 20. Since Table 22 is based on an overall measures and on a somewhat different procedure of calculation, the figures differ somewhat for individual years. Differences may also arise due to averaging. Yet on a broad view the two calculated series are not wide apart. For example, the cumulated evasion figure from 1952 to 1961 comes to 139.2 percent in Table 22, which is not far wide of the figure of 132.6 percent obtained for 1961 in Table 20. The evasion rate continues rising and reaches a peak of 181 percent in 1962, after which it starts to decline. It touches a bottom of 61.66 percent in 1965, after which it starts to stride up once again. The cumulated tax evasion of 1968 was about 121 percent of actual collections. It must be noted further, however, that the cumulated evasions (column 10 in cruzeiros and column 11 in percentages) are in terms of annual changes or flow measures. That is, starting from 1952 the rate of tax evasion cumulated to Cr$4.66 million (in 1949 prices), about 121 percent of the actual annual tax collections of Cr$3.86 million (in 1949 prices) in 1968. The cumulated stockof tax evasion in 1968 can be found by aggregating the figures of column 10 for the cruzeiro measure and column 11 for the percentage measure; they come to Cr$38.41 million (in 1949 prices), or about ten times the actual collections in 1968. Thus, were the government to reduce the current rate of evasion to zero, it would have to have the taxpayers declare their current hidden incomes so as to pay 121 percent additional taxes. Were it to capture all the past evasions since 1952, it would have to force the taxpayers to declare their past hidden stocks so as to collect (even applying zero rate of interest) a lump-sum amount of about ten times the annual income-tax revenues.

The foregoing estimates of the evasion of personal income taxes were made under the assumption of a constant distribution of income. As the analysis of the earlier part of this chapter, particularly the section on exemptions and deductions, has revealed, however, there are definite indications that incomes have been steadily redistributed in favor of high-income groups. On that account the tax buoyancy ought to have been higher than was observed. Ipso facto, the foregoing estimate of tax evasion must be considered underestimation. It might also be observed that more than 70 percent of the calculated tax evasion comes from the three uppermost income brackets.

Capital-Gains Tax?

So far, we have measured the extent of evasion by income brackets. We have not said which social groups of taxpayers evade taxes most

TABLE 22

Estimation of Overall Personal-Income-Tax Evasion, 1953-68

Year (1)	Non-withheld Tax Collection (2)	Withheld at Source (3)	2 + 3 = Actual Total Tax Collections (4)	Rate of Growth, GNP (5)	$(\text{Col. }5)s^{*}_{t}$ = Required Percent Increase in Tax Yield (6)	$(\text{Col. }4)_{t-1}\left(\frac{100+\text{Col. }6}{100}\right)_{t}$ = Required Tax Yield (7)	Col. 7 - Col. 4 = Tax Evasion: Year-to-Year Rate (Cr$) (8)	Year-to-Year Rate (Percent)[b] (9)	Cumulated Rate (Cr$)[c] (10)	Cumulated Rate (Percent)[d] (11)
1952	2.34	.10	2.44	5.6	12.32	..	..	..	..	..
1953	2.20	.14	2.34	3.2	7.12	2.61	.27	8.12	.27	8.12
1954	2.27	.21	2.48	7.7	16.47	2.73	.25	10.08	.52	20.97
1955	2.30	.34	2.74	6.8	14.89	2.85	.11	4.01	.63	23.00
1956	2.46	.43	2.89	1.9	4.16	2.87	-.04	-1.39	.59	20.34
1957	2.36	.76	3.12	6.9	15.11	3.34	.22	7.05	.81	26.00
1958	2.50	1.06	3.56	6.6	14.45	3.57	.01	.28	.82	23.03
1959	2.04	1.03	3.07	7.3	16.07	4.43	1.36	44.30	2.18	71.01
1960	1.85	1.11	2.96	6.7	14.56	3.52	.56	18.91	2.74	92.23
1961	1.18	1.40	2.58	7.3	15.90	3.43	.85	32.98	3.59	139.16
1962	.69	1.60	2.29	5.4	11.81	2.87	.58	25.30	4.17	181.30
1963	1.28	1.30	2.58	1.6	3.52	2.37	-.21	-8.14	3.96	153.48
1964	1.49	1.98	3.47	3.1	6.85	2.72	-.75	-21.61	3.21	92.51
1965	1.49	2.89	4.30	4.7	10.29	3.83	-.55	-12.80	2.66	61.66
1966	..	..	3.84	5.1	11.22	4.78	.94	24.5	3.60	93.75
1967	..	..	3.62	4.8	10.56	4.25	.63	17.4	4.23	116.85
1968	..	..	3.86	8.4	18.48	4.29	.43	11.15	4.66	120.73

Notes: Cruzeiro figures are in millions of 1949 cruzeiros. Personal-income tax includes personal taxes not withheld at source as well as those withheld (i.e., *impôsto de renda:* (i) *impôsto lançado das pessoas físicas,* (ii) *impôsto de renda retido nas fonte,* and (iii) *adicional de proteção a família*). The relation for the calculation of tax evasion here can be put into symbols as follows:

$$[(1 + g_t)Y_{t-1}]s_t^* - T_t^{pi}$$

where Y_{t-1} is the overall taxable income of the past year, g_t is the current rate of growth of income, T^{pi} is the actual tax collections, and s_t^* is the legal elasticity, as defined earlier. The coefficient of s_t^* was calculated for each year except for post-1961 period, for which the 1961 s_t^* was employed.

[a] The symbol s_t^* was explained in footnote a of Table 21.

[b] Percent evasion is calculated in reference to current-year collections —e.g., percent evasion for the year 1953 = 100(0.27)/2.34 = 8.1 percent.

[c] Figures of this column signify the (cumulated) *flows* of evasion. Between 1953 and 1965 the *stock* of evasion increased to Cr$26.15 million (in 1949 prices), which is over 600 percent of the tax collections in 1965.

[d] The cumulated percent evasion is calculated with reference to the current-year tax collections. For example, the cumulative percent evasion for the year 1958 = 100(0.82)/3.56 = 23.03 percent.

Figure 10: Influence of Various Provisions, Including a Relatively Lower Tax Rate, on Capital Gains, Effective Rates, and Taxable Returns, United States, 1960

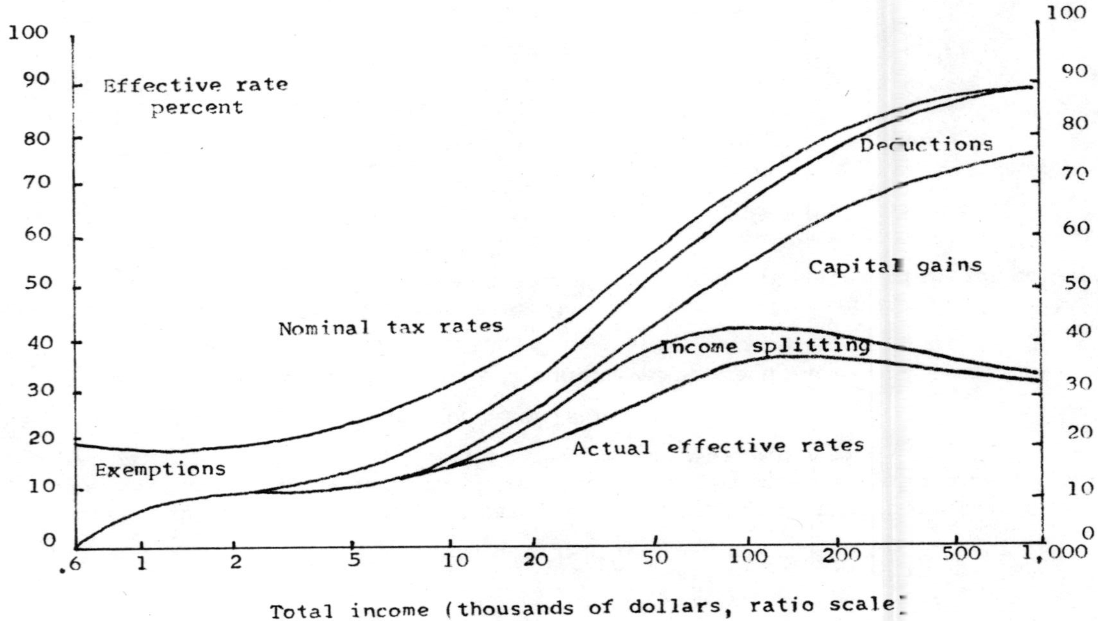

Source: Reproduced from Richard Goode (1964, Table A-10 and Chart 3).

nor how they do it. Apart from conjecturing, it is not easy to find all the answers to the "which" and "how," and we shall not try to do that in this study.* One loophole of tax avoidance almost suggests itself: it is the absence, for all practical purposes, of a capital-gains tax in Brazil. An idea of the extent to which income-tax payments can be avoided via capital gains can be had from the U. S. experience. The United States has a capital-gains tax, but the tax rate on capital gains (the bulk of which are earned by high-income people) is much lower than that on high incomes. The tax rate has revolved around 25 percent of realized, and not accrued, capital gains, while the income tax rates on the upper-income brackets have been as high as 80 percent or so, up to 1964. The U.S. picture for 1960 is sketched in Figure 10. It will be observed that even though the marginal income-tax rates in the United States went up to above 80 percent in 1960 and the effective tax rate on "earned incomes" went up to about 53 percent, the effective tax rates on the "earned and unearned" incomes did not exceed about 33 percent of total incomes. The major cause of low effective rates on the "earned and unearned" incomes is evidently a low rate of tax on capital gains.

In Brazil, the nontaxation of capital gains might very well be one of the major sources of tax avoidance, though the nonreporting of incomes, in the author's judgment, remains the foremost cause of outright tax evasion. People can convert their current incomes into one-or-the-other type of property to earn capital gains and avoid and evade income taxation. During high inflation periods, this kind of investment can be a substantial source of private wealth. It is true that most of the apparent sources of capital gains are spurious in that they are fictitious price rises instead of real gains. It is also a fact, however, that real capital gains do accrue in many sectors of the economy, and more so during inflation periods, when any kind of material property, in general, gets a premium price over personal labor. Changes in relative prices and income distribution resulting from general inflation add to the possibility of increased capital gains. There is no justification, thus, for not taxing capital gains, especially when the marginal tax rates on high incomes, as seen earlier, are low in comparison with those prevailing in several other countries, such as the United States, the United Kingdom, and India.

There remains the question of feasibility and economic effects. For instance, it is not very practicable to tax capital gains on an accrual basis. Outside of the stock market, the realized capital gains are usually a very small proportion of overall accrued capital gains.

---

*Insofar as the analysis of evasion is concerned, the scope of the present study is confined merely to making quantitative estimates of it. For an examination of the nature of income-tax evasion and some of the methods of reducing the same, see Oldman (1965).

Moreover, the stock exchange market (which is the major source of capital-gains tax in the advanced countries) is still in its infancy in Brazil, and in fact its development needs to be (and is being) encouraged by tax concessions and similar incentives. Political and administrative feasibility is different from economic rationality.

An impression is held in some circles that the absence of a capital-gains tax is conducive to economic growth. This may come through several ways, such as through increased private saving and development of the capital market. Going into the economics and ethics of a capital-gains tax in Brazil, however, is beyond the scope of this section. It is mentioned here only to point out a possible source of the avoidance of personal income tax, which in turn might, in large part, explain the calculated inelasticity of this tax in Brazil.

Apart from plugging loopholes, a typical case of which has just been discussed, one other important remedial measure that should be apparent from the discussion in this section is the gearing-up of tax administration. It is not possible to find out the cost of collection of income taxes from the available data. In the United States it is estimated to be less than 2 percent of total tax collections. In that country, however, the degree of compliance is surprisingly very high even as compared with the European countries, not to speak of the underdeveloped countries. In view of the bad effects that tax evasion has on the government and on the economy as a whole, even a ten-cruzeiro collection cost for every hundred cruzeiros of tax collected from potential evaders will not be a high price, perhaps. Economy in the taxation staff of a high caliber is often a good example of a penny-wise-pound-foolish policy.

It might be argued that the computed high rate of evasion is a blessing in disguise, in that the sources of the high rates of saving and investment, observed in the beginning of this study, may very well, in part, be the result of tax evasion. But if that is so, the proper way to earn the blessing is to remit or reduce the rates of certain taxes on all citizens, or to give subsidies to the deserving investors and consumers. Widespread tax evasion leads to misallocation and waste or resources (out of concealed incomes); it corrupts the potential tax-paying morality of the public; it impairs the spirit of the people of participation in the government and weakens democratic traditions; it penalizes honest taxpayers and condones the cheats; it thwarts the efforts of honest tax officers and tempts others to tolerance and even complicity; it frustrates the economic, social, and political aims of the government, insofar as these are embodied in federal finance.

The Brazilian government has, indeed, been aware of the problem, even without having any quantitative estimates of tax evasion. As can be observed from Table 22, there was a definite reversal of the rate of increase of tax evasion during 1962-65 when the rate of personal income tax evasion declined sharply from about 181 percent in 1962 to about 62 percent in 1965. Surprisingly enough, the trend of the early

1960s was reversed after the 1964 Revolution, although prima facie evidence suggests that the effort to check tax evasion was intensified. For instance, since 1967 the government has even shown the audacity, not very common in the Latin American tradition, to send a number of big tax evaders to jail and to order the confiscation of their property. Perhaps our reference period ended short of the effects of this campaign.

Tax evasion is a common feature of the underdeveloped countries; it is not peculiar to Brazil. To tackle this problem, the sources of tax evasion need to be identified. In this study, quantitative estimates of tax evasion have been made by income brackets only. For a policy aimed at reducing tax evasion, similar quantitative estimates, or at least some qualitative indicators, of tax evasion by groups of people, by occupations, by the nature of the sources of income, by geographic regions, and similar categories might be equally useful. Research in these directions for the personal income tax and estimates of tax evasion on the present lines in respect of other taxes remains to be begun.

CHAPTER

11

PUBLIC INSTRUMENTS AND SIZE-INCOME DISTRIBUTION

Functional income distribution and size-income distribution are likely to be related in any economy, but they are known to be more closely related in developing than in developed economies. In Brazil and other developing countries, therefore, the conclusions concerning the one distribution may, in general (though not necessarily) be expected to apply to the other distribution.* Yet the contexts in which the two distributions have conventionally been discussed have generally been different. Functional income distribution is discussed largely in macroeconomic parlance; size distribution has been studied primarily in the field of public finance. Functional distribution is associated with factor shares, whoever may own them; size distribution concerns personal incomes, whatever their sources. Theories of functional distribution are more elegantly developed than those of size distribution. The theoretic underpinnings of the former are derived from relative demand and supply theories of factors, production techniques, elasticity of substitution, product exhaustion, and responses to such aggregate categories as inflation, growth, unemployment, and so on. Size-income distribution does not have similarly developed and equally rigorous theoretical bases and response relationships. Even the measurement parameters so far developed for this distribution—e.g., Pareto's Law of Income Distribution, the Gini coefficient, the personal income tax progressivity norms, and so on—are not so precise or sensitive to changes in distribution. Despite these weaknesses, the size distribution of income, by virtue of its being available for a larger

*The recent rapid growth of human capital formation and the diffusion of stock among all income classes, including workers and the lower-middle class, in the industrialized countries, seem to have blurred the relationship between functional income distribution and size-income distribution somewhat but not entirely.

disaggregation of income groups (usually eight to ten income classes) than functional distribution (usually two or three factors), has been equally useful for several policy matters, such as tax policy.

In principle, there is no reason why size distribution of income cannot be analyzed as an integral part of a multiequation econometric model. The difficulty lies in the lack of appropriate data. Size distributions are usually derived from cross-sections, such as family-budget surveys or population census reports, while economy-wide econometric models are invariably estimated from time-series data. Another difficulty pertains to the functional form. Functional income distribution equations can usually be described and estimated by simple and well-behaved relations. Examples are a) a wage-bill equation and a nonwage equation, or simply a W/Q equation as in this study, and b) a production function with factor-demand equations derived from marginal conditions, or similar macro relations. Size distribution often cannot be reduced to one- or two-parameter relations. For example, the distribution of income-tax incidence scarcely lends itself in any country to simple and well-behaved functional relations, and any one-parameter measure usually conceals more than it reveals.

Considering these and other caveats, size-income distribution and one of its important direct determinants—namely, public income distribution—are analyzed separately from the main econometric model of this study. An attempt is then made to relate the three distributions—functional, size, and public—and thus combine the direct and indirect effects of different policy instruments on overall income distribution. Regional income distribution, which could not be analyzed in the econometric model, is also analyzed in this chapter.

Basic public income distribution is computed in two main parts: tax-burden distribution and expenditure-benefit distribution. All federal, state, and local taxes and expenditures are included. To this basic distribution are added the effects of some of the other policy variables. Thus, the effects of foreign-trade policy (or import-substitution policy) are incorporated to the extent that tariffs represent such a policy. The distributional effects of monetary policy are expected to be partially captured in an analysis of the distribution effects of what is called here the "inflation tax." The indirect effects of tax evasion and tax concessions under various investment incentive schemes are duly recognized. The effects of all these policy instruments are measured for nine income classes by regions and sectors as follows: eight different urban areas (state capitals, to be called cities), nine groups of towns or interior cities representing the major regions of the country (to be called town groups or simply towns), and seven typical rural areas.

The basic data come from the household survey of 1961-63 (FGV, 1967). The relative expenditure and income patterns of this sample are used to calculate tax burdens for 1968. The family (in contrast to the person or the household) is used as the unit of study. Family incomes reported in household surveys are, by and large, equivalent to personal incomes and are gross of direct taxes. The national-accounts category that comes close to this concept is net national product at market prices. Hence, one does not have to be apologetic about the income concept imposed by the available data.

The merits and demerits and conceptual weaknesses of incidence studies have been widely discussed in public finance literature and will not be reviewed here.[1] Our view is that incidence analysis complements our study and is in order in its own right for Brazil. In the 1960s tax-incidence studies for developing countries enjoyed a high priority.[2] In recent years, the incidence of overall government policy—taxes, expenditures, and other policy measures bearing on distribution—has become the focus of analysis.[3] The present analysis is in keeping with this trend.

SELECTION OF REGIONS

Brazil is a vast country with 47.3 percent of the area of South America. It is as unevenly developed as the cross-section of the countries of this continent. For example, the state of São Paulo with 2.9 percent of the area has about 19 percent of Brazil's population and produces approximately one-third of the nation's net domestic product. The city of São Paulo alone, with about 5 percent of the country's population, is estimated to produce nearly 10 percent of the nation's net domestic product. The state of Ceará, with 1.75 percent of the area and 4.4 percent of the population, on the other side, contributes only 2.1 percent to the country's net domestic product. Mato Grasso, with nearly 15 percent of the area, has only 1.5 percent of the population and less than 1 percent of net domestic product. Due to such large disparities in development and standards of living, the national averages of the distribution of tax-expenditure incidence may not represent (even approximately) the true picture of more than a couple of the 22 states. Moreover, an average state of Brazil is larger in area than, for example, an average state of the United States; in view of this, several typical regions and sectors are studied separately.

The selection of regions is determined basically by the availability of data. Nevertheless, the regions and sectors available from the family-budget surveys in themselves provide a fairly representative sample of the typical econographic entities of Brazil. The study is done for eight capital cities of eight different states (two cities of which, Rio de Janeiro and São Paulo, contain about 10 percent of the total population); nine town groups in five different (mainly southern) states; and seven rural areas in seven different states. These econographic areas, which extend in all over 11 of the 22 states of Brazil, are shown in Figure 11. To provide an idea of the level of development of each of

Figure 11: Econographic Regions of Brazil Included in This Study

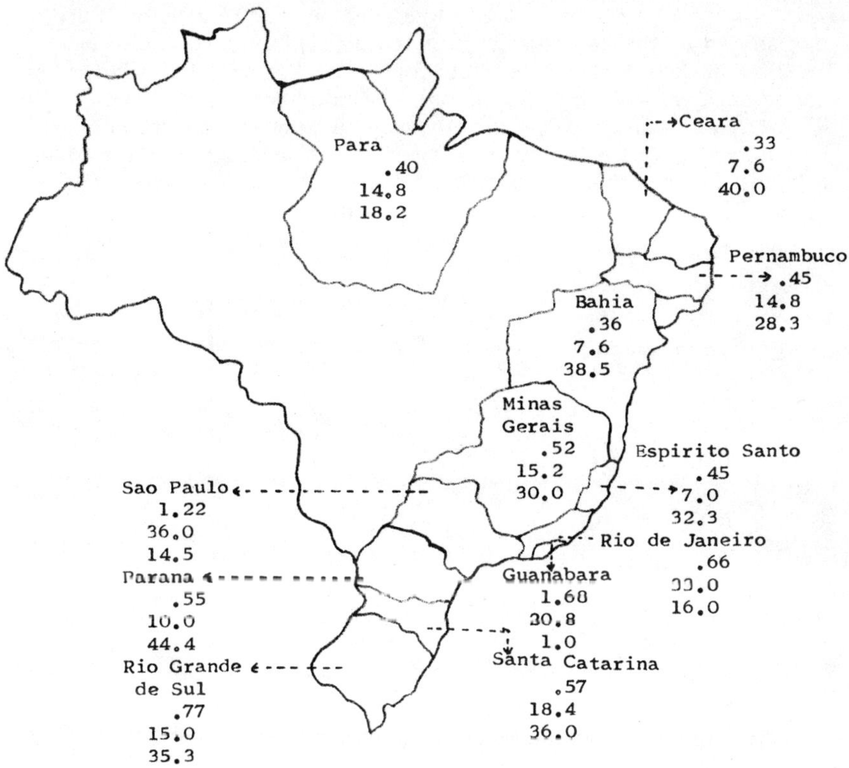

Note: The top figure for a state stands for per capita income in thousands of current cruzeiros in 1967. The middle figure denotes income originating in industry as a percentage of net domestic income of the state concerned. The bottom figure gives the same percentage for income originating in agriculture. The corresponding average figures for Brazil as a whole are 0.66, 26.2, and 19.2, respectively. The U.S.\$/Cr\$ ratio in 1967 was approximately 1:3.

the states, a few economic indicators are also given in the illustration. The names of cities and states may be of no particular interest to the non-Brazilian reader; however, when it is realized that a) three different sectors (urban city sector, urban town sector, and rural sector) are represented by the 24 econographic areas of this study, and b) most areas for each of the three sectors represent distinguishably different stages of development, the regional analysis becomes of special interest to students of economic development in general. Accordingly, on the basis of the data in Appendix A and the economic indicators given in Figure 11 and other unreported information the econographic areas studied here may be fitted into certain patterns of development levels as follows:

Urban City Sector	Para (N)	Underdeveloped Industrial	Underdeveloped
	Pernambuco (NE)		
	Ceará (NE)	Underdeveloped Agricultural	
	Bahia (East)		
	São Paulo (S)	Developed Industrial	Developed
	Guanabara (ES)		
	Minas Gerais (ES)	Modern Agricultural	
	Paraná (S)		

There are three town groups in each of the following:

Urban Town Sector	Espirito Santo (ES) and Rio de Janeiro (ES)	Median	Underdeveloped
	São Paulo (S)	Developed	Developed
	Paraná (S) and Santa Catarina (S)	Modern Agricultural	

Ceará (NE)	Underdeveloped	Underdeveloped
Pernambuco (NE)		
Espirito Santo (ES)	Median	
Minas Gerais (ES)	Developed	Developed
São Paulo (S)		
Santa Catarina (S)	Developed	
Rio Grande do Sul (S)		

THE ESTIMATING PROCEDURE FOR TAX-BURDEN DISTRIBUTION

The main steps involved in the procedure for estimating tax-burden distribution can be explained briefly. They include a) the choice of the dimensions of classes for which tax-burden distribution study may be desired and the selection of taxes to be allocated among various classes of the chosen dimensions, b) the calculation of average (percentage) indirect tax rates for the consumption categories available in the family-budget data, and the average direct tax rates after allowing for various exemptions and deductions, c) the allocation of the burdens of individual taxes by classes and the translation of this allocation into schedules of tax payments, and d) the calculation of "differential burden" in accordance with the proportionality and/or progressivity norms chosen.

Choice of Classes of Economic Units and Selection of Taxes

In selecting taxes for the problem at hand, distinction needs be made among taxes as such against which there is no direct quid pro quo such as personal income tax, sales tax, etc.; fees and prices that are charged in return for specific service rendered or goods sold, such as fees and sales proceeds from public enterprises; and certain special-purpose charges and levies against which some quid pro quo can be established but not precisely, the most important examples of which are social security charges, fuel tax (and road fund), and electricity tax. For the purpose of this study the second category of government income, that is, fees and prices, was ignored. Likewise, fuel and electricity taxes were also neglected, as they, too, are of the nature of fees and prices. Insofar as social security charges are concerned, tax burden was estimated with and without these charges. Tax structures of all levels of government (federal, state, and local) were included.

Averaging Tax Rates

Consumption Taxes

The calculation of average tax rates turned out to be arduous and yet perhaps not very satisfactory. This is especially so in the cases of the consumption tax,* import duties, and inflation tax. The legal tax rates are available for scores of commodities. The questionnaire on family budgets included approximately 150 expenditure items. When these data were compiled for publication by Fundação Getúlio Vargas (FGV) in 1967 the 150 or so items were aggregated into 20 nondurable-consumption items and three durable-good items. The problem, therefore, was to calculate average excise-tax rates for these 23 items of expenditure.

The procedure followed here is, first, to calculate average tax rates for the raw data available from the questionnaire files on the shelves of FGV. Going through the huge stack of these files would have been a stupendous job beyond the time and resources available to me (the FGV studies themselves came out with about five years' lag). In view of this, it was decided to use one typical rural area and one representative urban center for calculating the weights for the 150 or so expenditure items. Rural Espirito Santo was chosen as a representative rural area, and Guanabara as a representative urban area. The term "representative" is used here to indicate the relative weights and not the absolute levels of detailed consumption items. The average tax rates for the 23 expenditure items of the published reports for

---

*The Brazilian name for this tax is imposto de consumo, which, literally translated, is "consumption tax." This group of taxes actually includes what in common terminology are called "selective excise taxes," though the taxes in Brazil are of the nature of value-added taxes. However, the bulk of excise revenue in Brazil is drawn not from a few items of mass consumption, such as tobacco, liquor, etc., as is the case in several other developing countries, but from a large number of consumption goods. It is a fairly general tax. Since the name "consumption tax" appears to be more representative and appropriate than the term "excise tax," in this study these taxes will sometimes be referred to by the former name.

various cities, towns, and rural areas were then computed from these weights and legal tax rates.

A further problem in calculating the average rates of these taxes is that while the consumption taxes are levied at the production level, the family expenditures are naturally shown in retail prices, which include the mark-ups of the wholesaler, trader, and retailer, costs of transportation, and so on. Moreover, the mark-ups, after their final pyramiding, cannot be assumed to be uniform for different commodities. Hence, no iteration could be applied. In view of these difficulties, the average tax rates in terms of retail prices were worked out (for the 150 or so commodities of the questionnaire) in consultation with experts and knowledgeable people in the field and using whatever evidence could be obtained.

Import Duties

The calculation of average tariff rates for the available consumption items was an equally complex job. In addition to calculating average tariff rates from a large number of items, one has to go to the input taxes in this case. A large proportion of imported stuff is for industrial uses: raw materials, machinery, intermediate goods of various kinds, and so on. The products of the industries in which these production goods are used have to be allocated to the consumers, unless the tax is assumed to be borne by the producers. In keeping with the other outlay taxes, however, it was assumed that the incidence of import taxes also falls on the consumers.

Accordingly, the commodities subject to tariffs were, first, allocated to the input-output categories. (The input-output table used in this study is the one prepared by Rijckeghem [1967].) The ratios of imports to domestic production were calculated in terms of percentage. The input-output coefficients were adjusted by these percentages to indicate the import proportions of inputs. The next step was to calculate the average tariff rates on the inputs of particular products. The final tariff rates were calculated by taking the weighted averages of the above (input-tariff) rates and those on the respective final products. Finally, the input-output categories were translated in terms of the family budget items (again taking weighted averages). Symbolically, the average tariff rates on the consumer expenditures were calculated by the following relation:

$$w_1 m_j t_j + w_2 \left(\sum_i a_{ij} m_{ij} t_i \right) \qquad \text{(xlvii)}$$

where the a_{ij} are the input-output coefficients, the m_{ij} are the ratios of imports to domestic supply in the ijth cell of the input-output table, m_j is the ratio of imports of product to the domestic supply of product, t_i is the tariff rate on input i, t_j is the tariff rate on the final product j;

and w_1 and w_2 are the weights for direct consumption of the imports of product j and the intermediate use of the same, respectively. For lack of other information, w_1 is simply the ratio of total direct-consumption imports to total imports. Naturally, $w_2 = 1 - w_1$.

Sales Tax

This tax is levied nearly uniformly all over the country. Thus, it is easy to calculate the average sales-tax rates. They were obtained by taking the ratio, expressed as percentage, of the total revenues from this tax in the state in which a given city or rural area happened to be situated to the income originating in that state, adjusted for exempt items. Usually this rate was around 6.5 percent, but it differed slightly from state to state. The effective sales tax rate also differed from class to class, as the tax rate is applied to expenditures while the denominator is income. The difference is accounted for by the proportion of saving of various groups of families.

Personal Income Tax

Among the remaining taxes, the average rates of personal income tax are particularly necessary. The relevant tax rates in the context of this study should be total tax payments to total incomes from all sources—what are usually called the effective rates for different income brackets. In this study, four sets of rates are considered.

1. One set of tax payments is available from the family budget data. It is an item called imposto sobre a renda (that is, income taxes). The effective tax rates can be calculated by taking the percentages of these tax payments to the incomes reported in the same source. Unfortunately, the tax rates calculated from this source are extremely low for all of the states. The rates rarely exceed a small fraction of 1 percent of incomes. Referring to the nine income brackets of the family budgets within which the incomes of the present education classes fall, we find that choosing at random for Fortaleza, for instance, there are no entries in the first six brackets and in the ninth bracket. The entries in the seventh and eighth brackets indicate ratios of income tax payments to incomes of only 0.0003 (or 0.03 percent) and 0.0007 (or 0.07 percent), respectively. In view of this, these tax payments were ignored.

2. The second set of rates used in this study is available from the standard published sources (for example, Anuário Estatístico do Brasil), at least up to 1962. These rates are the actual tax payments as percentages of "reported" incomes. They are net of all relevant exemptions and deductions. Actually, these rates are the legal average (or effective) rates. The snag about these rates, however, is that they are calculated in relation to "reported" incomes only. In practice, not all incomes are reported, and different income brackets have different

propensities to evade personal income taxes. In this study, these rates are referred to as "legal" rates, and the tax burden calculated from these rates is called the "legal" tax burden.

The calculations give zero effective rates for the first four income brackets and the following approximate rates for the remaining five brackets, in ascending order: bracket 5, 3.35 percent; bracket 6, 7.25 percent; bracket 7, 12.24 percent; bracket 8, 21.00 percent (both for personal income tax and for personal and company income tax combined); and bracket 9, 36.17 percent (when only personal income tax is considered) and 30.2 percent (when company income tax is combined with personal income tax). Since the definition of income used here includes income from dividends also, no further adjustment was considered necessary.

3. A third set of rates was calculated by taking into consideration the probable tax evasion as estimated in Chapter 9. A rough average of tax evasion from incremental incomes, as translated from the income tax brackets to the family budget income classes, comes to 30 percent, 17 percent, 23 percent, and 72 percent, respectively, for the sixth, seventh, eighth, and ninth brackets. In other words, roughly speaking the reported incremental money incomes are 100/130, 100/117, 100/123, and 100/172 of actual money-income increments, respectively for the four upper brackets. The remainder of the increments in incomes are not reported.

However, because a major part of the incomes of the uppermost (ninth) income bracket and a part of the incomes of the eighth bracket are derived from dividend and other company income sources, the legal effective rates, as can be verified from the rates stated one above, are already close to the company income-tax rate of about 30 percent. Since the company tax rate is only mildly progressive, it is practically uniform for all brackets of company earnings, and although there is as much scope in Brazil as anywhere else for evading this tax, for lack of data we assumed the actual tax rate also close to 30 percent. In view of the situation, the difference between the second and third measure for the upper two brackets (which are assumed to enjoy investment incomes) is supposed to be slight. This leaves the sixth and seventh brackets as having some propensity to evade taxes. According to the estimates of Chapter 9, the incremental evasion rates are roughly about 30 percent and 17 percent, respectively. In the overall tax payments, this comes to roughly 2 percent of reported incomes. The second and third measures discussed above are thus practically the same.

4. The tax rates just discussed still do not reflect the actual tax burden, as the aforementioned rates of evasion pertain to the incremental incomes only. That is, these are the evasion rates from the additional flows of incomes rather than the existing cumulated levels of evasion from total incomes. Accordingly, a fourth measure was approximated to reflect the actual tax rates, which were calculated by taking the ratio of actual total tax collections to the total tax yield as

indicated by the calculations according to the rates stated in the second measure above and the income distribution patterns from the census data. Minor adjustments were also made to incorporate the relative evasion rates as stated in the third measure above. The last described measure is supposed to give the "actual" or "realized" burden of personal income tax. This is the measure used here for empirical work.

Distribution of Tax Payments by Classes

The allocation of tax payments by taxpaying units grouped according to income and other classes involves two steps. First, the final locale of a tax among broad economic categories, as distinct from the detailed distribution of taxpaying units, was determined from the general theory of tax-shifting. However, empirical evidence where available was profitably used. Second, the allocation by broad economic categories was translated in terms of allocation by taxpayers grouped according to various classes. The allocation process at this stage enters a strictly empirical base, and the calculations are pure measurements from data. This is the main part of the study. A brief discussion follows of the ways in which the shifting and allocation of various taxes were handled.

Consumption Taxes

In this study, as in other incidence studies, consumption taxes are assumed to fall on the consumers of the commodity concerned. Apparently, this will be so only if the supply schedules are perfectly elastic (or the demand schedules are completely inelastic). It might not be a very unreasonable assumption if only marginal changes are considered, but this study is not confined merely to marginal changes. If the supply schedule concerned is sloping upward, the price of the commodity taxed will rise by less than the tax; part of the tax will thus be borne by the factors of production. Unless, therefore, the consumers of the commodity concerned share equally in the production of that commodity, part of the burden of the tax may be shifted to the nonconsumers of the commodity. The overall effect of such shifting will be reduced if the consumption taxes are fairly general as, indeed, they are in Brazil. In any case, the errors that may be introduced in accounting for the slopes of the supply schedules of all the taxed commodities may result in greater distortion of tax-burden estimates than the possible distortion due to the errors in making the simple assumption of horizontal supply curves. Following Musgrave (1965) and others, possible changes on the income-source side due to consumption taxes are, therefore, disregarded. The entire tax is assumed to be borne by the consumers of the taxed commodity.

Once this assumption is made, the allocation of consumption taxes to various groups of taxpayers is straightforward and can be obtained by taking the sum of the products of consumption outlays on individual commodities and the respective average tax rates.

Export Revenue

The most important source of export revenue in Brazil is coffee, though the revenues from coffee export are not called export duties but are referred to, frequently, as "exchange confiscation." Brazil is a member of an international coffee agreement among exporters and importers. The other two most important suppliers of coffee are Colombia and the African countries as a whole. The quotas are fixed; the prices are left free for the exporting country concerned to fix as it likes. Although the qualities of the coffee supplies of the three big suppliers differ somewhat, the rate of substitution of one country's coffee for another's is extremely high, and the international price is believed to be determined by price leadership. (The quotas are revised periodically in accordance with the recent past sales.) The government of Brazil thus maintains two prices: the international (oligopolistic) price and the domestic guaranteed-minimum price to the coffee growers, at which the government is the residual buyer. Though the bulk of international trade in coffee is in the government's hands, anybody can export. The excess profits go to the government's coffers. This is what is called "exchange confiscation." The government's share in the foreign-exchange earnings from coffee currently runs around 60 percent of the export price. The domestic free-market price is only a fraction of the foreign market price. But given a favorable price, Brazilian coffee growers can more than meet the entire world and domestic demand. For instance, when coffee prices soared in the world markets in 1955, Brazilian farmers increased the plantation of coffee trees, so that in 1958 Brazil produced nearly twice the amount of coffee consumed all over the world, including that consumed within Brazil. The coffee price came crashing down, so that large stocks of coffee had to be burned. In view of this situation of an inelastic foreign demand market and a very elastic supply curve, it is difficult to say whether in the long run Brazilian farmers would gain or lose in the absence of the government's control of the coffee trade and in the absence of the "exchange confiscation" or profits on coffee exports. Part of the revenues from this source are applied for the improvement of the quality of Brazilian coffee and as subsidies for the eradication of coffee trees. A fair assumption, in respect to this source, according to the pseudo supply and demand curves, seems to be that the entire revenue from coffee is borne by foreigners. The duties on commodities other than coffee are negligible and can be ignored or assumed to be treated in the same way as those for coffee.

Import Duties

Import duties are treated in the same way as excise duties. That is, the incidence of import duties is assumed to be borne by the consumers of the imported goods. The allocation to various groups of taxpayers is also on the same lines.

General Sales Tax

The general sales tax is likewise assumed to fall on the consumer and is allocated in proportion to the expenditures. Certain food items are usually exempt from this tax. But certain other expenditure items, such as rent, though seemingly exempt from the sales tax, are not really so. The house owner, for example, has paid a sales tax on the construction material and does so on current maintenance and utility services. That part of the construction cost can reasonably be assumed to be passed on to the tenant. Only savers, in general, can escape this tax, and only as long as they continue accumulating.

Personal Income Tax

The burden of this tax is assumed to stay put with the initial payer. This assumption is valid only if total factor supplies remain unaffected by the tax. Textbook theorizing tells us that work effort, capital accumulation, and risk taking may on the whole, though not necessarily, be reduced if income tax is levied at progressive rates. Empirical evidence with regard to work effort, investment, risk taking, and productivity is indecisive, especially when compared with feasible alternative sources of revenues (and we are for the time being neglecting the effects of government expenditures). Saving is usually not considered to be invariant to tax burden, especially when it is levied at progressive rates, though the evidence in this case, too, is rather lean. On the whole, the least objectionable assumption would perhaps be to surmise that insofar as groups of taxpayers (as distinct from individual taxpayers) are concerned, all such effects are cancelled out, so that the burden of individual income tax rests with the initial recipient of income.

Once the above assumption is granted, the allocation of income tax is simple. Since it is a tax on incomes, the average tax rates discussed above directly yield the required tax burden. Mean incomes are available from the family budget data for all the classifications used here.

Company Income (Profit) Tax

More work, empirical as well as theoretical, has been done on the shifting of this tax than on many other taxes.[4] Yet the verdict remains undetermined. According to the classical theory, if the producers are assumed to maximize profits, the entire tax on profits will stay with the producers, since neither marginal revenues nor costs are then affected. According to some pragmatic approaches—e.g., that producers maximize sales rather than profits (for example, in Baumol [1965]), that producers follow average-cost-plus-a-margin practices (for example, in Hall and Hitch [1939]), and so on—profit taxes are passed on to the consumers.

In the present study, the starting point is the profit-maximizing approach, but certain modifications are introduced for the following reasons. First, in the underdeveloped countries, during the incipient growth stage, industrialists enjoy some monopoly power. Hence, the existence of monopolistic elements cannot be excepted in Brazil. It is true that a tax on monopoly profits will largely be borne by the producer. Since profit is a return to factor capital, the incidence will fall on capital. Second, the reference period has been a high-inflation period. During this period, the government has been imposing some regulation of prices. The price increases have been permitted on the basis of rise in costs, and taxes have invariably been included as costs. Accordingly, it is possible that, in the process, the company tax was partially passed on to the consumer. Third, in the long run, capital will shift out of high-tax investments into low- or nontax ventures, so there will be a tendency for rates net of return to equalize. This will affect long-run output and prices. Consequently, profit taxes may get shifted.

In view of these factors, allowing partial shifting of the company tax appears in order. Certain empirical studies of the incidence of corporation tax also indicate partial, or even full, shifting forward.[5] Accordingly, for tho purpose of this study, it was assumed that one-third of the tax is shifted forward to the consumer while the remainder falls on profits.[6]

The former one-third is then allocated according to the sales tax basis, as follows:

$$t_i\left(\frac{T_h}{T_s}\right) \qquad \text{(xlviii)}$$

where t_i denotes the calculated per capita sales tax payments (or rates) by the ith class, T_h stands for the total collection of the tax that is to be allocated, which in the present case is one-third of the company profit tax, and T_s indicates the total sales-tax collections. Thus, the sales tax basis is employed to allocate that part of company tax that is assumed to be shifted forward. For some of the other (relatively minor) taxes, a relation similar to xlviii is used with one modification. That is, the sales tax basis is replaced by the consumption tax basis, and the relation is:

$$t_i\left(\frac{T_h}{T_c}\right) \qquad \text{(xlix)}$$

where T_h stands for total collections of the _h_th tax under investigation, T_C is total consumption-tax collections, and the t_i's pertain to the burden of consumption tax on various classes i.

The remaining two-thirds of company tax (whether retained or paid as dividends) are considered part of the incomes of dividend receivers. The income concept used here is understood to include retained profits also. Accordingly, the personal-income-tax rates, discussed earlier, were revised to allow for profit incomes as well as corporation-tax rates.

The data on profits by size brackets and such are not available. In the absence of data it is assumed that the lowest seven brackets receive practically no income from profits, that 20 percent of the profits accrues to the taxpayers who fall in the eighth size bracket of this study, and that 80 percent accrues to those in the uppermost (ninth) bracket. Since this tax was levied at a nearly uniform rate (or approximately 22 percent) in the reference year, the above assumption implies that 20 percent of the tax was paid by the eighth bracket and 80 percent by the ninth bracket. The resulting tax rates are applied to the other classifications also, using the respective mean incomes as the base.

Company tax was aggregated from both the deductions at source and payments by annual returns. The effect of adding the corporation profit income (taxed at 30 percent) to other incomes (taxed at higher rates) is to reduce the effective (legal) tax rate on the ninth bracket from about 36 to about 31 percent.

Social Security Contributions

It is assumed that, insofar as the payment by the income earner is concerned, the incidence of social security payments rests with the initial payer, as in the case of personal income tax. The part contributed by the employer is assumed to be shifted either forward to the consumer of the product in which the income earner participates or backward to the income-earning unit itself.

The measure of social security contributions used here is taken directly from the family budgets. Evidently, the social security contributions as reported in this source are only the income receiver's share. These are simply doubled to include the employers' part of contributions. Since there is no definite agreement among economists as to whether social security contributions should be treated with or without a quid pro quo, two estimates of tax-burden distribution are made: one with social security contributions and the other without these. The former measure is more relevant when government expenditures and transfers are also considered. The latter measure is more relevant when only the revenue side is taken into account. To avoid multiple tables, however, we report only one measure, namely, the one _with_ social security contributions.

Other Taxes

The foregoing taxes account for nearly 90 percent of the total federal, state, and local revenues. The remaining 10 percent or so include stamp and registration duties of federal and state governments, rural land tax, inheritance tax, gift tax, local taxes, and a few very minor taxes. Fuel and electricity taxes, as stated earlier, are ignored as being more in the nature of fees and prices rather than of taxes.

Although these taxes form a small proportion of the total revenues, their incidence may not be as general as that of the major taxes. In the present study, gift tax is allocated almost entirely to the highest income bracket considered here. Local taxes are assumed to be paid mainly by the local residents. Rural land tax is assumed to fall on rural incomes. The stamp duties are allocated according to the formula of Equation xlix.

Inflation Tax

Government's deficit finance (through the printing press or the creation of central bank credit, rather than through public borrowing), as a percentage of GNP in the reference year, is assumed to represent what has here been called the inflation tax. This tax has special significance in Brazil. The effect of inflation, insofar as it is caused by deficit finance, is not merely to transfer real resources from the private sector to the government sector but also to cause shifts within the private sector—that is, among different groups of population. The redistribution of incomes due to inflation arises from the "income-use" side as well as "income-source" side.

On the "income-use" side, this tax was allocated according to the relative rise in the prices of various consumption items in the family budgets. (Actually, only half of these relative price changes were taken into account; the other half were assumed to occur in spite of inflation, due, for example, to differential productivity changes and factors other than deficit finance. See Chapter 7.) The allocation of the distribution effects of the inflation tax from the "income-source" side was done simply by a quantification of deductions from the theory of inflation. Specifically, it was assumed that the entire tax falls on the urban sector's first seven brackets in proportion to their incomes, implying that the net effects of inflation from the "income-source" side are neutral for the two upper-income brackets and the agricultural sector.

THE ESTIMATING PROCEDURE FOR EXPENDITURE BENEFITS

Ideally, as emphasized by Samuelson (1969), the allocation and the distribution aspects of the government budget ought to be solved simultaneously. An implication of this is that the income-distribution effects of the government budget (discussed in this chapter) cannot be realistically studied in abstraction apart from the budget's allocation effects (studied in the econometric model). However, Samuelson's approach is hardly workable in practice. Accordingly, we resorted to the Musgravian real-world approach (1969) in solving the two parts separately and then bringing them together.

Allocation among Income Classes

Because of lack of public-expenditure-incidence theory, the allocation of public expenditures among various income classes is rather arbitrary. For example, instead of measuring the distribution of the incidence of public expenditure on education, health, and other functions, the allocation is done under various assumptions as regards their incidence. The main interest therefore dwells on the distribution of overall public expenditure (in various sectors and areas) rather than on the distribution of its components. The detailed breakdown of Brazilian public accounts is available by administrative classification. Economic classification and functional classifications are sketchy and incomplete. Fortunately, the classification by administrative units is quite detailed and so is expected to come close to functional classification. Accordingly, public expenditures were first grouped into administrative categories—e.g., by ministries. Allocation was then done for each category on the bases of family characteristics, number of families in different income classes, business income, wage and salary income, and other similar factors.

Any assumption with regard to the incidence allocation of public expenditures can be controverted. To minimize disagreement and to maintain comparability the allocation of this study generally follows the traditional pattern, though special economic and institutional characteristics of Brazil are taken into account by making minor but appropriate departures from the traditional pattern. The allocation procedure, moreover, is by and large similar to that used in his aggregative study for Brazil by Aaron (1968). However, since that study and this one are not coextensive, the allocation procedures here differ in several respects from those of Aaron. These differences arise largely due to a disparity in the scopes of the studies: while Aaron estimates expenditure benefits at the national level (among income classes) only, this study makes the same estimates at the state level and by urban and rural sectors (interclass, interregion, and intersector estimates).

Broadly, the social-good types of public expenditure were allocated among various income classes on the basis of income. These include federal, state, local, and autarquia expenditures on such functions as the legislative branch (poder legislativo, orgaos auxiliares, etc.), the judiciary, various general-purpose commissions, and various executive functions of this nature (namely, administrative, defense, police, national security, finance, foreign relations, federal district, etc.). Public expenditure on education (at all levels of government) was allocated on the basis of an index prepared from the number of households, average family size, and the number of years of schooling of the head of the family. Public-health expenditure was allocated on the basis of the number of households in each income bracket and the average family size in each bracket of the family-budget survey data. The expenditures of the Ministries of Industries and Commerce, Mines and Power, Telecommunications (half—the other half being on the basis of consumption outlays), and similar services were allocated on the basis of business income. Half of the expenditures of the Ministries of Merchant Navy, Federal Roadways, Industrial Estates, and similar others were allocated by business income and half by consumption outlays. Following Aaron, the expenditures of the Ministry of Agriculture and Allied Products (referring always to all-level governments) were assumed generally not to have provided any benefits to the two lowest income brackets "on the assumption that any significant benefits would have moved family income above those levels" (Aaron, 1968, p. 7). Aviation and public-works outlays were allocated equally on three bases: income, business income, and expenditure on transportation. Social security benefits were allocated one-fourth by pension incomes and three-fourths by wage and salary incomes, assuming that this relatively recent system in Brazil benefits retired and active workers in that ratio. Development expenditure was allocated according to income, assuming that in the long run everyone (in the region concerned) benefits from development expenditure.

Federal government expenditures by function are available from Revista Brasileira da Economia (annual) and in greater detail from Balanco dos Investimentos Publicos since 1965 (Brazil, 1965). For states a functional breakdown has been prepared at the Ministry of Finance by Silva (1967). For local bodies, a functional breakdown is available since 1961 in Revista de Financas Publicas (annual). Income and expenditure patterns were taken from the household survey data compiled by FGV (1967).

Allocation between Rural and Urban Sectors

It was assumed that rural and urban families of equal size and having equivalent incomes get benefits in the ratio of 1:3 from expenditure on public health (totaling about 4 percent of total public expenditure in 1968), public education (totaling approximately 10 percent of public expenditure in 1968), and labor and social affairs (amounting to roughly 3 percent of federal expenditure in 1968). A family that has lived in both rural and urban areas knows well that ambulance and emergency medical services and even general hospital services are very unevenly distributed between the two sectors. Expenditure on malaria control and similar measures that benefit the rural population is not a very large fraction of the overall public-health outlays. Likewise, the availability and the quality of schooling favors the urban sector. Similar conditions apply to labor and social-welfare measures. The indicated ratio is a rough approximation of the division of such expenditures. The impact of expenditures by the Ministry of Agriculture (about 3.2 percent of federal budget in 1968) is apparently on the rural sector, but the ultimate incidence in the form of lower agricultural prices, etc., is assumed to fall 50-50 on the equivalent rural-urban families. The federal government incurs approximately 2.5 percent of revenue expenditure on the administration and secretarial work for the development of the northeast. In addition, a good proportion of corporate tax money (nearly 3 to 4 percent of the federal budget) currently goes to the northeast. However, while the money comes from the urban sectors of the south and southeast, it is almost entirely spent on the urban industrial sector of the northeast, so that the transfer to the rural sector is negligible. The social security system is almost entirely confined to the urban areas. Hence, the entire expenditure on _previdencia social_ and 75 percent of the expenditure of other autarquias is allocated to the urban sector. The whole of municipal expenditure is allocated to the urban sector, as were all municipal taxes.

All public-good types of expenditure (nearly two-fifths of the total) are assumed to benefit both sectors in proportion to their incomes.

Allocation among States

This is done on the basis of wages and salaries in the public sector (_salarios e ordenados no setor publico_) for the year 1968, as given in _Conjuntura_, 24, no. 11 (1970). This source gives this information according to states and by broad classifications for federal, state, and municipal governments. Federal expenditures of the public-good nature, such as defense, legislative, secretarial, federal district, presidential, and part of the central executive and judiciary, are shown in the states where federal capitals are located—namely, Guanabara (which consists of the city of Rio de Janeiro alone) and the federal district (which chiefly consists of Brasilia). These federal expenditures were therefore reallocated in proportion to the domestic incomes of the various states. The concentration of public wage payments in state capitals does not bias the results, as the unit of study is now the state or the urban and rural sectors of the state and not individual cities as in the tax-burden chapters. In using wages and salaries (rather than the budget or state incomes or any other basis) as an approximate index of public services it is assumed that by and large equivalent bundles of services are rendered against these payments. To the extent that services rendered do not match wages paid or that the ratios of services rendered to wages paid differ in different states, the indicated basis will give erroneous results. It would not be unrealistic to assume, however, that loafing at work (if any), multiple-job holding by public employees (not uncommon in Brazil), or any other cause of a wedge between services rendered and wages paid is fairly evenly spread over the country. The unevenness of nominal military, executive, and other federal expenditures has already been taken care of. The transfer of public funds to the northeast, as represented by SUDENE investment-incentive schemes, were assumed to come in proportion to the industrial incomes of the southern states and to be distributed in proportion to industrial expansion in the various states of the northeast.

SUBSTANTIVE RESULTS

The results are presented in Tables 23 (tax burden), 24 (expenditure benefits), and 25 (budget incidence) and are sketched in Figures 12 and 13. Among the broad conclusions that emerge from these calculations, the following may be noted.

The expenditure benefits are, by and large, pro-poor, except that the uppermost (ninth) bracket lies off the progression line and tips the upper end of the curve upward to produce pro-richness at that end. This observation is generally true for both urban and rural sectors. Tax-burden distribution, on the other hand, is broadly bowl-shaped. Because of the upturn of the curve at the end in general, the shallow bowl-like shape of the urban-tax-burden distribution is almost in reverse of the expenditure-benefit distribution.

The rural curves open up leftwards (towards lower-income classes) like a pair of shears, the upper blade representing the benefits. The gap between burdens and benefits is, in general, lowest at the level of the eighth bracket and starts to widen again above that bracket (for the rich class). The uppermost income class seems to have managed to reverse the trend of the rigors of progression to some extent, but, ignoring that, the budget incidence in the rural areas is more consistent with the general progressivity norm than is its incidence in urban areas. The urban middle class is squeezed out from both sides, being subjected to an inequitable tax burden and to inequitable expenditure benefits.

The rural sector as a whole is the net gainer from the public activity. (Compare the urban and rural panels of Table 25.) Hardly any

TABLE 23

Tax-Burden Distribution, Including "Actual" Tax Payment, Social Security Contributions, and Inflation Tax, by Size-Brackets of Income, Different Econographic Areas of Brazil, 1968

Area	Size-Bracket of Income in 1962-63 Cruzeiros										
	Up to 99	100 - 149	150 - 249	250 - 349	350 - 499	500 - 799	800 - 1,199	1,200 - 2,499	2,500 and Over	Average	National Average
(1)	(2)	(3)	(4)	(5)	(6)	(7)	(8)	(9)	(10)	(11)	(12)
Capital Cities											
Belem	23.15	30.50	34.73	34.61	34.19	33.30	32.78	29.23	21.65	32.80	27.75
Fortaleza	30.36	38.64	36.00	38.75	36.54	36.93	35.90	36.59	25.42	35.95	30.27
Recife	25.88	29.66	32.60	36.15	34.70	35.95	37.35	30.83	31.40	32.48	27.34
Salvador	26.13	36.99	39.86	39.95	39.77	39.69	42.00	35.96	35.90	37.47	31.33
Belo Horizonte	26.78	33.88	34.90	35.57	35.48	36.93	39.06	32.32	30.18	34.98	29.72
Rio de Janeiro	25.73	35.34	35.78	35.84	39.32	36.50	35.71	30.46	25.40	36.48	29.68
São Paulo	31.65	33.51	36.12	34.73	38.93	38.90	35.78	32.06	29.01	36.12	30.19
Curitiba	29.00	39.36	36.00	43.00	40.65	36.42	36.92	27.66	32.21	37.25	31.18
Average	26.33	34.23	35.74	37.32	37.32	36.82	36.93	31.88	28.89	35.81	30.32
Towns											
Espirito Santo I	26.73	26.45	30.63	32.51	35.76	36.10	38.07	34.82	34.82	35.35	28.81
Espirito Santo II	20.86	23.64	27.68	26.51	36.25	31.35	38.49	33.99	35.62	33.13	28.81
Espirito Santo III	22.59	33.74	28.77	33.53	34.75	37.19	37.45	30.20	31.70	34.69	28.81
São Paulo I	19.10	16.89	32.63	39.40	39.53	39.40	40.23	27.77	22.95	36.86	30.19
São Paulo II	23.76	24.92	36.00	31.10	31.48	36.77	44.70	27.05	..	33.60	30.19
São Paulo III	33.85	27.12	30.56	34.95	35.00	37.73	35.54	33.96	26.21	35.08	30.19
Paraná I	23.26	26.81	31.42	41.93	33.46	36.35	38.99	33.06	27.04	34.48	31.18
Paraná II	..	35.85	32.90	39.38	41.30	43.79	36.02	27.93	29.02	36.16	31.18
Paraná III	26.30	31.39	37.88	43.08	41.36	45.02	40.67	30.31	35.48	40.09	31.18
Average	24.29	26.59	32.04	35.28	36.25	37.30	37.61	31.00	30.24	35.49	30.32
Rural Areas											
Ceará	7.41	10.72	7.78	7.03	7.87	6.64	7.90	10.09	..	7.56	30.27
Pernambuco	5.40	7.78	6.13	7.00	7.00	6.13	10.00	8.76	..	6.78	27.34
Minas Gerais	6.58	7.15	6.25	8.23	6.01	10.03	8.24	10.44	10.81	8.67	29.72
Espirito Santo	4.00	6.94	10.09	6.21	7.00	7.95	9.24	10.51	10.12	6.47	28.81
São Paulo	7.60	6.97	10.57	7.86	8.68	8.70	11.82	10.41	10.03	8.17	30.19
Santa Catarina	6.14	5.59	7.41	7.41	8.22	9.22	7.89	13.56	10.45	7.48	31.10
Rio Grande do Sul	7.45	8.47	7.56	6.80	7.54	6.67	7.93	10.12	12.30	7.38	31.08
Average	6.25	7.46	7.97	7.22	7.47	7.91	9.00	10.56	10.74	8.04	30.32
National Average	21.49	25.82	28.67	30.48	30.92	31.26	31.60	27.26	25.78	30.32	30.32

TABLE 24

Distribution of Benefits of Public Expenditure, 1968

	Income Class in 1962-63 Cruzeiros											
State (1)	Up to 99 (2)	100 - 149 (3)	150 - 249 (4)	250 - 349 (5)	350 - 499 (6)	500 - 799 (7)	800 - 1,199 (8)	1,200 - 2,499 (9)	Over 2,499 (10)	Average (11)	National Average (12)	Per Capita Income, 1967 (13)
Urban												
Para	45.43	42.53	39.94	40.44	33.83	35.59	33.64	29.98	30.53	37.00	33.88	759
Ceará	44.84	43.52	39.28	40.39	34.93	34.39	34.24	29.68	30.63	38.86	35.43	505
Pernambuco	45.63	44.33	39.95	41.39	35.43	34.28	34.93	30.11	32.59	38.06	35.07	620
Bahia	44.35	43.40	40.32	42.14	33.90	34.68	33.93	31.59	33.69	38.28	35.24	551
Minas Gerais	36.62	33.83	33.10	33.16	28.15	28.00	27.13	25.99	29.43	28.77	26.15	793
Espirito Santo	36.61	35.24	32.12	32.45	27.08	27.72	27.23	24.67	25.73	28.09	25.96	885
Rio de Janeiro	53.44	47.46	42.99	45.39	35.52	35.39	33.76	30.18	34.02	36.34	34.06	891
São Paulo	32.45	28.53	28.70	29.24	24.63	23.62	22.87	20.33	20.75	25.11	23.70	1582
Paraná	29.28	28.52	26.40	27.32	22.90	23.72	22.97	21.39	22.95	23.73	22.71	870
Santa Catarina	31.77	30.83	28.44	29.44	24.57	25.38	24.58	22.81	24.48	25.43	23.32	1046
Rio Grande do Sul	39.95	43.32	40.14	41.87	34.40	29.03	32.23	29.53	30.39	35.52	32.73	1011
Urban average	43.46	42.00	37.45	38.10	30.91	29.97	29.23	26.50	23.51	33.91	..	895
Rural												
Para	25.72	23.56	21.45	21.85	17.24	17.34	16.46	14.72	15.85	19.69	33.88	132
Ceará	25.97	23.68	21.57	21.96	17.33	17.43	16.53	14.78	15.90	19.80	35.43	194
Pernambuco	26.40	24.14	22.08	22.38	17.64	18.00	17.03	15.22	16.35	21.49	35.07	227
Bahia	26.11	23.93	21.78	22.17	17.50	17.59	16.70	14.88	16.05	21.41	35.24	209
Minas Gerais	20.48	18.80	17.12	17.61	13.72	13.80	13.10	11.71	12.60	14.22	26.15	262
Espirito Santo	20.39	18.71	17.04	17.53	13.62	13.70	13.00	11.63	12.52	16.24	25.96	226
Rio de Janeiro	29.24	26.78	24.35	24.79	24.03	19.65	18.67	16.69	17.96	23.70	34.06	300
São Paulo	17.43	23.01	20.92	21.33	16.83	16.92	16.05	14.36	15.45	17.27	23.70	524
Paraná	16.68	22.09	20.12	20.49	16.20	16.30	15.41	13.80	14.85	18.07	22.71	370
Santa Catarina	17.93	16.43	14.95	15.23	12.01	12.08	11.46	10.25	11.03	13.68	23.22	328
Rio Grande de Sul	26.51	24.25	22.17	22.47	17.74	18.11	17.12	15.23	16.39	20.01	32.73	507
Rural average	24.97	22.86	18.36	22.13	18.48	17.61	16.86	14.29	14.47	19.14	..	312
National average	40.13	38.55	35.01	35.23	29.67	28.75	28.00	25.30	25.98	..	30.16	672

Note: Calculated values are expressed as percentages of income.

TABLE 25

Distribution of Budget Incidence: Expenditures Minus Taxes to Income, 1968

	Size-Bracket of Income in 1962-63 Prices									
State (1)	Up to 99 (2)	100 - 149 (3)	150 - 249 (4)	250 - 349 (5)	350 - 499 (6)	500 - 799 (7)	800 - 1,199 (8)	1,200 - 2,499 (9)	Over 2,499 (10)	Average (11)
Urban										
Para	22.28	12.03	5.21	5.83	-0.36	2.29	0.86	0.75	8.88	4.20
Ceará	14.48	4.88	3.28	1.64	-1.61	-2.54	-1.66	-6.91	5.21	2.91
Pernambuco	19.75	14.67	3.35	5.26	0.73	-1.67	-2.42	-0.72	1.19	5.58
Bahia	18.22	6.41	0.46	2.28	-5.87	-5.10	-8.07	-4.37	-2.21	-2.23
Minas Gerais	9.86	-0.05	-1.80	-2.41	-7.33	-8.93	-11.94	-6.33	-0.75	-8.83
Espirito Santo	13.21	7.24	3.12	1.65	-8.62	-7.18	-10.77	-8.33	-8.27	-6.31
Rio de Janeiro	27.77	12.12	7.21	9.55	-3.80	-1.11	-1.96	-0.28	8.62	-2.42
São Paulo	0.80	-2.98	-7.42	-5.49	-14.30	-15.28	-12.91	-12.27	-8.26	-12.42
Paraná	0.28	-10.84	-9.60	-15.68	-17.75	-12.70	-13.96	-6.27	-9.26	-14.54
Santa Catarina	6.77	-0.77	-5.66	-11.86	-13.43	-16.00	-13.12	-7.69	-6.02	-12.37
Urban average	18.16	11.60	3.55	1.80	-5.89	-7.13	-8.03	-4.94	-5.55	-1.74
Rural										
Ceará	18.31	13.42	13.80	14.93	9.46	10.87	8.63	4.69	..	12.24
Pernambuco	20.57	16.38	15.95	15.38	10.64	11.87	7.03	6.46	..	14.71
Minas Gerais	13.90	11.65	10.79	9.30	7.71	3.77	4.76	1.31	1.79	5.54
Espirito Santo	16.39	11.77	6.96	11.32	6.62	5.65	5.00	1.12	2.50	9.77
São Paulo	9.83	16.04	10.35	13.47	8.15	8.22	4.23	3.95	5.42	9.10
Santa Catarina	11.79	10.84	7.54	7.82	3.79	2.86	3.57	-3.31	0.58	6.20
Rio Grande do Sul	19.06	15.78	14.61	10.94	10.20	10.44	9.19	5.11	4.09	12.63
Rural average	18.72	15.40	10.39	14.91	11.01	9.70	7.86	3.72	3.73	10.10
National average	18.64	12.73	6.34	4.75	-1.25	-2.51	-3.60	-1.94	0.20	-0.16

Figure 12: Distribution of Tax Burden and Expenditure Benefits in Urban Sector, 1968

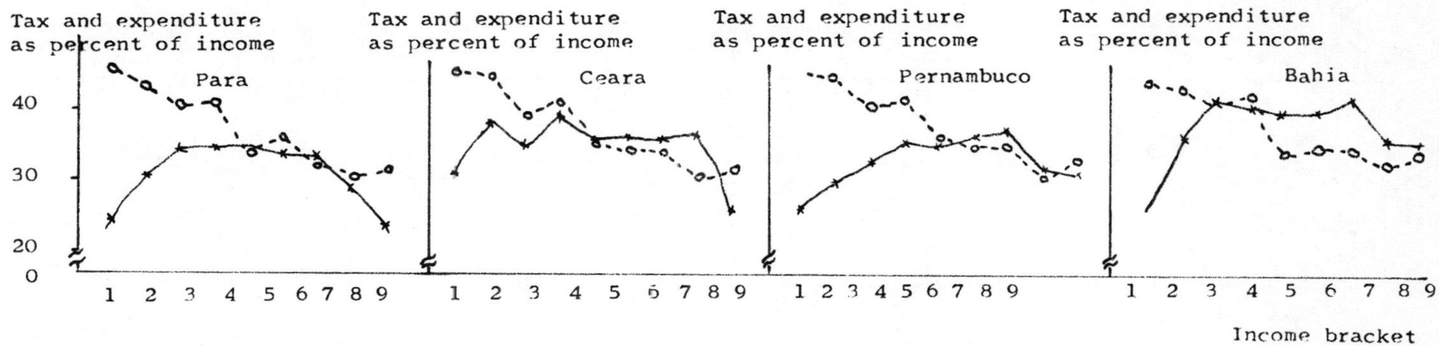

(continued)

Figure 12, continued

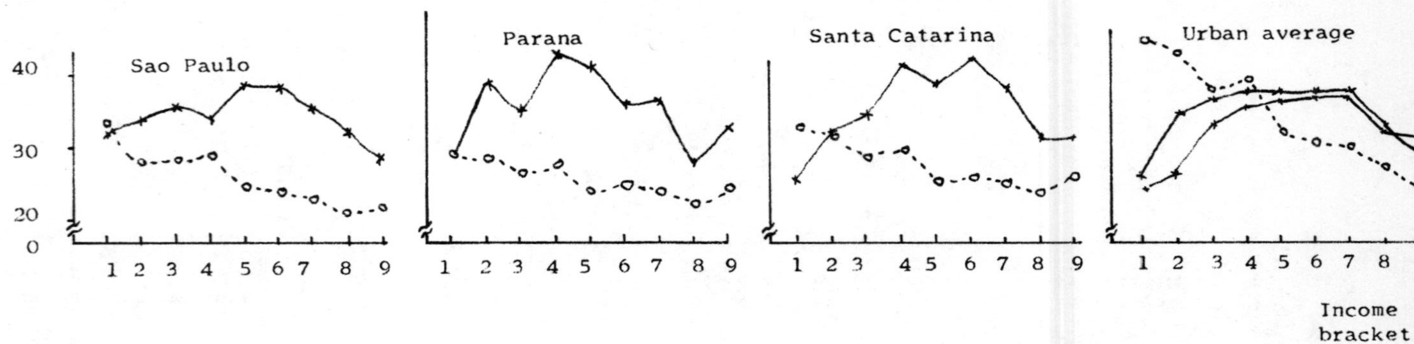

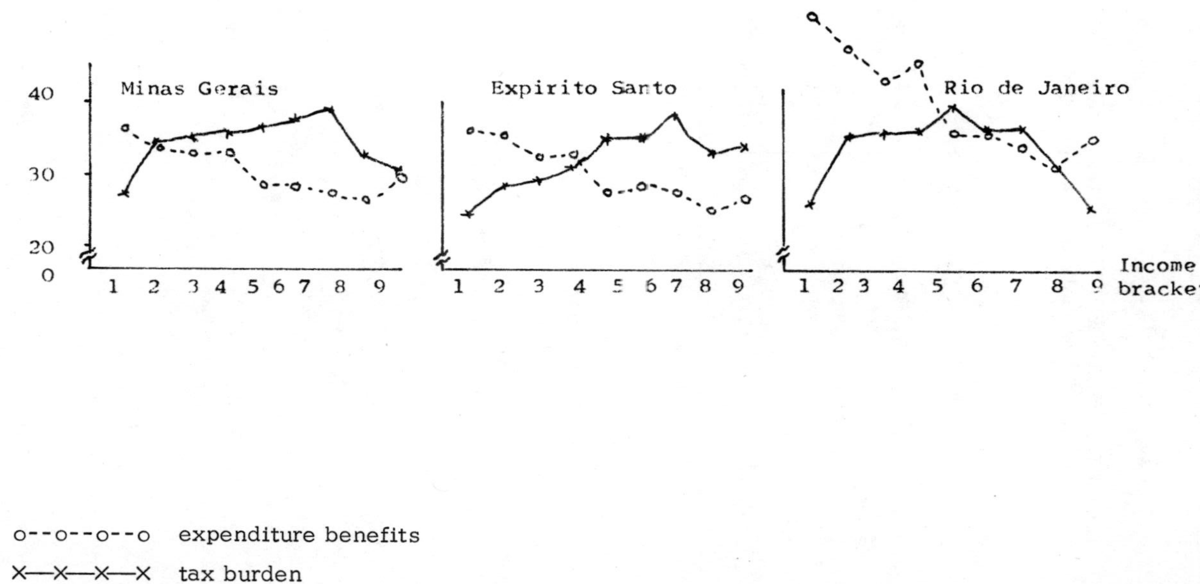
Minas Gerais
Expirito Santo
Rio de Janeiro
40
30
20
0
1 2 3 4 5 6 7 8 9
Income bracket
o--o--o--o expenditure benefits
X—X—X—X tax burden

Figure 13: Distribution of Tax Burden and Expenditure Benefits in Rural Sector, 1968

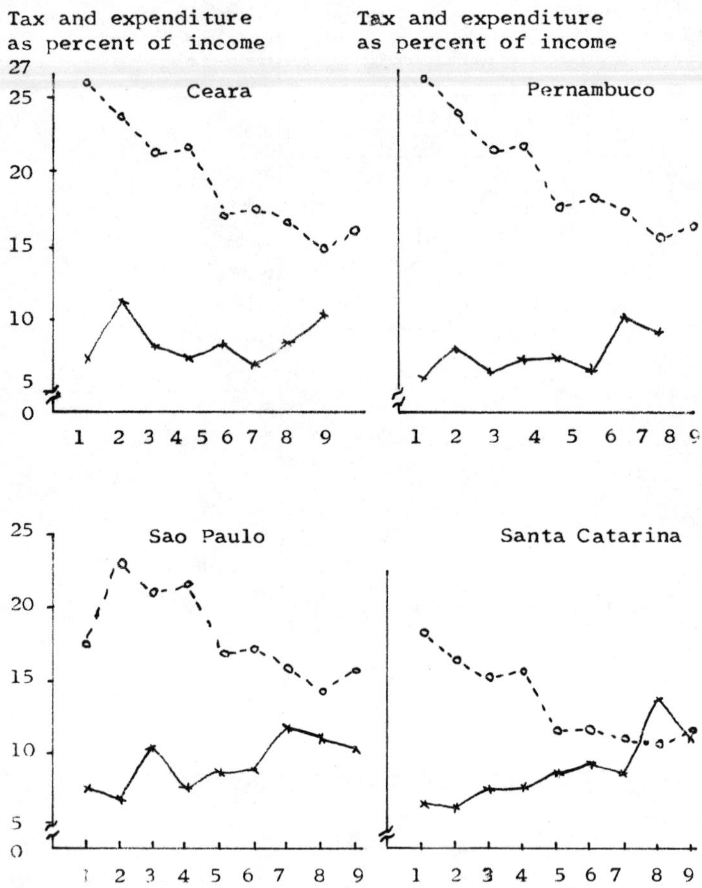

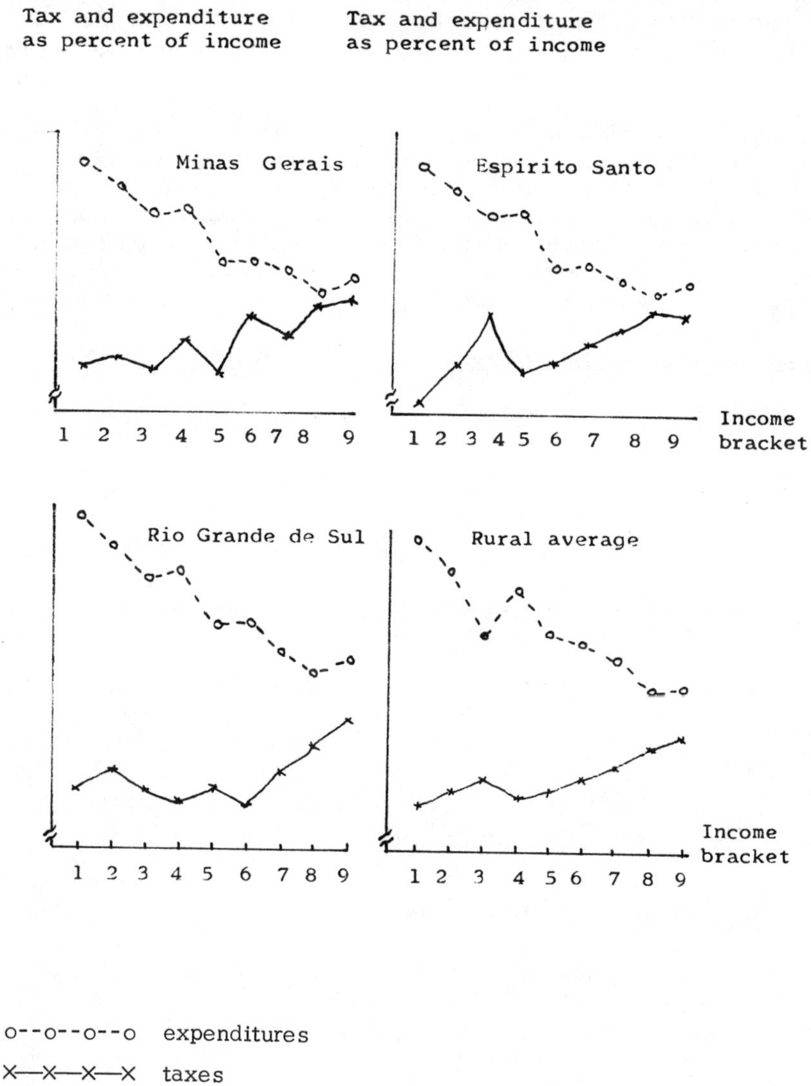

Tax and expenditure as percent of income
Tax and expenditure as percent of income
Minas Gerais
Espirito Santo
Rio Grande de Sul
Rural average
1 2 3 4 5 6 7 8 9
Income bracket
o--o--o--o expenditures
x—x—x—x taxes

income class in the rural areas of any state is a loser. (Observe the positive signs of the values for rural areas in Table 25.)

Within the urban sector, the less developed northern and northeastern states gain at the cost of the developed south. A cursory look at the urban sector of Table 25 will indicate that the high positive values lie to the northwest, and increasingly negative values lie to the southeast of the array. The overall budget incidence, thus, is redistributive of incomes from the upper-income families to low-income families and from more to less well-off regions. This is so both for the urban and the rural sectors. The combined pro-poorness of taxes and expenditures is of a small order, however, and will not stand the test of a reasonably-determined progressivity norm as distinguished from the proportionality norm (i.e., as distinguished from the benchmark of the status quo in income distribution) used in this chapter and commonly employed in tax-burden studies. The calculated aggregate budget incidence for Brazil can be compared with that estimated for the United States by Gillespie (1965).* The U.S. estimates for seven family-income classes are arrayed against the Brazilian estimates for nine income classes as follows:

<table>
<tr><th>Bracket</th><th>Brazil, 1968</th><th>United States, 1960</th></tr>
<tr><td>1</td><td>18.64</td><td>55.1</td></tr>
<tr><td>2</td><td>12.73</td><td>44.4</td></tr>
<tr><td>3</td><td>6.34</td><td rowspan="2">18.5</td></tr>
<tr><td>4</td><td>4.75</td></tr>
<tr><td>5</td><td>-1.25</td><td>-1.3</td></tr>
<tr><td>6</td><td>-2.51</td><td>-2.9</td></tr>
<tr><td>7</td><td>-3.60</td><td rowspan="2">1.7</td></tr>
<tr><td>8</td><td>-1.94</td></tr>
<tr><td>9</td><td>0.20</td><td>-13.2</td></tr>
</table>

The distributions of the U.S. and Brazilian budget incidence present somewhat dissimilar pictures. While the U.S. budget incidence (benefits minus taxes) curve is of the shape of an inverted tilde, the Brazilian budget incidence curve is of the shape of a leftward tilted L or a Chinese soup spoon. If Gillespie's estimates represent the real picture, the low-income classes in the United States are highly favored by public

---

*It should be noted, however, that Gillespie's estimates for the United States do not allow for tax avoidance under various tax concessions and tax loopholes by upper-income classes, whereas the same have, at least partially, been accounted for in the present study for Brazil.

finance.* The uppermost income class in the United States is highly disfavored by public finance. The middle-income classes of the two countries end up, more or less, with similar adverse budget incidence.[7]

It may, thus, be concluded that, insofar as the aggregate estimates are concerned, on the whole the urban-tax-burden distribution bulges upward from the middle and the urban-expenditure-benefit distribution dips downward in the middle. The middle-income classes, thus, are hollowed out from both sides: they are losers on both sides of the budget. The low-income classes are the net gainers. The gap of net losses closes towards the upper end. Combining these segments and subject to the observation made with regard to the middle-income classes, the overall budget incidence on the urban sector comes out, by and large, regressive. The same conclusion applies to rich and poor states.

The rural sector is the net gainer from the budget incidence in two respects: a) expenditure benefits exceed tax burdens substantially for all classes and b) the tax burden is broadly pro-poor, as is expenditure distribution (except between the eighth and ninth brackets, which do not conform to the said pattern in all states). Relative to other classes, Brazilian low-income families gain much less from the budget operation than U.S. low-income families; Brazilian upper-income families lose much less than their U.S. counterparts; Brazilian middle-income families end up roughly in the same (losing) position as U.S. middle families.

The foregoing are the overall results. Individual tax and expenditure instruments (not reported in detail in this chapter) differ in their effects on distribution. In general, the following instruments are redistributive: personal income tax (in spite of evasion), excise tax, gift and inheritance taxes, public expenditures on health, education, and social welfare, direct or indirect transfer payments among both people and regions, and other similar measures. In general, the

---

*Consider a tax function with the desired parameter a^*:

$$t = by^{a^*}$$

where t is tax payment and y is family income. The parameter b is expected to be less than zero and a^* is to be viewed as the desired value (norm) and not the computed value of this parameter. The following conditions on a^* define the indicated norm:

$a^* = 1$ defines the proportionality norm (status quo)
$a^* < 1$ defines regressivity norm (pro-rich tax)
$a^* > 1$ defines progressivity norm (pro-poor tax)
$a^* = 1.5$ is a commonly-used value of progressivity norm which is also consistent with Brazil's income distribution. (See Sahota [1974].)

following instruments exacerbate inequalities: the sales tax, the inflation, municipal taxes to some extent, general public expenditures, and similar others.

PERSONAL-INCOME-DISTRIBUTION FUNCTION AND A SYNTHESIS

In the foregoing sections I have attempted to measure public income distribution and to establish relationships between it and size-income distribution. The next step is to see whether the inequality measures based on size-income distribution—e.g., the Gini coefficient—can be related to functional income distribution and the relevant macro variables of the econometric model. For this purpose, using the Gini coefficients for 20 regions (ten urban and ten rural) as calculated from the 1970 census by Langoni (1973, Table 7.5), the following regression equation was computed:

$$\begin{array}{lllllll}
G = & = .4196 & - .0459y & - .0022ILL & + .0085g & + .0031IND & + .1105DU \\
[.47] & (4.33) & (-1.84) & (-2.06) & (1.68) & (1.65) & (2.90) \\
 & & [2.29] & [33.49] & [5.08] & [11.57] &
\end{array}$$

$$\bar{R}^2 = .7504$$

Standard error of estimate = .0415

where the numbers in parentheses are t values and the figures in square brackets are the means of the respective variable. All variables, excepting g, are measured for 1970 and are defined as follows: G = Gini coefficient; y = per capita income in hundreds of (new) cruzeiros; ILL = illiteracy, in percentages; g = annual rate of growth of state real internal income from 1961 to 1967; IND = the degree of industrialization measured as value added in the secondary sector as a percentage of GDP; and DU = a dummy variable for urban areas.

The fit is pretty good and most of the coefficients are tolerably significant. There are two intercorrelations—namely, those between ILL and y (r = -.88) and IND and y (r = .76)—that raise the problem of multicollinearity. The relevant individual coefficients, therefore, are vulnerable to probable indeterminacies. Subject to that, it may be seen that only per capita income levels are equalizing of personal incomes. Education, growth, urbanization, and industrialization seem to exacerbate income inequalities.

The reasons for the negative sign of variable g are not immediately clear. A possible explanation is that in the cross-section employed here the short-run effects of income expansion are captured by growth (g) and favor the rich, while the long-run effects across states are captured by per capita income (y) and are income-equalizing. Another reason that may or may not be incidental is that growth rates have been lower in high per capita income states, such that the interstate gaps have been tending to close somewhat. For instance, between 1954 and 1967 the relative shares in the total internal income of such developed states as São Paulo, Guanabara, and Rio Grande do Sul have declined from 36.5 to 33.92 percent, from 13.8 to 10.58 percent, and from 9.40 to 8.45 percent, respectively. Naturally, if the Gini coefficient is positively related to y, it can pick up a negative sign with respect to g. A third reason may be that g is a surrogate for left-out forces. Since the growth-rate variable is measured over the preceding eight years, and thus pertains to changes of the preceding decade, while the remaining variables of the regression are for 1970, g may proxy for left-out changes of the corresponding period even when the influences of industrialization and urbanization have been accounted for. Moreover, it is only a partial model estimated by ordinary least squares. One should note that the coefficient of g can barely be stretched to be significant at the conventional levels. As a matter of fact, when a sensitivity analysis is carried out by permutation and combination of variables, g acquires positive signs as frequently as negative ones, but in no case significantly so. However, since the coefficient of g may stand for something, we may not ignore it in the regression equation reported here.

Taking the coefficients at their face values, it may be seen that a near-doubling of the rate of economic growth from an average of approximately 5 percent in the period 1960-67 to an average of approximately 10 percent in the early 1970s would worsen personal-income distribution in terms of the Gini coefficient by approximately 5 points—to be precise, .04234 = .008468(10-5). On the other side, a sustained rate of growth of 10 percent would double per capita incomes in eight and two-thirds years. This will reduce the Gini coefficient by approximately ten points (.0459 × 2.29 = .1051, where 2.29 is the mean per capita income in hundreds of new cruzeiros in 1970), giving a net improvement in distribution of more than five percentage points.

The Gini coefficient is on the average 11 points higher in urban areas over the general constant of .42, due to factors other than those considered here. Since economic growth naturally leads to increased urbanization, as the weight of urbanization in overall population rises, the overall Gini coefficient converges toward a value of .53 and away from .42. Similar conclusions apply to industrialization and education: Both factors are associated with economic progress; both worsen income distribution.

It is to be noted that we have not included the effects of the income-redistributing policy instruments in the model estimated in this chapter. In other words, only (some of) the automatic economic forces built into the Brazilian economic system have been analyzed. Insofar as these forces are concerned, under plausible assumptions about the changes in relevant variables, Brazilian income distribution is expected to taper off at its present state.*

The distribution relationship in the macroeconomic model of the earlier chapters is given by W/Q. It will be seen from Table 4 that most of the coefficients of Q and W differ only by a constant multiple, given approximately by the sample means of these two variables. The opposite influences are found only for g_{-1}, e^{ex}, and w^{min}/y, which are redistributive. Policy manipulation in terms of the system determined by the econometric model can, thus, be done only in terms of two instruments: e^{ex} and w^{min}. Interestingly enough, while labor's share was negatively related to minimum-wage rate in the structural equation of Chapter 4, the two variables turn out to be positively related when all economy-wide direct and indirect effects are accounted for in impact multipliers (Table 4). Likewise, while distribution and and growth were not found to be positively related in the simple regression equation of this chapter, in the general equilibrium impact multipliers they are. The influences of the tax and expenditure instruments of the two computations (those of the econometric model and those of the present chapter) are not inconsistent with each other when the following three provisos are kept in view. First, functional and size distributions are broadly related but cannot be expected to depict

---

*Assuming a sustained annual rate of growth of national income of 9 percent during the decade of the 1970s (thereby doubling per capita income by the end of the decade) and annual rates of growth of one percentage point in industrialization, urbanization, and literacy, the resulting changes in the Gini coefficient over the decade can be calculated as follows:

Due to a sustained rate of growth of 9 percent (implying an increase of roughly four percentage points over the mean of the reference period)	= (.0085)(9 - 5)	= .0340;
Due to a ten-percentage-point reduction in illiteracy over the decade	= -10(-.0022)	= .0220;
Due to a ten-percentage-point increase in industrialization over the decade	= 10(.0031)	= .0310;
Due to a ten-percentage-point increase in urbanization over the decade	= [.8(.42) + .2(.53)] - [.7(.42) + .3(.53)]	= .0110;
Due to a doubling of the 1970 per capita income (of 2.29 units of hundreds of cruzeiros)	= -.0459(2.29)	= .1052

Net result = -.0072, which is less than a one-percentage-point change.

precise relationships even in developing countries. Second, there was no way of incorporating the graduation or progression of different tax rates across different income brackets and across the income and expenditure characteristics of different subgroups in the econometric model, whereas such disaggregation has been duly incorporated into the analysis of the present chapter. Three, the output and other similar effects of different taxes are taken into account in the econometric model but not in the present size-distribution model, which stops at the stage of the final locale of a tax or expenditure. Thus, both models have their own merits and demerits and are useful in their own ways. Not until the analysis by income brackets of the type in the present chapter can be incorporated into the full model of the earlier chapters can we learn more precisely about the ultimate size-distribution effects of taxes and expenditures. For this one needs much more detailed data than are at present available in Brazil.

NOTES

1. For a discussion of various conceptual and other difficulties of incidence measurement, see Shoup (1969), Musgrave (1965; 1969) and Prest (1962; 1968). While Musgrave favors such studies, Prest is highly critical of them.

2. The following excerpts from a recent work (Southworth and Johnston, 1967) underscore the statement of the text:

> The 1954 conference at Harvard pointed out that there existed virtually no detailed empirical studies on tax incidence in developing countries, even of a relatively simple sort [Wald and Froomkin, 1954, especially pp. 51-56]. In the decade since the conference, little has been done to remedy this situation except for several studies on India. Fortunately a major piece of work was recently completed [Gandhi, 1966] summarizing and improving on the estimates of tax incidence in India. . . . The lack of empirical studies, however, remains a major problem. (Lewis, 1967, p. 458.)

> The Harvard Conference pointed out in particular that tax incidence studies even of a simple sort were not available in the developing countries. They recommended that one priority of research was to prepare such studies to indicate, under simple assumptions, the relative taxes falling on various groups within the economy. Such studies have not come forth. Due [1963] has presented computations for some of the African countries, but only India has had comprehensive treatment. (Lewis, 1967, p. 481.)

The priority assigned to research in tax-burden distribution in the developing countries by the Harvard conference in 1954 and noted by

Lewis in his 1967 survey article excerpted above, continues to remain a priority even in the 1970s.

3. See, for example, Sahota (1972). In this connection, the following statement by Robert McNamara, President of the World Bank Group (1972, p. 17), which beacons a change in the Bank's erstwhile policy, is relevant:

> Shifts in the patterns of public expenditure represent one of the most effective techniques a government possesses to improve the conditions of the poor. . . . Governments can best begin . . . by initiating surveys on the effects of their current patterns of disbursement. The bank will assist in such surveys and, based on them, will help to design programs, to be financed by it and others, which will improve the distribution of public services.

4. For an excellent discussion of the traditional and nontraditional views on the shifting of profit tax, see Musgrave (1959). For a survey of the current discussion on the topic, see Slitor (1963). For a rigorous theoretical analysis of the long-run effects of corporation tax, see Harberger (1962).

5. Most work in this field has been done either by Musgrave himself or by his advisees or the advisees of his advisees. In particular, see Musgrave and Krzyzaniak (1963) and Spencer (1967). See also Kilpatrick (1965).

6. The same assumption was made by Musgrave in his tax-burden study for the United States and some of the Latin American countries referred to earlier. See Musgrave (1965).

7. This result is somewhat at variance with the inference that Stigler has drawn from what he calls Director's Law of public income distribution, according to which the middle class is the gainer from the public budget. (Stigler, 1970).

CHAPTER

12

AN OPTIMIZATION OF ECONOMIC POLICY

In Chapters 3 through 11 attention focused on the response mechanism of various aspects of the Brazilian economy to changes in different policy instruments and exogenous variables. In this chapter we reverse the focus of inquiry and look at changes in policy variables that are required by specified changes in different objective variables of the economy. Comparisons are made of the effectiveness of the major instruments of policy studied here.

Some of the dynamic aspects of the computed relationships between objectives and instruments have already been analyzed, under impact and dynamic multipliers and the dynamic properties of the system. In the present chapter we apply the theory of ends and means proper, as can be done only after quantitative relationships between objectives and instruments have been established. The main focus of this chapter is on tradeoffs, among both multiple instruments and multiple objectives. The analysis is carried out in three steps. In the first section a very brief summary of the effects of each of the instruments on each of the objectives is given. There is a more general analysis in which an attempt is made to bring together in a tabular form the relationships of instruments and objectives estimated in various earlier chapters of this study, including those done outside the econometric model in Chapter 11.

The second section presents simultaneous solutions of a few sets of equations in which instruments are so selected as to keep the number of objectives (knowns) equal to the number of instruments (unknowns) without specifying a social-welfare function. This is a system of absolute targets with no tradeoff among them. Since this is a form of the mathematical theory of economic policy that is authored and extensively discussed by Tinbergen (see Chapter 1 above), we denote it, for short, as the Tinbergen approach. In the last section, we start with Theil's approach in formulating a quadratic social-welfare function. Employing the optimal-control theory to determine appropriate sets of

TABLE 26

Summary of Some Computed Results

Line Number	Primary and Derived Objectives	T^{pl} (1)	$T^{c'}$ (2)	T^{exc} (3)	T^{sle} (4)	e^{im} (5)	e^{ex} (6)	T^{inf} (7)	T^{def} (8)
	1.1 Growth								
1	Y one year	- .24	- .25	- 3.69	- 7.82	.002	-.015	2.26	-.002
2	Y one cycle	2.46	5.34	50.0	17.3	-.01		2.17	.003
3	g one year	- .03	- .03	- .47	- .99	.001	-.002	.45	-.0002
4	g one cycle	.07	.18	.77	2.16	-.001		.006	.001
	1.2 Resource Allocation Between future and present consumption								
5	I one year	- .56	- .59	- 8.65	-18.4	.007	-.04	-33.3	-.01
6	I one cycle	1.49	3.16	8.28	41.3	-.03	-.04	-34.0	.002
7	I^{pl}	- .56	- .59	- 8.65	-18.4	.007	-.04	-33.3	-.01
8	Y^{ind}	- .57	- .60	- 8.8	-18.6	.005	-.04	6.1	-.004
	2. Income Distribution								
	1. Functional								
9	Q one year	-1.33	-1.4	-20.5	43.5	.02	3.5	-19.8	-.008
10	Q one cycle	1.13	2.76	24.9	24.9	-.03	3.5	-12.0	.002
11	W one year	- .32	- .33	- 4.9	-10.3	.005	-.02	- 4.7	-.002
12	W one cycle	.40	.88	- 2.17	9.08	-.01	-.02	- 1.77	.001
	2. Personal								
	A. Progressivity of taxes								
13	a. Dev. U.	np	np	rp	rr	rn		nr	nr
14	Dev. R.	np	np	nr	rr	nn		nn	nn
15	b. Undev. U.	np	np	rp	rr	rn		nr	nr
16	Undev. R.	np	np	nr	rr	nn		nn	nn
	B. Pro-poor effects of expenditures								
	3. Regional								
17	Dev.-ind(A)								
18	Undev.-ag(B)								
	4. Sectoral (by brackets)								
19	U								
20	R								
	3. Stabilization								
21	1. F one year	.27	.28	4.1	8.8	-.003	4.8	15.9	.007
22	F one cycle	.78	.27	.96	2.01	-.024		- 1.54	.006
23	2. DEF one year	-2.29	-2.40	-35.1	-76.5	.03	-.15	4.59	.002
24	DEF one cycle	2.12	5.36	- 8.5	34.8	-.05		26.0	.0004
25	3. P one year	.04	.04	.59	1.26	.4	.003	4.3	.0002
26	P one cycle	- .14	- .08	- .43	- 4.19	.54	.003	3.33	-.0003
27	4. Built-in flexibility	.5	.9	1.3	.3	.9			
28	5. (1-Z) one year	- .01	- .01	- .17	- .35	.0001	-.001	.12	-.000
	4. Revenue								
29	1. Tax elasticity	.89	.65	.86		.85			
30	2. Tax buoyancy	1.56	1.46	1.47	1.15	1.58			
31	3. (T) one year	2.29	2.40	35.1	76.5	-.03	.15	- 4.6	-.002

Line Number	$\eta^{a}_{-1}-\eta^{c}$ (9)	w^{min} (10)	W^g (11)	C^g (12)	I^g (13)	H^g (14)	E^g (15)	D (16)	E (17)	I^{pe} (18)	T (19)	G (20)
1	.01	.0001	-.02	-.02	.14	.17	.117	.02	-.02	.14		
2	.02	.00	.27	.27		1.61	1.61	.27	.27			
3	-.001	.000	-.002	-.002	.02	.003	.003	.002	-.002	.02		
4	.002	.00	.025	.025		.0005	.021	.025	.025			
5	-.06	.0002	-.04	-.04	1.34	.06	.06	-.04	-.04	1.34		
6	.06	.0001	.045	.045		.23	.23	.05	.05			
7	-.06	.0002	-.04	-.04	.34	.06	.06	-.04	-.04	.34		
8	-.02	.0002	-.04	-.04	.34	.06	.06	-.04	-.04	.34		
9	-.04	-.02	-.09	-.09	.80	.14	.14	-.09	-.09	.80		
10	.09	-.008	.44	.44		.47	.47	.44	.44			
11	-.009	.0001	-.02	-.02	.19	.034	.03	-.02	-.02	.19		
12	.016	.00	.025	.025		.12	.12	.025	.025			
13											.99	..
14											.96	..
15											1.01	..
16											.88	..
17			nn	nn	nn	rr	rr	pp	pp	nn	nn	pp
18			nn	nn	np	rr	rr	pp	pp	nn	nn	pp
19			nn	nn	nn	rr	rr	pp	pp	nn	pr	pp
20			nn	nn	np	rr	rr	pp	pp	nn	pn	pp
21	.03	-.0001	.018	.018	-.64	-.03	-.03	.018	.018	-.64		
22	.08	-.0001	-.13	-.13		.11	.11	-.13	-.13			
23	.008	.001	1.02	1.02	.82	.24	.24	1.02	1.02	.82		
24	.004	.0001	.89	.89		.59	.59	.89	.89			
25	.001	-.000	.003	.003	-.02	-.004	-.004	.003	.003	-.02		
26	-.003	-.000	-.13	-.13		.02	.02	-.13	-.13			
27											.12	..
28	.0003	.000	-.001	-.001	.007	.001	.001	-.001	-.001	.007		
29											1.03	..
30												
31	-.008	-.001	.1	.1	.18	-.24	-.24	.10	.10	.18		

(continued)

NOTES TO TABLE 26

The row values marked "one year" and "one cycle" pertain to one-year impact multipliers and one-cycle dynamic multipliers, respectively. The two period values of multipliers were considered useful for the reader, as the signs of one-year multiplier might otherwise have given a wrong impression, since they are usually reversed over a period of one cycle. A cycle is completed normally in five to six years. All other rows—i.e., those not specifically marked for "one-cycle" periods—define one year effects. However, not all coefficients are multipliers. Built-in flexibilities, tax-yield elasticities, tax-buoyancy elasticities, and so on, have to be interpreted accordingly. These are taken from the structural estimates (see Chapter 4). Blank cells indicate that appraisal is difficult, that the entry is not relevant, that the coefficient is not computed quantitatively, or that it is accounted for in alternative ways.

Since the distribution measures of taxes and expenditures cannot easily be captured quantitatively by single numbers, the same are indicated by rather broadly defined symbols for the lower and the upper halves of income brackets, as follows: n = neutral or proportional, p = progressive or pro-poor, r = regressive or pro-rich; nr = neutral in the lower half and regressive in the upper half of the nine brackets; and so on.

For the actual, detailed distribution the reader is referred to the relevant tables. The impact multipliers for C^g, W^g, I^g, and H^g were calculated under the assumption that these cover the typical categories that, by an large, cover all other categories of public expenditure. The values for E^g, D, A, and I^{pe} are also reported in this table by making a plausible assumption that the marginal effects of the components A and D are the same as those of their aggregates C^g or W^g, that the marginal effect of E^g is, more or less, equal to that of H^g, and that the marginal effect of I^{pe} is roughly equivalent to that of I^g.

alternative instruments, we minimize the deviations of the actual values in relevant intermediate and final periods from the desired values of specified objectives under the constraints specified by the mechanism of the economy. Theil's essentially static approach is further fertilized by a dynamic control-theory technique that analyzes the paths followed by different variables as well as their final outcomes. To identify the author of this theory, we shall refer to it as the Theil-Pontryagin approach.

A SUMMARY OF ESTIMATES AND PARTIAL POLICY TRADEOFFS

A partial analysis of policy tradeoffs has already appeared in Chapters 4 through 11. In the present chapter the emphasis is on the comparative aspects and relative effectiveness of policy instruments in the achievement of policy objectives. As a first step, numerical results obtained in the earlier chapters are summarized in Table 26. The summary is a general and truncated one. The values reported are described in the notes to this table. The effects of each of the instruments noted at the top on different objectives noted on the left-hand side can be read down the respective columns. A detailed discussion of this table is not warranted at this stage because it would amount to a repetition of the conclusions already discussed in earlier chapters. Only a few of the salient features of the estimates will be briefly noted. The conclusions will be stated in qualitative terms (directions of change). For quantitative magnitudes the reader should refer to the table.

Among other statistics the estimates collected in Table 26 include a) one-year impact multipliers, which shall be defined here as "short-run" effects, and b) the cumulated effects of dynamic multipliers over one complete cycle (usually five to six years), which for the purpose of distinction will be referred to as "long-run" effects. It so happens that in the model of this study the long-run effects of most of the instruments are opposite to the short-run effects. Consequently, unless both are presented side by side, one may get misleading conclusions and contradictory policy implications. Let us first look at the short-run effects.

Short-Run Effects

The two major direct taxes of Brazil—namely, personal and corporation income taxes—have negative short-run effects on growth and investment. They are neutral as regards functional income distribution, in that their negative effects on wage income and on nonwage income in the industrial sector are almost proportionate. However, their partial impact effect on personal income is redistributive. As expected, they are elastic with respect to income and have positive effects on revenue collection and, interestingly enough, on foreign balances. The personal income tax has positive (favorable) effect, and the corporation income tax has negative effect on capacity utilization; both induce cost inflation. To summarize, the short-run effects of income taxes in Brazil are unfavorable to growth, inflation, and tax compliance (not reported in Table 26); favorable to balancing the government budget and the foreign account budget; and from favorable to neutral in regard to income distribution. In the short run, thus, these taxes basically serve a revenue-gathering function. Their redistributive function is uncertain. By implication their unfavorable effects on growth and inflation can be reduced by tax concessions. It follows, therefore, that Brazil's extensive investment-incentive schemes through corporation-tax concessions are growth-promoting, at least in the short run.

Interestingly enough, the direction of the effects of the two major indirect taxes—namely, central excise duties and sales tax—on stabilization, capacity utilization, growth, and investment is similar to that of direct taxes. On the other hand, these taxes have adverse effects on personal income distribution.

Long-Run Effects

Contrary to their short-run effects, the two major direct taxes have positive long-run effects on growth, investment, resource allocation, revenue collection, foreign balances, personal income distribution, and, surprisingly enough, functional income distribution. They have unfavorable effects on capacity utilization but favorable effects on the other two aspects of the stabilization objective—namely, budget surplus and control of inflation. Broadly speaking, therefore, insofar as the long-run effects are concerned, these taxes have many of the desirable properties of a good tax. (However, they are bad taxes from the criteria of tax evasion and capacity utilization.) The corporation tax has a marginal effect on growth, resource allocation, and income distribution roughly twice as high as the personal income tax. Remembering that all cruzeiro figures are expressed in 1963 prices, an increase (in 1968) of the personal income tax rate of one percentage point in terms of GDP leads in a period of five years to an increase of Cr$1.49 million of overall investment and Cr$2.46 million of GDP. In the process it augments nonwage incomes by Cr$1.13 million and wage incomes by Cr$0.40 million. The tax is redistributive of personal incomes among the income classes having a minimum annual income of Cr$350 but neutral among the income classes below the Cr$350 limit.

Interestingly enough, the direction of the effects of central excise duties and the sales tax on stabilization, capacity utilization, growth, and investment is similar to that of direct taxes. On the other hand, these taxes are bad from the viewpoint of functional income distribution, though insofar as personal income distribution in the rural areas is concerned the excise tax does not seem to be inequitable at all.

Although the directions of some of the effects of the two major direct taxes (personal and corporate) and the two major indirect taxes (excise and sales) are similar, their efficiency levels differ significantly. The efficiency of an instrument was defined in terms of the magnitude of the change in the instrument concerned relative to its effects on the given objective variable—i. e., the smaller the proportionate changes, the more efficient the instrument under this definition. According to this criterion, Brazil's direct taxes are less efficient than her indirect taxes. Thus the income-tax rate has to be raised about 15 times the increase in the excise tax rate to obtain an equivalent increase in growth.

Tariffs have adverse influence on growth and investment but are, by and large, distributionally neutral among income classes. A foreign-exchange-rate policy that raises the exchange rate for imports has, broadly speaking, similar effects on stabilization of foreign balances, capacity utilization, budget deficit, and revenues to a policy that lowers the exchange rate for exports.

When excess capacity exists in the economy, inflation might promote growth or prevent growth from slowing down, but not when capacity is fully utilized, since its influence on capacity utilization is positive but negative on investment. A million-cruzeiro increase in GDP in a period of five years might be brought about either by a Cr$0.6 million once-and-for-all sustained increase in public-health (or education) expenditure or a Cr$3.5 million once-and-for-all sustained increase in what is categorized as public consumption. The former category is conducive of foreign balances and capacity utilization; the latter is not. Investment in physical capital is a drag on foreign balances. Investment in public health and education, as just noted, improves foreign balances. The other effects of the two forms of investment seem to be similar. Thus, public health expenditure and public education come out creditably from almost all aspects.

STATIC POLICY TRADEOFFS WITH ABSOLUTE TARGETS: THE TINBERGEN APPROACH

The number of instruments of policy in our system exceeds the number of "final" objectives to be achieved; hence, there are numerous solutions. For example, the number of combinations and permutations available from 18 instruments and four ultimate objectives of this study rises to 3, 060. Apparently, there is little point in trying to obtain all the solutions for this study. At the same time, not all of the instruments analyzed here are distinct, nor are all the "derived" objectives entirely independent of one another. For the purpose of this section, therefore, we selected a few alternative sets of algebraically independent objectives and the equivalent number of distinct instruments from the econometric model of this study, to obtain unique solutions as possible policy programs.

In view of the constraint that only a subset of the econometric model can be worked with, the appropriate model from which the relevant equations have to be selected is the reduced-form model rather than the structural system. Accordingly, the solution we get is of the following form:

$$\tau = [B^{-1}D]^{-1}Y - [B^{-1}D]^{-1}B^{-1}AX \tag{4'}$$

instead of the solution obtained in Equation 4—i.e., instead of:

$$\tau = D^{-1}BY - D^{-1}AX \tag{4}$$

The results of five experiments of Equation 4' are presented in Table 27. For comparison purposes the results of an experiment from Equation 4 are also reported, namely Experiment #6 of Table 27. That experiment, as noted earlier, gives only partial effects of instruments, as distinguished from the total effects of the other experiments. Only a few of the cells of the 11 × 11 matrix (comprising 11 targets and 11 instruments) of Experiment #6 have nonzero elements. On the other hand, Relation 4' used for the first five experiments is arithmetically weaker (due to larger rounding errors) than Relation 4. The coefficients of this table indicate the values that ought to be given the instruments (noted in the columns) to achieve a unit increase in the objectives (noted in the rows on the left) in the course of one year.

In this model all targets have equal weights. Such a weighting scheme, though outwardly simple, precludes easy simulation for optimization and practical planning, as will be evident from the ensuing discussion.

In interpreting these results it must be remembered that once the objectives to be achieved simultaneously have been chosen, the estimation of the values of one target variable requires the values of the other objectives to be held fixed at their target values simultaneously. The condition is that in determining the instrument levels for one objective, the target values of the other objective variables in the program have also to be achieved or maintained at preattained levels. This is an imperative that is at the same time an extremely difficult condition, except when the instruments happen to be specialized—i.e., when one instrument is related to a single objective, another to another objective, etc. This is scarcely, if ever, the case in economics. Generally, a change in instruments required to achieve the target value of one objective disturbs the values at which the other target variables are to be held constant. As a consequence, to achieve the

TABLE 27

Some Solutions of the Tinbergen Model

Exper- iment	Objective	Instrument													
		τ^{pi}	τ^{exc}	e^{im}	e^{ex}	τ^{inf}	τ^{def}	η	H	I^{g}	$\tau^{c'}$	τ^{sle}	w^{min}	W^{g}	NTR
#1	C^{p}, I^{pl}, Y^{na}, W	..	3.0788	686.46	-613.13	-.0166									
#2	C^{p}, I^{pl}, Y^{na}, W, F	..	.3013	198.6669	1.1322	-.0098	- 737.85								
#3	C^{p}, I^{pl}, Y^{na}, W, F, P	..	.2307	3.0617	1.5859	-.0456	-1149.62	41.1213							
#4	C^{p}, I^{pl}, Y^{na}, W, F, P, T	..	8.8508	2.1134	1.2940	.0655	- 514.71	6.9660	-1411.8						
#5	C^{p}, I^{pl}, Y^{na}, W, F, P, T	232.1	..	.3434	1.3080	.0395	- 685.00	-1.0400	1788.9						
#6	Y', C^{p}, I^{pl}, W, Ex, T^{pi}, T^{c}, T^{exc}, T^{sle}, T, Y^{na}	.7813	.0510	..	.2080	-.0711	..	..	.5988	1.00	.3574	.0231	-46.40	-1.7887	3.7900
Observed mean values		8.18	14.3	70	78	1.4	1.7	1.47	89	388	22.2	10.7	64	698	200

target values of objective variables simultaneously, especially when they may be incompatible or conflicting (without being contradictory), in some cases extremely large changes in particular instruments may be required.*

Moreover, it turns out to be rather difficult to predict even approximately the relative or absolute values of different instruments from the solution of a given set of targets when the number of targets is altered. The addition of an objective (and a corresponding number of instruments) sometimes changes the values of the previously included instruments by big margins. To see this one has simply to compare the values of the corresponding instruments in the successive sets of solutions in Table 27. In general, the availability of an additional instrument contributes to the solution only to a small extent. The addition of an objective, on the other hand, increases the constraints in a big way.

Thus, the target values of a unit-increase in each of the objectives create serious problems. The selected objectives are in general competitive. Some may even be incompatible. A somewhat less implausible set of targets would apparently consist of varying target values instead of uniform one-unit increases of each target. Any such set of target values would, however, be an exercise in trials and errors. Yet this kind of simulation may be a practical way of obtaining plausible solutions. An alternative procedure is to allow departures from target values but with penalties of varying degrees—i.e., to present solutions by assigning alternative sets of weights to different objectives. This forms a part of the analysis of the following section of this chapter.

A number of experiments were made. We selected various sets of variables without distinguishing between "final" and "derived" objectives. These include T for the revenue objective, P for the price stability objective, F for foreign balances, W indicating the redistributional aspects, Y^{na} for long-run structural changes concomitant to growth, I^{pi} as a factor of growth, and private consumption (CP) for its welfare interest. Apparently, these are not independent objectives. The final and the derived objectives are intermixed; this was done simply to get some unique solutions from an equal number of objectives and instruments.

Looking across and down the instrument columns, we find that for the given changes in the objective variables, the personal income tax (τ^{pi}) and deficit finance (τ^{def}) are the least efficient instruments. In some experiments, public expenditure on health (H^g), minimum-wage policy (w^{min}), and the excess-reserve ratio (η) are also extremely

---

*A drawback of all analysis based on derived reduced forms, whether as impact or as longer-period dynamic multipliers, is the lack of standard errors of individual multipliers. Hence estimates based on derived multipliers can only be deterministic. This deficiency applies to all solutions in this chapter.

inefficient.* The results for other instruments are not so discouraging. For example, the changes required in instrument τ^{exc} for all of the six experiments lie between 2 percent (Experiment #3) and 60 percent (Experiment #4). But one should not be misled by plausible values for some instruments in an experiment in which some other instruments are unstable. For example, Experiment #2 looks quite good if one neglects the required change in τ^{def}. The latter change is astronomical, however, and takes away the heat entirely from other instruments in the experiment. It may also be mentioned that the low values for Experiment #6 are also misleading, in that they include only partial effects and not the total effects of the entire system.

The experiments of Table 27 were carried out for one-unit arbitrary change in each objective (making sure, of course, that the changes are in the desirable directions). Thus, a one-percentage-point decline in P is as important in this model as a million-cruzeiro increase (in 1963 prices) of industrial output. Moreover, in this model, for the reasons mentioned earlier, the derived and final objectives are intermixed. With a view to setting more desirable target values, we carried out the following experiment in which, as far as possible, only the final objectives (as defined in this study) are used. For this experiment, we chose five objectives and five instruments as follows: Y, W, F, P, T; τ^{pi}, τ^{exc}, τ^{inf}, I^g, H^g. The target values of the objective variables were fixed, as usual, for a one-year plan. Since most of the data for the year subsequent to the sampling period are as yet unavailable, we applied the Tinbergen model to the 1964 set of exogenous variables. The one-year target values for the objective variables for the year 1964 were specified as follows:

Y: One half of 1 percent of additional growth in GDP in 1964—i.e., g = 3.4 percent, in place of actual g of

---

*The variable η is defined as $\eta^a_{-1} - \eta^c$. The interpretation of η is somewhat involved but can be approached as follows. The actual reserve ratio (η^a) depends, in part, upon the state of demand for bank loans but can be considered a better approximation to desired reserve ratio and thus a more stable one than the compulsory reserve ratio (η^c). The latter may be changed pursuant to the monetary policy of the time; it is a change in η^c that represents monetary policy. A positive change in η can be interpreted to mean a monetary expansionist policy: the central bank lowers η^c and widens the gap (i.e., η). This interpretation is consistent with the computed positive coefficient of P and negative coefficient of W with respect to η. However, this interpretation can be confused if unwanted reserves get accumulated and banks cannot, for one reason or another, invest them, even in government or similar other securities.

2.9 percent or Y^* = 10283 instead of 10071 (actual) or 10479 (predicted)

W: An increase in W of 1 percent, over and above the constant ratio of W/Y in 1964, or W^* = 943 in place of actual W = 914 or predicted W = 961

F: No change from the actual value for 1964

P: No change in the rate of inflation from its previous-year level—i.e., ΔP^* in 1964 = 78 percent, against the actual price rise of 88 percent and predicted price change of 69 percent.

T: An increase of 0.1 percent in T/Y, or T^* = 2425 instead of the actual T = 2373 and predicted T = 2135.

The results are noted below:

Instrument	Actual	Solved
τ^{pi}	6.87%	Over 100%
τ^{exc}	18.0%	Over 100%
τ^{inf}	3.2%	73.3%
I^g	Cr$513	Cr$2217
H^g	Cr$ 99	Cr$6245

It will be seen that the instrument values required to achieve the specified target values are extremely high and make no sense for practical policy purposes. Two conclusions, among others, can be drawn from this analysis. The first is that Tinbergen's static theory of economic policy is inapplicable to a dynamic system of the type of the present study. One-year impact multipliers or the reduced-form estimates are the appropriate coefficients to use for the Tinbergen type of policy tradeoffs. The use of dynamic multipliers would be inappropriate as they are calculated from a particular (prespecified) change in given instruments, which are precisely the unknowns of the present model. Accordingly, all solutions of this section were obtained from impact multipliers. The lacuna of this approach is a serious one in the present context, since the cumulated effects of most instruments over a cycle are, in general, the reverse of their impact effects.

A second conclusion is that for a policy to be practical, flexibility of targets is essential. Sometimes an extremely high price may have to be paid in terms of the opportunity cost of alternative goals if attempts are made to achieve a set of fixed targets simultaneously without allowing tradeoffs among them. One can arrive at a feasible solution, of course, by trials and errors or by pure coincidence. There are more scientific methods of obtaining not only feasible but also optimal solutions available, however, and to these methods we turn next.

DYNAMIC POLICY TRADEOFFS WITH FLEXIBLE TARGETS AND A SOCIAL-WELFARE FUNCTION: THE THEIL-PONTRYAGIN APPROACH

The solutions of simultaneous (third-degree) difference equations were described in Chapter 8; the dynamic optimal-control solutions involve (among other things) similar solutions of difference equations. The basic difference between the dynamic optimal-control problem of this chapter and the problem of conventional solutions of difference equations of Chapter 8 is essentially the same as that between simulation and optional control delineated in Chapter 1. Mathematical derivations of optimal-control solutions have recently appeared in the economic literature and will not be reproduced here. They usually employ either the method of dynamic Lagrangian multipliers (e.g., Chow [1972a]) or Bellman's dynamic programing approach (e.g., Chow [1972b]). Readers interested in the mathematical derivations of these methods are referred to Pindyck (1973, especially Chapter 3) for deterministic control solutions of the type of the present study and to Chow (1972b) for stochastic control solutions. The present study follows Pindyck's approach.*

Briefly, the dynamic control model used here starts from the terminal year, going backwards in time to the initial year. In this backward process of calculation it uses the coefficient matrix of the econometric model, the weights (cost functionals) for both target and instrument vectors (when the instrument vector is constrained not by boundary conditions but by penalties for deviations from its target values), the exogenous variables, and the target or ideal values of the state (objective) and control (instrument) variables. The transformations are such that even if the τ's are linearly related among themselves, they are unlikely to cause singularity. The computation of the optimal values of state and control variables then begins forwards in time, year by year, using (in addition to the information described above, once over again) the initial values of all current and lagged variables in the base year. The lagged values subsequent to the base-year values (where relevant) are not data but are those generated within the control solutions.

Before specifying the empirical model, the relationship of the analysis of this section with other recent studies on the topic is in order. When this research was undertaken at the onset of the 1970s, one could hardly find a reference on the economic theory of targets

---

*Pindyck's methodology was extended to handle additional characteristics of the present model, such as lagged-instrument effects (which did not appear in Pindyck's model). The computer program used for this study, which was developed at the Vanderbilt University Computer Center, is documented in Appendix C.

and instruments other than the decade-or-so-old studies of Scandinavian authors—for example, the studies of Tinbergen (1952; 1956), Theil (1956; 1961; 1965), Bogart and Theil (1959), Krischen (1964; 1965), and Van Eijk and Sandee (1959). These authors did not explicitly introduce dynamic factors in their models. Since 1970 the interest in this subject seems to have been revived and more than a half-dozen works have been published—notable among these the studies of Pindyck (1973), Chow (1970; 1972a; 1972b; 1973), and Poole (1971). Among the main aspects in which the present study differs from the studies just cited, the following might be mentioned:

1. Most control-theory models, including those by Pindyck and Chow, are linear; ours is linearized. Linearized models create difficulties for predictions for those periods that are not in the immediate neighborhood of the point of time at which a model is linearized. Since dynamic optimal-control solutions involve time periods that may lie several periods away from the period of linearization, point estimates will generally differ from expected values. Since it will be difficult to know how the differences between the point estimates and the expected values behave over time, the inefficiencies of instruments and the deviations of optimal values from the specified ideal values will be exaggerated. The optimal solutions of linearized nonlinear models will, thus, be inferior to those of linear models.

2. Most optimal control theory models use illustrative or small, aggregative econometric models. Thus, Chow (1972b) employs a nine-structural-equation model (including four behavioral equations); Pindyck (1973) uses a ten-equation model (including nine behavioral equations); and Poole (1971) selects a subset from the SSRC-MIT-Penn econometric model for income growth (as an objective variable) and monetary changes (as an instrument variable). The present study uses a relatively large econometric model consisting of 46 structural equations (including 34 behavioral equations); there are 17 exogenous variables, a score of instrument variables, and several lagged endogenous variables. In general, the larger the number of equations, the lower the precision of the reduced-form coefficients derived from solved structural equations; the larger the number of instruments, the larger the probability of errors in variables; the larger the number of exogenous variables that have to be predicted for (ex ante) planning periods, the greater the probability of error. All these factors can reduce the quality of the resulting estimates.

3. Almost all existing studies include lagged endogenous variables (this is what makes their models dynamic), but none except for Chow's study (1972b) includes lagged instrument variables. The presence of a number of lagged instrument variables in our model creates additional difficulties for obtaining cost minimization solutions.

4. The models published so far analyze no more than one to three instruments. For example, Chow (1973) and Poole (1971) analyze only one instrument; Chow (1972b) analyzes at most two instruments;

Pindyck (1973) analyzes three instruments. In the present model, we analyze 11 instruments (out of the 18 included in the econometric model).

5. It might also be mentioned that almost all existing studies select as their state variables mostly the derived objectives (defined in Chapter 1)—i.e., the usual endogenous variables of their econometric models, such as consumption, investment, and market yield on bonds. For the present study, we have selected (in the terminology of Chapter 1) the final objectives—namely, economic growth (efficiency), income distribution (equity), control of inflation and foreign balances (stability), and revenues (provision of public goods).* The objectives selected here are, by and large, commonly recognized in most developing countries. Hence, ours is an attempt to solve an empirical problem of current interest. In doing so, however, we also meet additional pitfalls. The probability of incompatibilities or inconsistencies of objectives increases. This is particularly the case when one has such competing goals in the planning model as we have—namely growth along with income distribution, expansion of social goods along with economic efficiency. An optimization of competitive objectives like these may require extremely large changes in the instrument variables or, when large changes in instrument variables are heavily penalized, wide deviations in the values of the objective variables from their targeted values.†

6. The present study is confined to optimal deterministic-control policy. Recent developments in optimal stochastic control hold promises of substantial improvement in gain of such an optimal policy over the optimal deterministic-control policy. For example, in the first study of its kind, Chow (1972b) found a gain of 30 to 80 percent by substituting optimal stochastic-control policy for optimal nonstochastic-control policy.

7. In addition to the quality of the econometric model, the plausibility of the solutions of the policy model of the present section especially depends upon the efficiency of instruments. The instrument

---

*Rigorously defined, the objective variables of this study (in particular, the revenue variable) are also "derived." However, the variables selected here as objectives conform to the conventionally-stated ultimate objectives in the planning documents of most developing countries. This point was discussed in Chapter 1.

†Chow's contributions and Pindyck's study must be categorized as pioneer works in the field, although their value lies more in theoretical and methodological refinements (tool-building) than in empirical insights (application), perhaps. The present study falls essentially in the application field. The differences underlined in the text are simply by way of putting the particular analysis of the present chapter in a proper perspective.

efficiency in the fully monetized and economically integrated developed economies is expected to be much higher than that in the dualistic economies of the developing countries (and Brazil, for that matter).

8. Finally, the Brazilian data series are not as reliable and complete as those for the United States. This applies particularly to the key data series for policy instruments: rates of interest and average wage rates are not available; numerous physical controls thwart the operation of free-market forces; the data on unemployment do not exist; many series are incomplete, and several others untrustworthy.

For all these and other reasons we do not expect that we can get as nice an orchestration of objectives and instruments in this study as was obtained in some of the studies reviewed above.

The multitarget multiinstrument policy model of this section has three components: a) a social welfare function, b) a set of technical relationship equations, and c) constraints or boundary conditions. Due to computation problems, component c) is handled as part of component a) in this study. For a specification of the social-welfare function we need a set (or alternative sets) of weights. For this purpose, as a starting point, a basic set of weights was prepared by the present author in consultation with a number of Brazilian experts, including and in particular, Delfim Neto, during whose eight-year term as finance minister the Brazilian economy experienced what Milton Friedman has called an "economic miracle" of the postwar era (Friedman, 1974). These weights are noted in Table 28. To provide perspective, the basic weights of the Dutch social-preference function as reported in van Eijk and Sandee (1959) and the weights for a U.S. model as used by Pindyck (1973) are also reported.*

For Brazil, initially the basic weights were determined for the early 1970s in terms of the magnitudes for 1968 as follows:

- One-percentage-point increase in the growth rate of GDP (which in 1968 in 1963 prices was Cr$117 million): $g = 1$
- One-percentage-point decline in the Gini coefficient for personal-income concentration or (as a rough-and-ready measure) a three-point increase in the ratio of W/Q: $\xi_1 = 3$
- One-percentage-point decline in the Gini-coefficient measure of inequalities between states†: $\xi_2 = 5$

---

*The weights used by Pindyck are not presumed to represent the desires of the U.S. public or its elected representatives. They were chosen simply to illustrate different aspects of the optimal solution as well as the dynamic properties of our econometric model. An analysis of the type of van Eijk and Sandee (1959) was done by Bogart and Theil (1959) for President Roosevelt's first administration. These authors avoided any value judgment about weights.

†As explained in Footnote b to Table 28, regional income distribution is not analyzed in the control theory model of this chapter.

TABLE 28

The Weighting Scheme

Objectives and Instruments (1)	Weighting Scheme for Brazil (2)	Weighting Scheme for Denmark[a] (3)	Weighting Scheme for United States[b] (4)
g	1.0	..	..
ξ_1	3.0	5.0	..
ξ_2	5.0	..	..
F equiv. lg.	0.1	1.0	..
P	-0.3	-7.5	-1.0
T equiv. lg.	0.1	0.5	..
C^p	..	..	1.0
I^p	..	.2	1.0
I^{stk}	..	..	0.0[d]
r_1	..	..	0.0[d]
r_2	..	..	0.0[d]
w	..	..	0.0[d]
Y^d	..	..	0.0[d]
UR	..	-0.2	-1.0
Instruments			
τ^{pi}	1	..	2.0
τ^c	1	..	..
τ^{exc}	1	..	..
τ^{sle}	1	..	..
τ^{inf}	1	..	..
e^{im}	1	..	..
e^{ex}	1	..	..
C^g	1	..	..
W^g	1	..	..
I^g	1	0.25	1.0
H^g	1	..	..
ΔM	..	..	1.0
η	1[e]	..	..
w^{min}	1[e]	..	..

Note: The new symbols of this table are defined as follows: UR = unemployment rate; r_1 = short-run rate of interest; r_2 = long-run rate of interest; w = wage rate; ξ_1 = Gini coefficient of national-income distribution among income classes; ξ_2 = Gini coefficient of regional-income distribution. All variables in this table are to be interpreted in terms of changes rather than levels.

[a] van Eyk and Dandee (1959).

[b] Pindyck (1973).

(continued)

TABLE 28, continued

[c] ξ_1 and ξ_2 are distribution variables —i.e., Gini coefficients. A change in ξ_1 is represented in the econometric model by an appropriate change in the ratio W/Q in the manufacturing sector. Our calculations indicate that a 3 percent improvement in W/Q is equivalent, roughly, to a one-percentage-point decline in Gini coefficient. Regional-income distribution (ξ_2) was not included in the econometric model —that is, the model was not estimated by regions. Personal-income distribution, which is a more representative measure of income inequalities than functional-income distribution as represented here by W/Q, was measured by regions, but outside the econometric model. It is analyzed in Chapter 11 and cannot be incorporated here.

[d] This is Pindyck's basic set of weights. He carried out several experiments, slightly altering a weight here and a weight there. Most of his zero weights are for those endogenous variables that are "intermediate" in terms of policy goals. With a view to avoiding double penalty on the whole (and a major component) of a variable, he assigns nonzero weight to C^p and zero weight to Y^d.

[e] The instruments of η and w^{min} were tried in some experiments but were dropped in other major experiments due to their insignificant efficiency levels.

- An increase in foreign balances equivalent to 1-percent rate of growth of GDP—i.e., Cr$117 million: F = 0.1
- 1-percent decline in inflation at a time (1973) when prices were rising at 15 percent per annum = 0.33
- The weight will be proportionately higher when the rate of inflation is higher. For the period of 1960-68, with an average inflation of 47 percent, therefore, the weight is .33 × 47/15: $\Delta P = 1.034$
- One tenth of one-percentage-point annual increase in T/Y: T = 0.1

All those instruments that are expressed in terms of percentages of their base (usually some component of national income or expenditure)—i.e., τ^{pi}, τ^{c}, τ^{exc}, τ^{sle}, and τ^{inf}—were given a weight of unity for each one-percentage-point increase. Those that are expressed in absolute units, such as government expenditures, were given a weight of unity for one-unit increase in their values (unit measures are given in Appendix A).*

In the actual solutions of our model, g is replaced by Y and ξ's are replaced by W. Since the new variables are scaled differently, the assigned weights, including those for instruments, also have to be scaled accordingly. The weights so scaled are as follows:

Objectives	Basic Weight	Scaled Weight
Y	1.0	1
W	3.0	347
F	0.1	11
P	0.33	35
T	0.1	4

*As defined earlier, two variables for national product are used by this study: Y', which is defined as GNP and comes from the identity equation of the model, and Y, frequently called "GDP" here, which is obtained from an aggregate national-production function. Accordingly, T is not a direct component of Y. Yet since both are closely related, there may be some element of excess penalization of Y and T when both are given individual weights. In this regard the low relative weight of 0.1 for T mitigates the effect of the indicated possibility. In addition, alternative runs are made by dropping out T.

Instruments	Basic Weight	Scaled Weight
τ^{pi}	1	1,448
τ^{c}	1	368
τ^{exc}	1	599
τ^{sle}	1	780
τ^{inf}	1	5,148
e^{in}	1	13
e^{ex}	1	10
W^{g}	1	1
C^{g}	1	2
I^{g}	1	2
H^{g}	1	9
W^{min}	1	15
η	1	2,500

It might be restated that the relative weights indicate the penalty for deviations from the respective target values. In addition, the boundary conditions (maximum-minimum limits) can be used to further circumscribe the physical constraints on these variables but are not used here.

Finally, it must be emphasized that the weights assigned in this study are not sacrosanct and are not meant to be valid for all time. As a matter of fact the relative importance or priority rating of different objectives (and instruments) is expected to change as economic conditions change. It may also be mentioned that the Dutch weighting scheme was unavailable to us when the Brazilian weights were determined in Rio de Janeiro in December 1973. It will, therefore, be useful to discuss some of the economic and other factors that may explain the variations of the Brazilian weighting scheme from that of the Dutch weighting scheme.

1. The Gini coefficient of income concentration for Brazil is approximately 0.55, against that of about 0.35 for Denmark.

2. The regional-income disparities in Brazil are considered to be even more striking than personal-income inequalities.

3. The Brazilian economy has been growing at a substantially faster rate, especially in recent years (exceeding an annual rate of 10 percent in the 1970s) than the Dutch economy. Such a high rate of economic growth might compensate for possible low achievement levels of other objectives.

4. The absolute minimum-wage rate fell during the 1960s. There is little evidence to indicate that the relative increase in average-wage rate has been consonant with the overall rate of economic growth. As was seen in Chapter 7, the ratio of wage bill to nonwage income in the manufacturing sector fell from approximately 45 percent in 1950-53 to approximately 33 percent in 1966-68. (See Appendix A, Table A.1.) The redistribution objective as reflected by the policy of the present government is not merely to keep the current W^m/Q^m ratio constant in a two-digit annual growth economy, but, in addition, to raise this ratio steadily, to enable wage labor to recoup its lost income position. It will, for instance, take more than a decade to reattain the relative shares of the early 1960s if the 1968 W^m/Q^m ratio should rise at a rate of one percentage point per year.

5. The foreign exchange objective has, by and large, been as important in Brazil as in Denmark. Only in the 1970s, as a result of an increasing quantum of manufactures exports, has Brazil's foreign balance situation improved to comfortable levels.

6. Although a reduction in inflation has been among the major objectives of the Brazilian government, Brazilians seem to have created a number of institutions—e.g., the monetary correction for bank deposits, and similar monetary claims—that to a large extent insulate the income positions of various groups against price rises. Brazil seems to have gotten used to (or, at least, reconciled to) mild rates of inflation much more than a North American or Western European country could be expected to. This does not mean that inflation is not a problem in Brazil. It surely remains an important problem, and its control is included in the set of national objectives. It only means that in spite of a high rate of inflation the preference-intensity coefficient for $\dot{P}$ might not be higher than that used for the Dutch model. The recent plateau of price increase at around 15 percent seems to have reduced the priority rating for inflation in this country. An important point emphasized by Delfim Neto in regard to the present inflationary situation is that to maintain the delicate confidence of businessmen and consumers in the Brazilian cruzeiro and in their holdings of cash and other balances, it is extremely important to keep inflation from rising at all. That is, the following constraint must be observed: $(\Delta P/P)_t \leq (\Delta P/P)_{t-1}$.

7. Even though the government budget still continues to be unbalanced, high economic growth and the buoyant tax structure have made the revenue position of the government quite comfortable. Accordingly, while some rise in the T/Y ratio is still desirable it gets a low priority relative to the other objectives listed above.

The solutions are obtained for two periods: a) The observation period, 1960-68, to be denoted the "historical period" and b) a seven-year planning period, 1969-75, to be called the "planning period." For the historical period we compare the calculated (optimal) values computed according to the optimal-control theory, with hypothetical ideal values on the one side, and with actual values and the model's predicted values on the other side. For the planning period, we compare only the optimal and the ideal or target values and compute the costs of alternative optima.

Optimal Control Theory Analysis: The Historical Period, 1954-68

Several optimal policy experiments, using alternative sets of weights, were made for a period of nine years, from 1960 to 1968. For this purpose the target values were specified according to straight-line trends between the base-year value and the terminal-year value (except for the price level, which was specified to rise at a constant rate of 30 percent per annum). The objective variables chosen for the analysis of the present section are those that we have defined as the final objectives—namely, Y, W, F, P, and T. With all the necessary information thus in hand, the empirical policy model can now be specified:

$$\text{Minimize} = \sum_{t=0}^{T} [(Y_t - Y_t^*)^2 + 347(W_t - W_t^*)^2 + 11(F_t - F_t^*)^2 + 4(P_t - P_t^*)^2 + 4(T_t - T_t^*)^2 + 1448(\tau_t^{pi} - \tau_t^{pi*})^2 + \ldots] \qquad (\ell)$$

subject to the economic mechanism as expressed by rows 9, 11, 17, 19, and 34 of Table 4: impact multipliers, after being expressed in terms of absolute-value variables as required by the model (see Appendix C), instead of first differences. For example:

$$Y = .0255Y_{-1} - .0255Y_{-2} + \ldots + \ldots - .2405\tau^{pi} + .3612\tau_{-1}^{pi} - .1207\tau_{-2}^{pi} + \ldots \text{ (78 terms)} + \text{Constant} \qquad (\ell i)$$

and similarly for W, F, P, and T.*

It will be seen that the boundary conditions of Equation viii have been softened and have been included in Equation ℓ. That is, all

---

*Computer programing for handling boundary conditions or physical restrictions expressed by inequalities in Equation viii of Chapter 1 was found to be too complex. The inclusion of instruments in the cost-minimization equation (ℓ) is a way of getting around the indicated physical restrictions. The same applies to the objective variables, if they also are subject to boundary conditions. In this sense, all empirical models so far derived from optimal-control theory, and the present policy model for that matter, give only a second-best solution.

objectives and all instruments are subjected to cost penalties rather than to physical-boundary constraints.

A number of alternative runs were made by altering the weights and the mix of objectives. We will report a couple of specimen solutions in detail and briefly discuss the outcomes of others. One of the runs is reported in Table 29 and is sketched in Figure 14. The results in general are not very cheering—in view of a number of caveats noted in the beginning of this section, they were not expected to be otherwise. Nevertheless, they throw ample light on the problem under investigation. To the extent that some of the caveats noted can be removed as more accurate and detailed data become available in the coming years, the present results hold a promise for improvement.

Let us have a look at the time paths of variables in Table 29. It will be seen that the program forces the objective variables to converge to their target values towards the end of the period. Only the prive-level variable behaves somewhat discordantly: instead of moving towards the end-year target in the terminal year, it streaks off from it. In the first couple of years all optimal values of the objective variables show deviations from their ideal values that are wider than both the observed values and the values predicted by holding all instruments at their historical levels. This result is consistent with the results of the dynamic multipliers in Chapter 7. It seems that early-period destabilization aids in long-period stabilization and in a minimization of overall costs. Similar early-period-divergence and late-period-convergence behavior is exhibited by instrument variables.

The long-run optimization objective that the system is supposed to follow begins with a higher-than-targeted growth rate of output. Perhaps because tax revenues are, on the whole, directly related to national product and form a substantial component of it, a corresponding upward bulge is apparent in the tax-revenue curve. Similar upward rise takes place in the case of the price-level curve. The expansion of these variables seems to come at the cost of a decline in wage-bill and foreign balances. After a period of two to four years things appear to fall in line for all objective variables, and the program no longer behaves strangely as it did in the early years.

The relative displacements in the values of the instrument set are much larger than those in the case of the objective set. Brazilian instruments seem, in general, to be woefully inefficient. For the 1961-68 plan reported in Table 29, the personal-income-tax rate has to be raised to 22 percent against the prevailing effective rate of 8 percent in 1961, and it has to be lowered to 5.3 percent against the prevailing effective rate of 8.4 percent in 1967. Since the end-year instrument values do not affect objectives, they are necessarily equal to their target values. Hence, for the purpose of instruments, 1967 is the end year. The corporation tax starts at a negative rate and ends at 21 percent (as against the actual rate of 29 percent) in 1967. The excise-tax rate starts at a high rate and ends up with a negative rate.

TABLE 29

Actual, Predicted, "Ideal," and "Optimal" Values of Objective Variables

	Y				W				F			
Year	Actual	Pre-dicted	Ideal	Optimal	Actual	Pre-dicted	Ideal	Optimal	Actual	Pre-dicted	Ideal	Optimal
1960	8, 140	7, 993	8, 140	8, 140	368	397	366	366	-132	-9	-132	-132
1961	9, 263	8, 592	8, 582	10, 478	403	454	453	149	-61	136	-128	-1, 219
1962	9, 723	9, 727	9, 024	10, 295	555	488	540	379	-191	-77	-124	-586
1963	9, 945	10, 292	9, 466	12, 284	873	610	627	512	-51	77	-120	-328
1964	10, 071	10, 479	20, 258	22, 144	914	961	714	618	131	80	-116	-217
1965	10, 114	10, 507			877	963	801	730	322	-76	-112	163
1966	10, 472	10, 364	10, 792	9, 686	950	758	888	824	110	230	-108	-143
1967	11, 094	10, 857	11, 234	11, 762	946	1, 091	975	916	-7	-96	-104	-133
1968	11, 673	11, 646	11, 676	12, 176	1, 004	994	1, 062	1, 021	-105	132	-100	-122

	P				T			
Year	Actual	Predicted	Ideal	Optimal	Actual	Predicted	Ideal	Optimal
1960	27	24	27	27	2, 028	2, 233	2, 028	2, 028
1961	338	36	107	103	2, 103	2, 178	2, 282	3, 289
1962	56	56	187	303	2, 092	2, 365	2, 536	3, 338
1963	100	89	267	418	2, 147	2, 365	2, 790	3, 376
1964	188	169	347	281	2, 373	2, 135	3, 044	3, 475
1965	292	290	427	388	2, 746	2, 455	3, 298	3, 545
1966	405	446	507	415	3, 201	3, 226	3, 552	3, 695
1967	515	547	587	418	3, 238	3, 193	3, 806	3, 856
1968	659	757	667	246	4, 026	3, 454	4, 060	4, 086

Figure 14: Actual, Predicted, "Ideal," and "Optimized" Values of Objectives

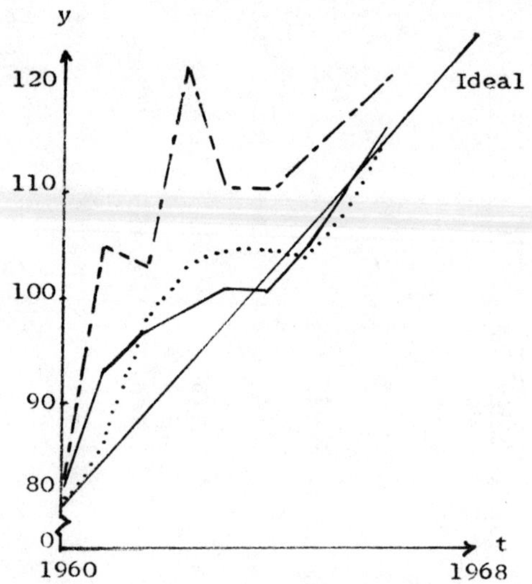

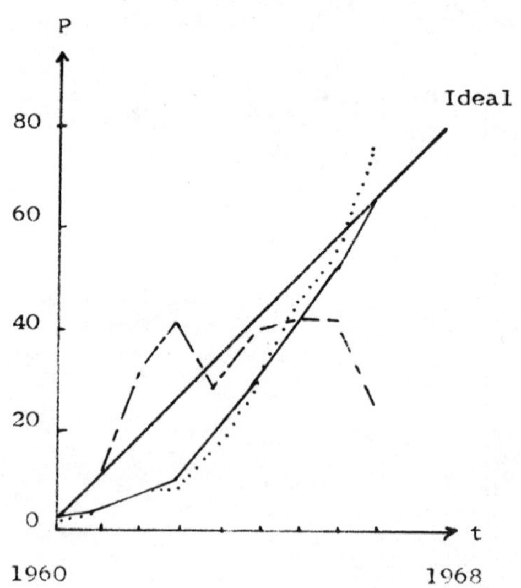

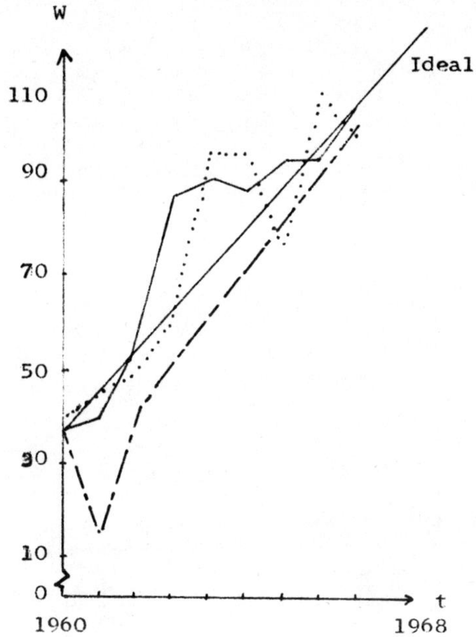

W
110
90
70
50
30
10
0
Ideal
t
1960
1968

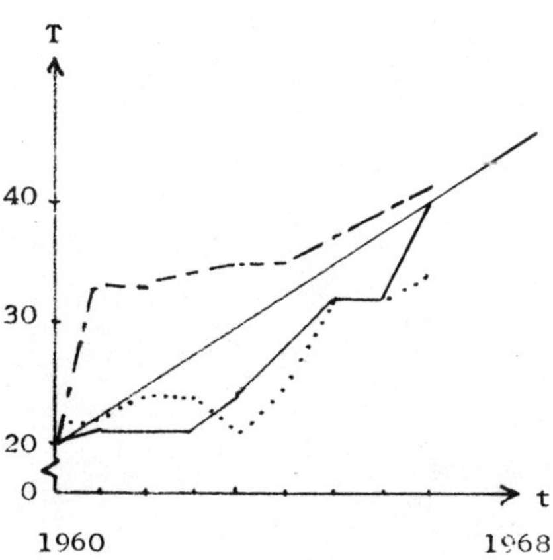

T
40
30
20
0
t
1960
1968

Figure 14, continued

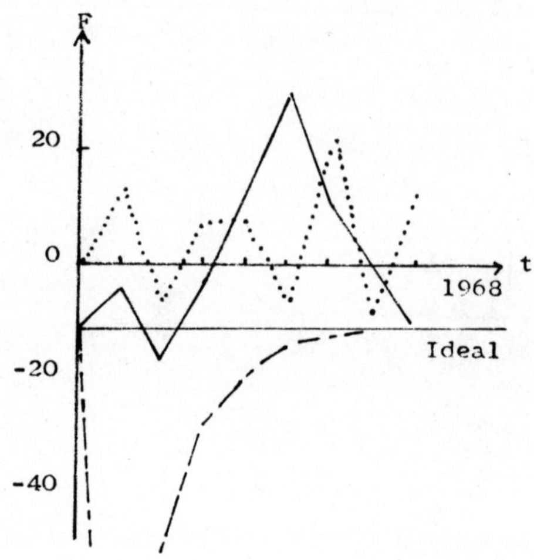

—— Actual Values

----- Predicted Values

—·—·— Optimal Values

The trend of the sales tax rate is the reverse of the excise tax rate and does not seem to make any sense. Similar implausible changes are indicated in the exchange rate and inflation tax. The changes required for public expenditures are within feasibility limits but still too large (roughly between 100 and 200 percent) for any legislature to undertake. The instruments of reserve ratio and minimum-wage rate show extremely low efficiencies.

The cost measured in Equation ℓ has a meaning only in a relative sense. While the cost functionals (the Ω and Θ weights of Appendix C) convert price indices, tax rates, cruzeiros of GDP, Gini coefficients, and other variables measured in their respective units, so to say, to a common denominator, it is difficult to conceptualize the absolute magnitude of the resulting hodgepodge. Whatever they may mean, the calculated costs of some of the runs (in millions of units) are noted below:

Run 1:	The program specified in model equations (ℓ) and (ℓi), with the weight of P divided by 10	879
Run 2:	Same program, except that the weight of P was restored and the weight of T was multiplied by 10 (Table 29)	791
Run 3:	Same program, except that the weight of Y was multiplied by 2 and the initial weight for P was halved and that for T was multiplied by 5	1, 308
Run 4:	The weights of all instruments doubled	1, 060
Run 5:	Using the Scandinavian weights for objectives	1, 379
Run 6:	The weights of all objectives of Program 1 raised by 10	10, 042
Run 7:	Only one objective P with its weight raised by 25 times	7
Run 8:	Same as run 7 except that the weights of τ^{exc}, τ^{sle}, and τ^{inf} were multiplied by 10	29

The differences of results from those of Run 1 are briefly summarized below.

Run 2

To correct for the poor performance of T we multiplied the weight of this variable by 10. This run gives better results than any other carried out here. Its cost is about 90 percent of the cost of the first run, as may be seen from the estimates given above. It must be remembered, however, that a lower cost is no basis for preferring a program over any other unless the alternative set of weights can also be justified from the point of view of the nation's desires and the feasibility of instrument values. Since this program elevates the importance of T by ten times the value assigned it in the Brazilian social-welfare function, we have no way of selecting Run 2 over Run 1. Nevertheless, it is interesting to note that the gain in a somewhat closer tracking of P^* and T^* is offset by somewhat wider tracking of Y^* and F^*, while the solution for W is not affected. Among the instrument set there is hardly any clear-cut gain or loss except for the inflation tax (τ^{inf}), which moves farther away from its ideal values. There are significant alterations, however, in the upward and downward deviations of indirect taxes (τ^{exc}, τ^{sle}), foreign-exchange instruments (e^{im}, e^{ex}), and welfare public expenditures (I^g and H^g).

Run 3

A doubling of the weight for Y and halving of the weights for P and T (as compared to Run 2) result in about 5 percent improvement in the tracking of Y^* and a corresponding deterioration in the time paths of P^* and T^*. These changes come at the cost of proportionately wider changes in the instruments of direct taxes (τ^{pi} and τ^c) and public outlays of the consumption type (W^g and C^g).

Run 4

A doubling of the instrument weights reduces the overall social cost of the program (as compared to Run 3, but not as compared to Runs 1 and 2) with minor loss in the tracking efficiency of the objective variables. The optimized values of several of the instruments for some of the years are still not controllable.

Run 5

An experiment with Scandinavian weights worsens the time paths of Y, W, P, and T and improves that of F. The improvement in F is natural, in that the Scandinavian weighting scheme assigns higher priority to foreign balances. The coefficient of variation of e^{ex} is higher and that of e^{im} is lower in this run than the corresponding values in the other runs. The reason the exchange rate for imports is stretched less than the exchange rate for exports is apparently that a reduction in the former is less effective than an increase in the latter (see Table 4, impact multipliers, row 17, columns 46 and 47).

In general, whatever weighting scheme is used for the objectives, relatively large changes and quick reversals are indicated in monetary policy (treasury's deficit finance). Equally large but less volatile changes are indicated in exchange-rate policy (e^{im}, e^{ex}) and indirect taxes (τ^{exc}, τ^{sle}). The changes required in fiscal policy (other than indirect taxes) are usually more gradual. One possible reason for this difference between monetary policy and fiscal policy is that the latter has shorter lags than the former.

Runs 6 and 7

The final runs for the historical period are those for inflation alone. In these runs we set a target for a straight-line decline of three percentage points each year so that P would come down to 9 percent in 1968. In Run 6, three of the four taxes (τ^{exc}, τ^{sle}, and τ^{inf}) show wide variations. Since all deviations cost equally, an inflation-correction policy would naturally use more extensively those instruments that are more effective. To reduce wide deviations in these instruments, however, we try ten-times higher weights for them in Run 7. The results are sketched in Figure 15. It may be seen that when several instrument handles are available, inflation can be controlled to whatever degree one may want. What happens to the objectives not included in this experiment is another matter. Since they cannot be ignored, Runs 6 and 7 are of analytical interest only.

Optimal Control Theory Analysis: The Planning Period, 1969-75

Considering a planning period of seven years, from 1969 through 1975 (i.e., t = 1, 2, ..., 7), the following ultimate (seventh-year) target values of the objective variables are assumed:

$$Y^*_{75} = Y_{68}\text{Exp}[.09 + .095 + .113 + .104 + .102 + .101 + .100] \quad \text{(growth)}$$

$$P^*_{75} = 10\% \quad \text{(inflation)}$$

$$F^*_{75} = \text{Exp} - \text{Imp} = \$250 \text{ million} \quad \text{(foreign trade balance)}$$

$$T^*_{75} = .35Y^*_{75} \quad \text{(government revenue)}$$

$$W^*_{75} = W_{68}\text{Exp}[.145 + .153 + .18 + .166 + .163 + .161 + .160] \quad \text{(redistribution of income)}$$

where the symbol "exp" stands for exponential. For those years since 1968 for which the data have become available we use the actual values. At the time of the final estimates of this chapter the rate of growth of Y was available up to 1972. The rate of economic growth for the terminal year was assumed to be 10 percent. The rates of

Figure 15: Optimal Control Runs for Inflation ($\dot{P}$)

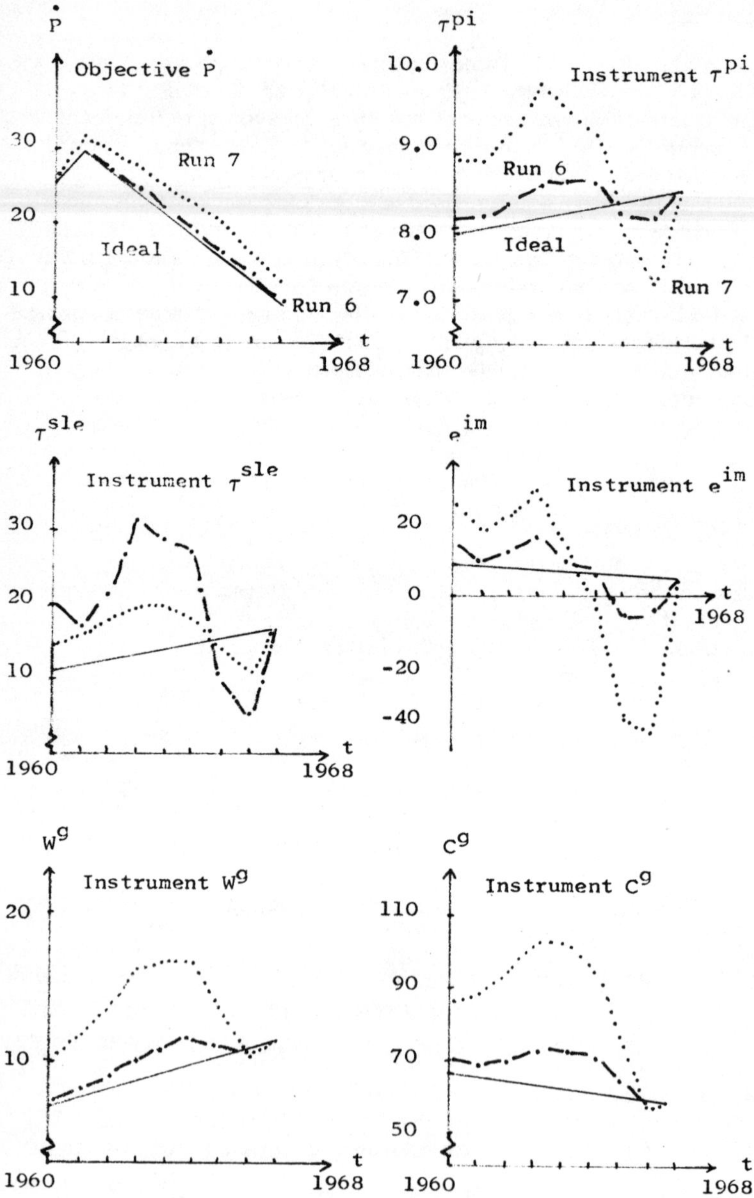

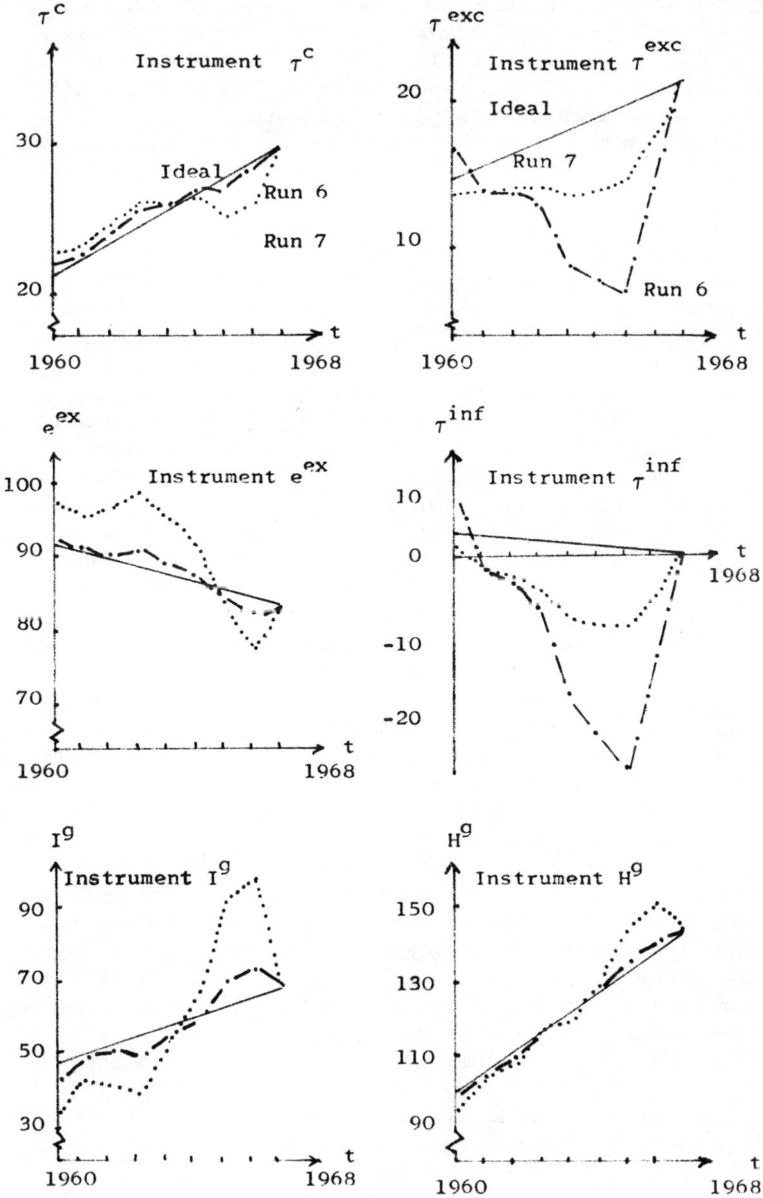

τ^c
Instrument τ^c
30
Ideal
Run 6
Run 7
20
t
1960
1968
τ^{exc}
Instrument τ^{exc}
20
Ideal
Run 7
10
Run 6
t
1960
1968
e^{ex}
Instrument e^{ex}
100
90
80
70
t
1960
1968
τ^{inf}
Instrument τ^{inf}
10
0
t
1968
-10
-20
I^g
Instrument I^g
90
70
50
30
t
1960
1968
H^g
Instrument H^g
150
130
110
90
t
1960
1968

growth for the intervening years were geometrically interpolated. The target values of W were fixed by raising W/Q by 1 percent per year. It may be recalled that W and Q are taken from the manufacturing sector. The rate of growth of this sector for the already-projected growth rate of GDP was estimated to be about 50 percent higher than that of GDP. The W^*_{75} target given above follows from these assumptions.

A T/Y of 35 percent was assumed for 1975. This is only slightly higher than the observed percentage of 34.5 percent for 1968. Foreign trade balances are targeted to improve slowly, by $50 million per year. For the intervening-year values of P, T, and F the Bogart-Theil procedure of linear interpolation was employed.

The target values for instruments were fixed as follows: the instruments τ^{pi} and τ^{inf} were fixed according to their linear trends; public expenditures (W^g, C^g, I^g, H^g) were fixed at a constant ratio of Y^*; the remaining instruments were fixed at their 1968 values. All exogenous variables were predicted to change according to their linear trends.

The results of a run with the standard weights of this study (except that the weight of P was halved) are sketched in Figure 16. In this run, instead of predicting the exogenous variables, we simply projected 80 percent of the objective variables. In other words, we want to see what it takes in terms of changes in the instrument set to effectuate 20 percent of the target values of the objective set. It was found that the variable P behaves quite strangely. It starts with a high positive value and ends with a high negative value. We could have forced it to fall in line by assigning higher penalties, but instead we chose to omit it as an objective. The other objective variables do not behave so strangely, and the model seems to force them over a time to approach the target values.

An instance of the basic weakness of the Theil model (the quadratic function), which was discussed in Chapter 1, note 7, is highlighted here. The model minimizes the deviations from the target values but does not distinguish between deviations to the desirable direction from those to the undesirable direction, as in the case of Y, where the solution starts with a 35 percent shortfall and ends with 19 percent shortfall (undesirable deviations). In this model there appears to be little conflict between growth (Y) and distribution (W); we have hardly any problem with meeting the W targets. The conflict seems to be between growth and inflation, although P behaves so erratically that this inference can be at best only a provisional one. The tax-revenue variable starts very low but catches up quickly.

On the instrument side, the program is executed with minor changes in direct taxes and in the sales tax. In other words, the program does not use these instruments extensively. This means that these instruments are not very effective. An equivalent program could be carried out at a lower cost by making necessary changes in the other, more effective, instruments. All other instruments undergo

Figure 16: Ideal and Optimized Values of Objective Variables, Planning Period (1969-75)

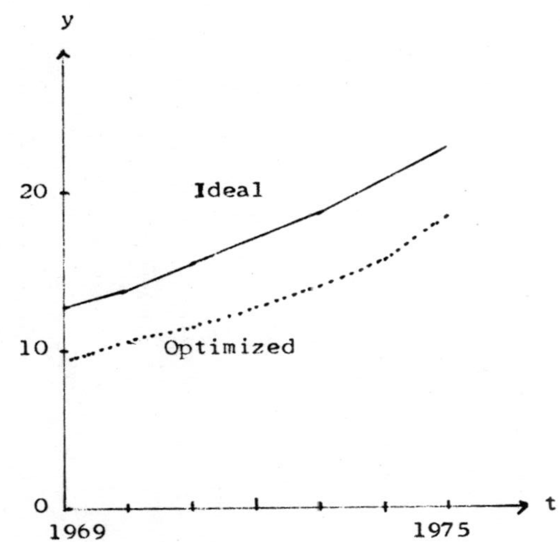

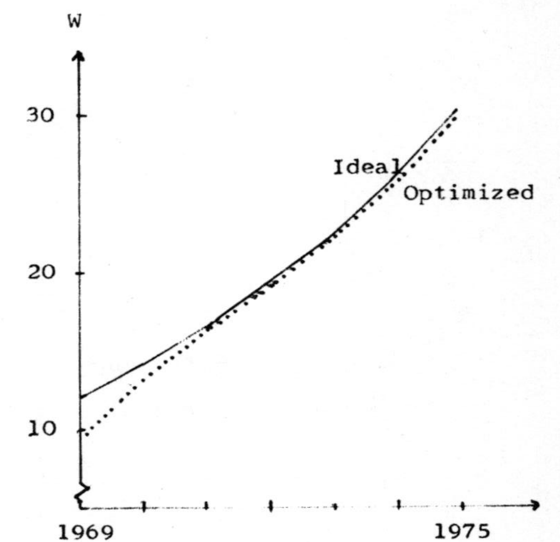

(continued)

Figure 16, continued

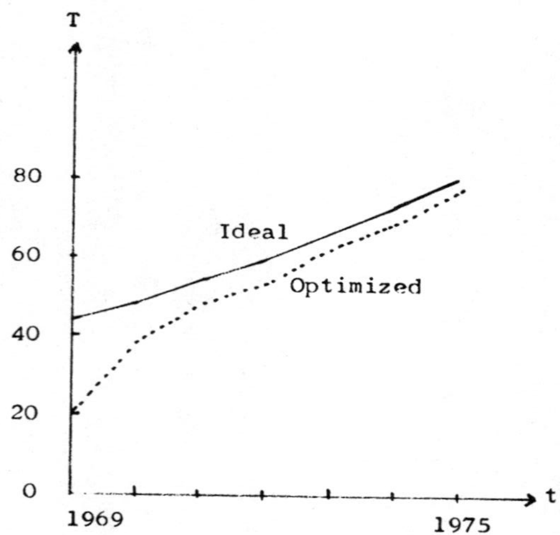

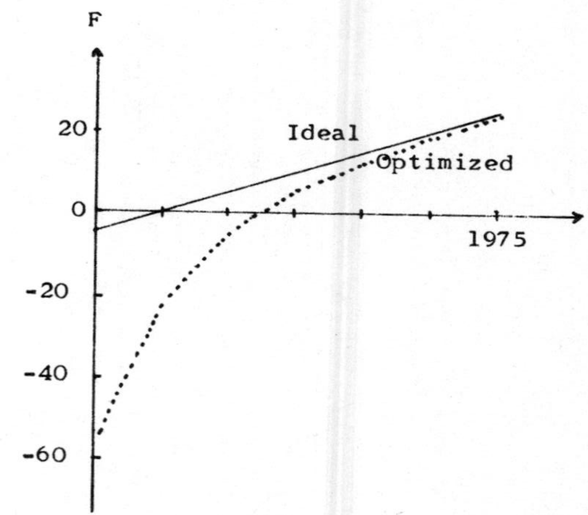

significant fluctuations. In general, it seems a matter of substituting one instrument for another. Governments' consumption expenditures (W^g and C^g) exceed and investment expenditures (I^g and H^g) fall short of the target values. A reduction in both of the exchange rates is indicated.

In the next set of experiments, we tried a few subsets of policy instruments and objectives. It has been seen that when P and T are dropped as objectives, the system becomes more controllable. When P and F are dropped, the system becomes even more compatible. Contrary to the prevailing belief about the relationship of growth and distribution, a set of targets consisting only of Y and W can be achieved without undue strains, provided the expenditure instruments are penalized lightly (i.e., if they are allowed to vary widely), which may, of course, render them uncontrollable. Similar results follow for growth (Y) and foreign balances (F) as a set of targets. In the latter case, however, the instrument of exchange rate for exports verges on instability.

We could go on with additional combinations of objectives and instruments and alternative sets of weights. Enough has, however, been done to squeeze information out of the present econometric and policy models. As it is said, "there appears to be no more oil left in these sesame seeds": this section has yielded no clear-cut results. Some of the lessons that can be pieced together from the positive and negative inferences of the analysis are mentioned in the following paragraphs.

If an econometric model is dynamically stable in its structural and reduced forms, that by itself is no guarantee that it will also be dynamically stable in its policy-specification form. In the former two cases one tests the stability of endogenous variables; in the policy-specification form (which has been the subject of analysis of the present chapter) one evaluates the stability of instruments. In the econometric analysis one analyzes the stability of the response mechanism of the system; in the policy model one examines the stability of the control mechanism. While the response mechanism of the model of the present study has been found to be stable, the controllability, if not long-run stability, of some of the instruments in achieving a set of targets of reasonable magnitudes is in question. This applies in particular to e^{im}, e^{ex}, τ^{inf}, τ^{def}, and τ^{sle}. Such an uncontrollability could, of course, be encountered even if an econometric model were free of inadequacies. The richer an econometric model, however, the greater the confidence one can put in the results (controllable or uncontrollable as the case may be) of optimal control analysis. Due to weaknesses in the data and other weaknesses noted at relevant places in this study, we cannot claim that there are no inadequacies in our model. It is also possible that various physical controls, relatively less predictable foreign investment/reinvestment, and rapid structural and other changes have frustrated our attempts to determine reliable relationships between instruments and objectives. Subject to this proviso, whatever facts have been brought to light by the present analysis give little suggestion that the Brazilian policy instruments as analyzed here are efficient enough to make reasonably ambitious long-run programs controllable. Efforts of future researchers will be better rewarded if more accurate data on instruments and their effects can be assembled.

A long-run stabilizing policy may require destabilizing policy measures in the early years of a long-run planning period. Since this will usually be unpalatable to politicians whose time horizon scarcely extends beyond the indicated destabilizing period, long-run optimal policies are unlikely to be popular.

Some of the objectives simultaneously pursued by developing countries may be incompatible. In the Brazilian case this may be observed from the following array of the directions of change in instruments required for impact increases in objectives:

	τ^{pi}, τ^{c} τ^{exc}, τ^{sle} NTR	e^{im}	e^{ex}	τ^{inf}	τ^{def}	W^g C^g	I^g	H^g	M_{-1}	w^{min}	η
Y	-	+	-	+	-	-	+	+	+	+	-
W/Q	0	0	-	0	0	0	0	0	+	+	0
F	+	-	+	+	+	+	-	-	-	-	+
P	+	+	+	+	+	-	-	-	+	-	+
T	+	-	+	-	-	-	+	-	-	-	-

For example, GNP can be expanded by lowering the four major direct and indirect tax rates. Such a policy is also antiinflationary, and it is neutral in regard to functional income distribution. However, it will have adverse effects on tax revenues and foreign balances. Exactly the opposite results follow from an expansion of public-health expenditure. At the same time, it is clear that, in general, if the number of instruments is sufficiently large, there may, in principle, exist feasible solutions. (Fifteen instruments are listed above; the sixteenth—i.e., M_{-1}—is not treated as an autonomous variable in this study and is listed here merely for academic interest.) In the present study, for example, if deviations from the target values are heavily penalized in the case of objectives and lightly or not at all penalized in the case of instruments, one can get close to the target values of objectives. In this respect, Samuelson's optimistic pronouncements of the 1950s concerning growth, employment, and distribution, were not without substance even though they may have underscored "fiscal dirigism."* What this study highlights is that policies matching the optimism of Samuelson are possible but, at least in the case of Brazil, not (as defined here in Chapter 1) "controllable."

---

*In 1955 Samuelson wrote that "with proper fiscal and monetary policies [the United States] can have full employment and whatever rate of capital formation and growth it wants. . . . [A]nd it can accomplish all this compatibly with the degree of income distribution it ethically desires" (1955, p. 234). [Emphasis added.]

See a discussion of "fiscal dirigism" in postwar Germany by Heuss (1963). See also NBER-Brookings (1966).

APPENDIX A: THE DATA AND THEIR SOURCES AND THE DEFINITIONS OF VARIABLES

THE DATA

The data used in the model of this study are given in Table A. 1. The symbols employed and the variables used in the study are defined below, followed by the sources of data and the procedure of preparing the tax rates and certain other variables.

Unless otherwise stated, the money-value variables are in millions of cruzeiros in 1963 prices. The physical quantities are also in millions, unless otherwise stated. Tax rates are in actual percentages or indexes, as the case may be. The base for price and other indexes is 1963 = 100. The currency unit used in this study is the new cruzeiro dollar of the late 1960s with the convertibility ratio of NCr$1 = Old Cr$1, 000. As time passed, the new cruzeiro came to be denoted simply as "cruzeiro"; in this study, the two terms are used interchangeably.

THE VARIABLES

1. Endogenous Variables

a. Tax Collections

T^{pi} = personal-income-tax collections, including those withheld at source

T^{c} = corporation-income-tax collections, including those withheld at source

T^{exc} = excise-tax collections, called imposto de consumo in Brazil

T^{im} = revenues from import-tariff duties

T^{stp} = federal stamp duties

T^{ss} = social security contributions

T^{fue} = fuel tax collections

T^{sle} = sales tax collections

T^{os} = other state taxes

T^{mun} = total municipal taxes

T^{fed} = total federal taxes

T^s = total state taxes

T^{dir} = total direct taxes

T^{ind} = total indirect taxes

T = total tax collections of all-level governments

Tr = government transfer payments, including subsidies, unless, when specifically stated, subsidies are treated separately

T^{disp} = government's disposable income—i.e., total revenues minus public transfers and subsidies

DEF = overall budget deficit

NTR = nontax revenue of the government

b. Other (Economy-Wide or Private Sector) Endogenous Variables

Y' = gross national product (GNP)

Y = gross domestic product (GDP) at factor cost

Y^d = disposable income of the private sector

Y^{na} = net domestic product at factor cost of the nonagricultural sector

Y^a = net domestic product at factor cost of the agricultural sector

$100(\Delta Y/Y)$ = g = annual charges of real GDP or rate of economic growth (expressed in percentages)

y = per capita GDP

C^p = private consumption

Q^m or simply Q = nonwage income, usually denoted in this study as "profits," calculated by subtracting the wage bill (W^m) from value added (VA^m) in manufacturing

W^m or simply W = wage bill in manufacturing industries

I^{pl} = private sector's gross fixed investment, excluding that of the public enterprises

I^{pe} = gross fixed investment of public enterprises

I^{stk} = variations in stocks and inventories

I^p = $I^{pl} + I^{pe}$

I = $I^p + I^g$

K = gross fixed capital stock in 1963 prices

Ex = exports of goods and services

Im = imports of goods and services

F = foreign surplus on current account

Z = excess capacity as a percentage of full-capacity K; note that the variable measure used here is usually (1-Z) or (1-Z)K

P = implicit price deflator

$\Delta P/P = \dot{P} = (P-P_{-1})/P_{-1}$ in percentages—i.e., annual percentage changes in implicit deflator

P^w = wholesale price index

$\Delta P^w/P^w = (P^w-P^w_{-1})/P^w_{-1}$ in percentages

P^{con} = consumer price index in Guanabara state

P^{im} = relative price of imports, calculated as real exchange rate times the weighted ratios of price index of imports to price index of comparable Brazilian goods

$\Delta P^{im}/P^{im} = (P^{im}-P^{im}_{-1})/P^{im}$ in percentages

M^n = nominal money supply, defined as the liabilities of the monetary system in currency and demand deposits to the domestic private sector

M_1 = currency in the hands of the public

M = real money supply—i.e., M^n deflated by wholesale price index, sometimes written as M^w; when M^n is deflated by GNP deflator, it is denoted by M^{gnp} or is indicated to be so in the relevant text

$\Delta M/M = (M-M_{-1})/M_{-1}$ in percentages

r = real rate of return on equities traded at the stock market

BL = real loans of the banking system to the private sector and public agencies other than the treasury

$\Delta BL/BL = (BL-BL_{-1})/BL_{-1}$ in percentages

2. Predetermined Variables

c. Exogenous Variables

Pop = population in hundred thousands

t = time trend, the serial number of years, with 1950 = 1

X_{-1} = capacity to import as defined in Brazilian national accounts, lagged by one year

$(\Sigma I/\Sigma K)_{-1} = \sum_{n=t-3}^{t-1} I_n \Big/ \sum_{n=t-3}^{t-1} K_n$, a measure of three-year moving sum of new vintages of capital normalized by the corresponding sum of capital stock

IL = international liquidity (gold holdings plus the reserve position in IMF plus foreign exchange—all in 1963 U.S. dollars, at the beginning of year

$\Delta IL/IL = (IL - IL_{-1})/IL_{-1}$, in percentages

N = active population, ten years and older, in hundred thousands, used as a proxy for labor force

M_1/M = ratio of currency held by the public to money supply

η^a_{-1} = actual reserve ratio lagged

θ = a parameter of the rate of mortality of capital stock

θ' = average straight-line depreciation rate

Fert = index of apparent fertilizer consumption, two-year moving average, 1952-53 = 100

A = cropped area in millions of hectares

GNP^{US} = GNP of the United States in billions of 1963 dollars

NRA = net income remitted abroad (renda liquida enviada às exterior)

Dummy 53 = a dummy variable with unity for the years 1954-68 and zero for earlier years

Dummy 59 = a dummy variable with unity for the years 1959-68 and zero for earlier years

d. Lagged Values of Endogenous Variables

$(\Delta M/M)_{-1} = \Delta M/M$ lagged by one year; $(\Delta M/M)_{-2} = \Delta M/M$ lagged by two years; $(\Delta M/M)_{-3} = \Delta M/M$ lagged by three years. Likewise there are lagged values of some other endogenous variables, including C^p_{-1}, I^p_{-1}, I^{pe}_{-1}, I_{-1}, $(\Delta P/P)_{-1}$, I^{stk}_{-1}, $(\Delta IL/IL)_{-1}$, T^{disp}_{-1}, T_{-1}, g_{-1}, $[(1-Z)K]_{-1}$, r_{-1}, Q_{-1}, $(\Delta BL/BL)_{-1}$, and Y'_{-1}.

e. Instrument Variables

i) Tax Rates

τ^{pi} = overall average effective rate of personal-income tax, with τ^{pi}_{-1} as its lagged value

τ^c = average effective corporate tax rate, with τ^c_{-1} as its lagged value; $\tau^{c'}$ is its measure when the SUDENE investment incentives have been translated in terms of reductions in tax rates

τ^{sle} = average ad valorem sales-tax rate

τ^{exc} = average ad valorem excise-tax rate

ii) Exchange Rates

e^{im} = indexes of average real (in the sense of having been deflated and not effective) exchange rate for imports deflated by wholesale price index, excluding coffee

e^{ex} = indexes of average real exchange rate for exports, including export premia, deflated as above

e^{nom} = index of nominal exchange rate corrected for changes in import duties

iii) Monetary Instruments

τ^{def} = overall deficit financing—i.e., treasury's deficit plus public borrowings, of all-level governments as a percentage of GDP calculated as (public saving - public investment)/GDP; entered as deficit (+) or as surplus (-)

τ^{inf} = treasury's deficit—i.e., deficits financed from printing press and the central bank's credit to the federal government as a percentage of GDP, referred to in this study as an inflation tax; entered as deficit (+) or as surplus (-)

η^c = compulsory reserve ratio; $\eta = \eta^a_{-1} - \eta^c$

EO = "foreign exchange operations," as called in Brazilian literature, including minimum support price for coffee; net outgoings (-) and net earnings (+)

iv) Wage Rate

w^{min1} = indexes of minimum-wage rate for Guanabara State (considered to be roughly equivalent to the country's average wage rate) deflated by Guanabara's cost-of-living index, assumed to be relevant for income-distribution and welfare purposes

w^{min2} or simply w^{min} = indexes of minimum wage rate as above, deflated by implicit deflator, assumed to be more relevant than the cost-of-living index for such rates and prices as are fixed in terms of minimum-wage rate—e.g., income tax brackets

v) Public Expenditures

W^g = government's current expenditure on wages and salaries

C^g = government's consumption expenditure on goods and services; $W^g + C^g$ = total consumption expenditure of the government

S^g = public savings

I^g = public investment in fixed capital

(The above expenditures are further classified, where relevant, into federal, federal autarquias, state, and municipal expenditures.)

E^g = expenditure on public education

H^g = expenditure on public health

D = defense expenditure

G = $C^g + W^g + I^g$

A = expenditure on general administration, police, and justice

SOURCES AND EXPLANATIONS

Some of the data sources were given in the footnotes to Tables 1 and 2 of Chapter 1. Additional sources are given below.

Tax Collections and Public Expenditures. Most of these series are available from Conjuntura Economica, 23, nos. 6, 10 (1970). Expenditures on public education, public health, and defense were taken from Instituto Brasileiro de Estatistica (IBGE), Anuario de Estatistica do Brasil, and IBGE, Atualidade Estatistica do Brasil (1970), Table 6. They are also reported (though complete series are not available) in UN, Statistical Yearbook (annual). The data for public education, public health, and defense are not entirely satisfactory, since they are available only by ministries for the federal government. For state and municipal governments, however, they are published under the respective functions.

Income-tax collections are available in three parts: a) personal-income-tax collections through annual returns (imposto de pessoas fisicas), b) corporate income tax collections through annual returns (imposto de pessoas juridicas), and c) a growing category of "deductions at source" of both personal and corporate taxes (arrecadacao nas fonte). The breakdown of the third category into the former two was done by the author from the ministry of finance sources. The background items for this breakdown are also assembled up to 1964 in about a half dozen subgroups in IPEA, Diagnostico Preliminar Setore de Finanças, mimeographed and bound (Rio de Janeiro, 1965). The withholdings at source were divided broadly on the following basis: those from remuneração do trabalho, outros rendimentos do trabalho, and capitais imobiliarios to personal income tax. The withholdings of juros da divida publica, juros diversos, and pessoas juridicas were added to the corporate income tax.

National Income Categories, Prices, Population, Capacity to Import, etc. From Conjuntura Economica, 24, no. 10 (1970), and Anuario de Estatistica do Brasil (various issues). Some of the series—e.g., consumer price index, minimum-wage rate, import prices, and so on—are available from secondary sources such as, for example, Ellis (1969). For fixed capital formation by public enterprises (I^{pe}), see Chapter 1, Table 2. The breakdown of I^p into I^{pl} and I^{pe} is not available from published sources after the year 1965. Accordingly, at some places we dropped the observations for 1966-68 for this variable. At other places the 1966-68 I^p series were broken down into I^{pl} and I^{pe} series on the basis of the 1965 ratio.

Capital Stock (K). For the years 1949 and 1950, from FGV, Centro das Contas Nacionais, Contas Nacionais (Rio de Janeiro, 1967). The initial-year (1949) figure for net fixed capital as reported in this source was subjected to a straight-line mortality-cum-obsolescence rate denoted by θ. For instance, depreciation as reported in national accounts is approximately 5.04 percent of GNP. GNP is about 56 percent of the estimated capital stock. This gives a θ of about 2.8 percent of capital stock. The estimates of depreciation of the national accounts, however, have lately come under expert attack. In this connection, for example, see Neto (1965). Including obsolescence, the true depreciation rate was assumed to lie somewhere between 2.8 and 5 percent. To be on the safe side, we used a mortality rate of 5 percent in this study. The 1950 base-year capital stock was subjected to 5 percent annual depreciation. To the surviving capital stock of 1950 was added annual gross fixed investment from 1951 onward to get the gross fixed capital series used in this study. The 1951 investment will remain intact until 1971, after which it will drop out entirely. This procedure amounts to the application of the one-horse-shay principle of depreciation with 20 years' life of the stock.

"Active Population Ten Years and Above" (N). Available in IBGE, Censo Demografico, Resultados Preliminar, 1960 (Rio de Janeiro, 1968) up to 1960. From 1961 onward this series was estimated from overall population by using the trend ratio of N/Pop of the 1950s.

Monetary Variables (Money Supply, Treasury Deficit, Bank Loans, Reserve Ratios, Rate of Return on Stock, International Liquidity, etc. From UN, Statistical Yearbook (various issues); Conjuntura Economica (various issues); Anuario de Estatistica do Brasil (various issues), Relatorio do Banco Central (various issues), and IMF, International Financial Statistics (monthly). The series for international liquidity after 1964 is not comparable to that of the earlier period as reported in Statistical Yearbook; both series, however, are available for 1962, 1963, and 1964. The overlapping values for these three years were used to adjust the 1964-68 series to the earlier series.

No representative series of interest rates has been prepared in Brazil. It is difficult to compute one. Due to inflation, some interest rates have been negative, while others have been exorbitantly high. Interest on government bonds hardly reflects the real market rate. Accordingly, as a provisional measure, the average prices of the most traded stocks, compiled by the National Investment Service or S. N. Index, deflated by the wholesale price level, was tried as an index of long-run real interest rates. Such a measure could, however, also reflect the rate of return on investment. No better series could be obtained when this manuscript went to press. The series used here are from Simonsen (1969, Table 7), Conjuntura Economica (1970, Appendix Table, column 250), and Conjuntura Economica (earlier issues).

Excess Capacity (Z). Z is defined here as the deviation in percentages of the actual output-capital ratio from the straight-line trend fitted to the consecutive peak-year ratios. The 1957 and the 1961 peaks were very nearly identical and have not been reached since. For the years 1962-68, therefore, the 1961 output-capital ratio was used as the full-capacity use ratio. For the purpose of this measure both Y^{na}/K and Y/K were tried. Although the magnitudes of the two ratios differ, the directions of change (and, by and large, the relative changes also) have been almost identical. The Y^{na}/K ratio, however, was considered to be more relevant and used here, as it is the shortfall in nonagricultural output that causes real excess capacity, while agricultural output also depends, to a large extent, on the vagaries of nature. The capital stock for the entire economy was used as the denominator, since separate series for the nonagricultural capital stock are not available. Note that as defined here $0 \leq Z \leq 1$, and the capital stock actually used is given by $K(1-Z)$.

The variable $(\Sigma I/\Sigma K)_{-1}$ equals:

$$\sum_{n=t-3}^{t-1} I_n \Big/ \sum_{n=t-3}^{t-1} K_n$$

expressed in percentages; it is the ratio of the corresponding three-year moving averages of I and K lagged by one year. It is used to separate out that part of technical change that might be embodied in the new vintages of physical capital. It would include learning-by-doing as well as technical knowledge acquired exogenously.

Area under Crops and Fertilizer Consumption. The series of hectareage under crops was prepared from the two benchmark figures for the years 1950 and 1960 from the Census of Agriculture (Censo Agricola), 1950 and 1960, by interpolation and extrapolation, employing an exponential trend. The series for apparent fertilizer consumption is available from Anuario Estatistico do Brasil (annual). The present series of fertilizer consumption was taken from Smith (1969, Table 6).

Tax Rates

Computing one average rate for such multiple-rate taxes as personal-income tax, excise taxes, import duties, and so on is not a simple task. In some cases the cost in terms of time and energy of the researcher in digging up information and calculations of all kinds to achieve high accuracy of these rates becomes extremely large in proportion to the gain in accurate estimates over the ones that can be obtained by approximate measures of the tax rates. Much depends on the data that are readily available. The line has to be drawn somewhere. The rates calculated in the present study are the result of such constraints and desirabilities.

Personal Income Tax Rate (τ^{pi}). The average effective rates of personal income tax were calculated as follows. Up to 1960 the income brackets and the tax rates were fixed in monetary units. Introduction of annual alterations in these rates did not start until 1957. Since 1961 the income brackets have been fixed as multiples of the statutory minimum wage rate, which is revised by the government periodically. Prior to 1961 the tax rates rose and the minimum exemption limit declined due to increases in price level, except when adjusted (since 1957) by the finance minister, though the adjustment seldom adequately matched the price changes. In 1960 and 1961, however, the income brackets were raised and the tax rates were lowered sharply. Thus, in 1960 the average effective tax rate was almost identical to that of 1950 and reached a low bottom in 1961. After 1961 the tax rates started rising again as the minimum nominal-wage rate was raised, usually at a rate that was significantly lower than the rate of inflation. This seems to have been so due to two main reasons. First, the government's predictions of price increases, which have formed the basis of the new year's minimum wages, invariably turned out to be lower than the actual price increases. For example, for the year 1966, the government planned a 15 percent expansion in money supply on the basis of which it predicted only a 10 percent rise in prices; as it turned out, prices actually rose by approximately 40 percent, and in the following year prices rose by approximately 30 percent, against the government's prediction of 15 percent. While the first reason was largely a consequence of wishful thinking and bad predictions, the second reason for the observed wage lag seems to be clearly the result of a deliberate policy to keep wages down. In its preoccupation with combating, according to its own diagnosis, the demand inflation of the early 1960s and the cost inflation of the late 1960s the government concentrated on controlling wages as the major demand component in the early 1960s and as the key cost price in terms of which many other prices were statutorily fixed in the late 1960s, including income tax brackets. The observed shortfall of the minimum wage from the general price level gave a measure of the degree by which tax brackets have become lower, and pari passu effective tax rates have become higher each year since 1961 (except for a year or so in the middle of the decade). Using the previous-year taxable incomes of different brackets as the revolving weights and the rates resulting from the interaction of inflation and the minimum-wage multiple rule, the average tax rates were computed from the (changing) effective rates of each bracket. For example, in 1964 the real minimum wage was about 6 percent lower than in 1963; accordingly, the tax brackets were lowered by 6 percent. Alternatively stated, 6 percent of the taxpayers of each bracket were pushed up to the next higher bracket rate in that year. In order to see what difference it would have made to the average tax rate, if the minimum real wage had been kept constant in absolute terms, even ignoring productivity or rising per capita income, 6 percent of the taxpayers of that year were shifted from each higher bracket to the next lower bracket. Apparently, 6 percent of the bottom bracket in this procedure now pay zero tax, while all along the line some taxpayers pay lower tax. The shortfall of the resulting average effective tax rate in comparison to the actual rate gave the wanted measure. The annual changes thus calculated were added to the base-year (1950) average effective tax rate.

In symbols, let τ_{jt}^{pi} be the jth bracket's income tax rate in year t, Y_{jt} the taxable income of the jth income bracket in year t, and $\bar{\tau}_0^{pi}$ the base year average effective tax rate. Then the current-year average effective tax rate, $\bar{\tau}_t^{pi}$, from 1961 onward, is given by:

$$\bar{\tau}_t^{pi} = \bar{\tau}_0^{pi} - \frac{\sum_{j=1}^{J} Y_{jt}\left(\frac{w_t^{min}}{w_{t-1}^{min}}\tau_{jt}^{pi} + \left[1 - \frac{w_t^{min}}{w_{t-1}^{min}}\right]\tau_{j-1,t}^{pi}\right) - \sum_{j=1}^{J} Y_{jt} - \tau_{jt}^{pi}}{\sum_{j=1}^{J} Y_{jt}} \qquad (A1)$$

where $\tau_{j-1}^{pi} = 0$ (i.e., the zero tax rate below the exemption limit), τ_1^{pi} = the first bracket rate, τ_2^{pi} = the second bracket rate, and so on, such that τ_J^{pi} = the top bracket rate. For withdrawals and accretions to the first bracket, it is assumed, for practical purposes, that there is an imaginary bracket below the exemption limit with taxable income Y_{0t} that is always equal to Y_{1t}. The last term in the numerator is the actual tax yield.

In addition to the changes in tax brackets and rates, various temporary concessions, such as tax postponements for short periods, were also taken into account. A major concession of this nature, for instance, was given in 1967. For several years, significant excess capacity had coexisted with high inflation. One cause of such a state of affairs was believed to be the shortage of working capital of the firms, as a sequel of the erstwhile tight-money and strict fiscal-policy measures of the government. Among other measures aimed at putting more working capital at the disposal of firms, the new finance minister, Delfim Neto, gave 40 to 45 days' postponement of various federal tax payments on industrial products. According to my calculations, the package of all such and other concessions in 1967 amounted to a reduction of the tax rate of approximately 5 percent for excise taxes and the personal income tax, 5.5 percent for the corporation tax, and 10 percent for tariffs and social-security contributions. The relevant tax rates for 1967 were, therefore, corrected by these proportions. Even though the concessions of the nature of the postponement of tax collections only impart a bubble effect, these concessions continued through 1968, the final year of our reference period. Hence, the estimates for 1968 were not affected.

Corporation-Tax Rate. Apart from various national sources, corporation-tax rates are also available from a number of (annual or periodical) English-language publications. Nearly complete sets of these publications are available at the Harvard Law School library. Among these the following might be mentioned:

1. Various publications of the Pan American Union (sometimes, but not always) done by different authors in different years; e.g., J.T. Nabuco, Laws of Brazil in Matters Affecting Business (Washington, D.C.: Pan American Union, 1955); and J.T. Nabuco and Isidoro Zanotti, Laws of Brazil in Matters Affecting Business (Washington, D.C.: Pan American Union, 1961).

2. Various publications of the Overseas Press and Consultants, usually done, again, by different authors; e.g., Walter H. Diamond, Digest on Brazilian Taxes and Business Organization (New York: Overseas Press, 1958).

3. Publications of Price Waterhouse & Co.; e.g., their Information Guide for Doing Business in Brazil (Boston: Price Waterhouse, 1962, 1964, 1965, 1968).

4. Publications of the U.S. Department of Documents; e.g., G.E. Heare, Brazil: Information for United States Business (Washington: the Department, 1961).

5. Henry J. Grumpel and R.G. de Souza, Taxation in Brazil, World Tax Series, Harvard Law School, International Program in Taxation (Cambridge, Mass.: The Riverside Press, 1957), gives tax rates as of 1957 and for some earlier years. Carl S. Shoup, The Tax System of Brazil, FGV, Comissao de Reforma do Ministerio da Fazenda (Rio de

Janeiro, 1965). Eurico Korff, *Imposto de Renda e Investimento*, Edição Financeiras, S.A. (Rio de Janeiro, 1965).

In the advanced countries, this tax is usually levied at a flat rate. In Brazil, this tax has somewhat complicated rates and is levied at mildly progressive rates. The basic tax rate has been between 22 and 27 percent. For a major part of the reference years there has existed an excess profits tax at a rate of about 5 percent, a surcharge, and a quasitax compulsory loan scheme at a rate of 10 to 15 percent of the tax paid (terminated in January 1965). On the other hand, there also has existed a "mask" of high nominal tax rates that were effectively reduced by about 5 percent of profits in the form of certain deductions on the previous year's profits. According to Shoup's estimates for the early 1960s, this mask nearly cancelled out the excess profits tax of those years (Shoup, 1965). In calculating the tax rates, these modifications, as far as information could be collected, were taken into account.

There are *various tax-incentive schemes.* Of these, the investment incentives for northeast Brazil under the development program of SUDENE have been the most significant. Under this program all Brazilian corporate entities can reduce their income-tax liabilities to half, provided they invest the other half in the northeast. In the first instance the saved half of tax payments is deposited in a blocked account at The Bank of the Northeast of Brazil. Only if these deposits are not committed to specific approved projects within three years do they revert to the federal treasury. Such deposits as a percentage of total corporate tax collections rose from 5.2 percent in 1963 to 50.3 percent in 1966. Since this is a major incentive scheme that affects corporate tax collections substantially, we have prepared two series of the rates of this tax: one that neglects the SUDENE tax incentive scheme (τ^{c}), and another that assimilates this scheme ($\tau^{c'}$). No allowance could be made for investment incentives in the form of enhanced depreciation allowances, tariff concessions, certain regional tax holidays, and similar other concessions; to that extent the incidence of this tax is overstated.

Tariff Rates, Exchange Rates, and Relative Prices of Imports. The main source of these series is the excellent work done in the field (for the years 1953-67) by the Berkeley Team at IPEA. The main sources of this information are: IPEA, *Diagnosto Preliminar, Setor de Comercio Internacional* (Rio de Janeiro: Ministerio da Planejamento, March 1967), also material published in Joel Bergsman and Pedro Malan, "The Structure of Industry Protection in Brazil," *Revista Brasileira da Economia*, 24, no. 2 (April/June 1970), Table 2; and Samuel A. Morley, "Import Demand and Import Substitution in Brazil," in Ellis (1969), pp. 283-315. The figure for tariffs for 1953 was calculated from Ellis (1969, p. 229), and that for 1968 was estimated. Prior to 1953 the Brazilian import policy worked largely through quantitative restrictions and import licenses. In 1953 this system was replaced by multiple exchange rates, price-based tariffs, and auction sales of exchange. It is, thus, difficult to calculate an average exchange rate for these earlier years of the study. It may be mentioned, however, that during 1950-52 the official exchange rate remained fixed at a constant value of Cr$18.7 to U.S.$1. The tariff structure as such also remained practically unchanged. Accordingly, the exchange rate for the years 1950-52 was assumed to be the same as that for 1953.

Once the e^{im} series was complete, the relative prices of imports (P^{im}) for the missing years in Morley's study were calculated by the same procedure as that employed by Morley—i.e., relative P^{im} = the price index of imports relative to the price index of comparable Brazilian goods, weighted by e^{im}.

Minimum-Wage Rate (w^{min}). Minimum-wage rates are published in standard sources, such as the Anuario Estatistico do Brasil.

Excise-Tax Rates (Imposto de Consumo) (τ^{exc}). The calculation of an average excise-tax rate is a laborious process. In Brazil, the excise duties are legally levied on the gross (price plus tax) price of the commodities concerned, presumably so that the rates will look low. The problem of calculating average rates on broad categories of commodities is rather acute as the data on the quantities consumed of different commodies are not available. Therefore, weighted average rates were computed from diverse sources for the main taxable commodity groups for one year. The indexes of production of about 17 commodity groups as closely corresponding to the tax categories as could possibly be obtained were then employed as the (changing) weights for the tax rates on the latter categories to compute the required series. About two-thirds of the excise-tax revenues in Brazil are secured from five consumer categories—namely, in 1965, tobacco (27 percent), metals and metal products (11 percent), textiles (11 percent), beverages (8 percent), and automobiles (7 percent). The 17 categories referred to above for which the production indexes are readily available are food, beverages, tobacco, textiles, clothing and footwear, paper and paper products, leather and furs, rubber products, nonmetallic mineral products, basic metals, metal products, electrical machinery and apparatus, and transport equipment.

Sales-Tax Rate. It was not found possible to compile tax rates for individual states and the exempted proportions of the sales in each state. A kind of multistage-sale pyramiding of this tax (prior to 1967) precluded any precise estimates of the actual incidence of these taxes in different states. In view of these difficulties, the average sales tax rate used here is simply the tax collections divided by a national product component prepared by adding the income of the secondary sector to an appropriately-calculated proportion—about half—of the tertiary sector.

Social-Security-Tax Rate. The contribution to this fund by the employee (about 8 percent), the employer (about 8 percent), and the government has remained practically constant in the reference period. What really is relevant here is not this constant rate but the system's expansion of coverage. The expansion of the system is not to any significant extent a state discretion, however; it is primarily an employee option. Yet the state can provide inducements and incentives to both employees and employers to promote this system, especially as the Brazilian government makes some contribution to this fund. Accordingly, the influence of this tax on the relevant variables will be tested from total contributions rather than its rate, presuming it to be in part subject to state policy.

The remaining tax rates were not compiled, mainly, because most of these rates hardly changed over the reference period and, secondarily, as these form relatively small proportions of the overall tax revenue.

TABLE A.1

The Data Series, 1948-68
(in 1963 prices)

Year	T^{pi}	T^{c}	T^{exc}	T^{im}	T^{stp}	T^{fed}	T^{ss}	T^{fue}	T^{sle}	T^{os}	T^{s}	T^{mun}	T^{dir}	T^{ind}	T	T^{disp}	NTR
1948																	
1949	21	81	119	36	34	291	77	26	151	66	217	36	179	562	741	760	
1950	19	77	115	38	38	346	77	19	173	77	231	77	269	519	788	750	175
1951	35	86	138	52	52	483	86	35	207	86	293	69	328	621	949	914	197
1952	45	106	136	45	45	424	106	45	212	91	288	76	348	621	969	894	195
1953	53	105	145	13	53	421	92	53	211	92	303	79	355	632	987	934	225
1954	54	109	163	22	54	565	98	43	228	109	337	76	359	804	1, 163	1, 065	199
1955	56	120	157	19	56	519	111	37	250	111	361	83	389	740	1, 129	1, 019	200
1956	60	120	173	15	60	602	135	30	271	113	391	75	436	834	1, 270	1, 060	180
1957	67	120	207	20	67	580	167	73	293	127	420	100	474	914	1, 388	1, 167	205
1958	78	120	240	72	72	828	186	84	342	138	480	114	528	1, 218	1, 746	1, 482	243
1959	65	157	250	88	83	880	204	106	431	130	565	97	556	1, 361	1, 917	1, 574	167
1960	61	172	307	81	95	922	212	102	490	146	637	95	575	1, 453	2, 028	1, 819	316
1961	55	185	339	99	99	795	290	149	544	149	693	102	668	1, 435	2, 103	1, 637	220
1962	50	167	363	103	114	810	283	119	587	116	703	103	616	1, 476	2, 092	1, 456	145
1963	55	188	408	87	107	855	278	121	581	113	695	131	622	1, 525	2, 147	1, 629	221
1964	74	183	468	66	100	919	308	128	707	120	827	122	663	1, 710	2, 373	1, 530	-62
1965	92	258	516	71	119	1, 046	389	231	750	96	846	154	852	1, 894	2, 746	1, 554	-212
1966	83	247	547	84	133	1, 199	503	221	847	118	966	174	1, 013	2, 188	3, 201	1, 935	213
1967	75	225	551	72	..	1, 034	527	201	1, 062	61	1, 123	112	1, 119	2, 119	3, 238	1, 720	-301
1968	83	260	771	124	..	1, 600	619	242	1, 289	82	1, 370	142	1, 326	2, 700	4, 026	2, 219	-425
Mean	61	158	315		86		246	108	499	109		104	605	1, 303		652	

Year	GNP	GDP	Y^d	Y^{na}	Y^a	Y^{na}/Y^a	g	y	C^p	I	I^{pe}	I^{pi}	K	Ex	Im	F	NRA	Z	P
1948	4,567	4,612	3,865	2,849	1,090	2.62	7.4	94											4.2
1949	4,852	4,894	4,041	3,065	1,100	2.79	6.6	96	4,568	754		532	8,239			4.7		0.00	4.7
1950	5,198	4,442	4,442	3,288	1,192	2.76	6.5	100	5,051	789	21	538	8,614	500	429	35	35	0.00	5.2
1951	5,536	4,638	4,620	3,448	1,224	2.82	6.0	102	5,442	1,017	20	793	9,238	584	707	28	28	1.39	5.8
1952	6,008	5,106	5,106	3,832	1,273	3.01	8.7	108	5,940	1,015	17	788	9,762	417	606	11	11	0.00	6.6
1953	6,131	5,198	5,198	3,869	1,368	2.83	2.5	108	5,426	947	30	711	10,369	434	379	47	47	2.09	7.6
1954	6,770	5,642	5,707	4,250	1,435	2.96	10.1	117	6,153	1,152	39	870	11,067	516	539	50	50	1.15	9.2
1955	7,201	6,102	6,176	4,621	1,546	2.99	6.9	121	6,483	991	54	751	11,637	566	521	54	54	0.00	10.8
1956	7,437	6,263	6,377	4,865	1,451	3.35	3.2	123	6,774	1,060	28	827	12,313	529	472	52	52	0.65	13.3
1957	8,080	6,797	6,923	5,283	1,561	3.38	8.1	128	6,913	1,154	30	754	13,006	495	546	44	44	0.00	15.0
1958	8,671	7,128	7,194	5,625	1,558	3.61	7.7	132	7,103	1,419	80	875	13,976	533	581	60	60	0.37	16.7
1959	9,142	7,413	7,570	5,783	1,695	3.41	5.6	136	6,660	1,704	131	1,176	15,272	621	707	68	68	2.75	21.6
1960	10,005	8,140	8,187	6,361	1,859	3.42	9.7	145	7,100	1,709	130	1,036	16,542	610	742	81	81	2.17	27.3
1961	11,119	9,263	9,483	7,353	1,976	3.72	10.3	153	7,251	1,923	194	1,156	18,107	767	828	65	65	0.01	36.2
1962	11,639	9,737	182	7,563	2,286	3.31	5.3	158	7,824	1,902	290	1,289	19,767	603	794	111	111	2.35	56.2
1963	11,857	9,945	228	8,036	1,981	4.06	1.5	155	8,154	2,099	283	1,360	21,436	1,156	1,207	17	17	3.12	100.0
1964	12,185	10,071	654	7,944	2,181	3.64	2.9	155	8,349	2,024	240	1,314	23,036	916	785	80	80	6.12	188.0
1965	12,457	10,114	903	8,016	2,294	3.49	2.7	153	8,224	1,848	197	1,009	24,457	1,110	788	135	135	7.83	292.0
1966	13,144	10,472	1,210	8,576	2,022	4.24	5.1	158	8,755	2,026	286	1,122	26,052	1,006	896	125	125	7.69	405.0
1967	13,715	11,094	1,996	9,083	2,164	4.20	4.8	159	9,073	1,955	375	1,038	27,670	919	928	153	153	7.78	515.0
1968	14,951	11,673	2,732	9,701	2,110	4.60	8.4	170	9,995	2,510	346	1,415	29,742	1,042	1,147	138	138	7.99	659.0
Mean	9,539	7,855	8,152	6,092	1,763	3.46	6.1	136	7,193	1,539	147	991	15,839	701	739			2.82	125.8

(continued)

TABLE A. 1, continued

Year	$\frac{\Delta P}{P}$	P^{im}	P^{w}	$\frac{\Delta P^{w}}{P^{w}}$	P^{Con}	M^{n}	$\frac{\Delta M^{n}}{M}$	M^{s}	M^{gnp}	r	η^{a}	BL	$\frac{\Delta BL}{BL}$	In 1963 prices Q^{m}	W^{m}	I^{stk}	Pop
1948	11.9		3.63			50	7	56.40	1,190								
1949	10.7	80.5	3.97	9.4		59	16	60.03	1,242			1,198		400		-43	
1950	11.2	76.0	4.52	13.9	3.9	78	35	70.64	1,500			1,402	17.0	442	211	-19	520
1951	12.0	75.8	5.41	19.7	4.3	91	17	68.35	1,569			1,626	16.0	552	224	-34	535
1952	13.2	64.7	5.96	10.2	5.0	104	14	71.29	1,576			1,797	10.5	530	227	30	550
1953	15.3	55.5	6.85	14.9	6.1	124	19	73.90	1,632			1,808	0.6	803	382	-26	567
1954	21.4	71.9	8.91	30.1	7.2	152	23	69.33	1,652	10.0		1,978	9.4	1,011	315	66	583
1955	16.8	81.9	10.07	13.0	8.7	178	17	72.06	1,648	10.3		1,970	-0.4	1,056	333	130	601
1956	23.2	80.4	12.06	16.6	10.6	217	22	73.45	1,632	10.1		1,901	-3.6	1,158	361	-8	618
1957	13.2	75.0	13.56	11.0	12.6	291	34	87.44	1,940	9.4		2,179	14.6	1,373	374	193	637
1958	11.1	100.0	15.21	12.5	14.5	353	21	94.62	2,114	9.7		2,085	-4.3	1,503	413	108	656
1959	29.2	94.7	20.89	37.3	20.0	501	42	97.80	2,319	11.7		2,249	7.9	1,509	417	277	676
1960	26.3	97.9	27.4	31.1	27.0	692	38	103.02	2,535	17.2	.180	2,269	0.9	1,596	366	150	697
1961	33.3	99.2	37.6	37.2	37.9	1,042	51	108.48	2,878	14.9	.184	2,340	0.0	1,761	403	251	718
1962	54.8	133.2	57	50.5	57.9	1,702	63	109.71	3,028	14.5	.162	2,763	18.1	2,483	555	288	739
1963	78.0	100.0	100	75.4	100	2,792	64	100.00	2,792	17.6	.165	2,310	-16.4	2,509	873	134	760
1964	87.8	111.9	190	90.0	187	5,191	86	97.15	2,761	9.6	.167	2,176	-5.8	2,662	914	261	782
1965	55.4	111.9	299	57.4	303	9,104	75	113.21	3,118	7.9	.165	2,051	-5.7	2,700	877	465	805
1966	38.8	57.8	412	37.8	444	10,470	15	113.25	2,585	7.4	.157	2,268	10.6	2,985	950	-3	828
1967	27.0	32.7	529	28.4	575	14,931	44	119.33	2,899	2.6	.150	2,590	14.1	2,970	946	102	852
1968	28.1	28.4	657	24.2	713	21,352	43	135.97	3,240	2.5	.106	3,569	37.8	3,341	1,064	108	876
Mean	31.4	81.5		48.0						12.4		2,123	6.4	1,734	537	130	

Year	X	$\Sigma I/\Sigma K$	IL	N	M_1/M	Fert	A	GNP^{US}	τ^{pi}	τ^{c}	$\tau^{c'}$	e^{nom}	e^{imp}	e^{exp}
1948	612				.3500									
1949	776				.3600				8.06	17.2				
1950	487	9.22	769	171	.3200		233	339	7.87	17.2	17.2	5	37	74
1951	527	9.84	533	173	.3100		234	362	8.14	17.2	17.2	6	37	61
1952	395	10.22	560	182	.3000		236	375	8.52	17.2	17.2	7	37	54
1953	435	10.14	650	184	.3000	100	238	390	8.78	17.2	17.2	8	37	57
1954	487	9.98	518	192	.3000	112	239	386	9.13	17.2	17.2	10	51	55
1955	461	9.34	529	199	.3100	127	242	419	9.24	20.0	20.0	15	64	71
1956	501	9.18	637	200	.3100	139	243	433	9.55	20.5	20.5	28	98	63
1957	474	8.70	482	209	.2900	152	245	440	9.73	20.5	20.5	28	85	65
1958	448	9.27	465	214	.2700	179	247	444	10.10	21.2	21.2	28	75	70
1959	514	10.12	365	221	.2600	178	248	481	10.22	21.2	21.2	48	88	86
1960	527	10.55	344	226	.2445	189	250	502	7.87	21.2	21.2	52	74	91
1961	579	10.69	470	231	.2440	187	252	518	5.47	21.2	21.2	100	100	100
1962	487	10.54	378	239	.2331	160	253	553	5.97	28.0	26.7	170	114	101
1963	553	10.33	318	246	.2445	168	255	585	6.87	28.0	27.3	273	104	85
1964	579	9.69	367	253	.2227	164	257	631	6.87	28.0	25.7	491	103	102
1965	645	8.66	736	260	.1901	156	258	684	6.87	28.0	22.1	643	87	103
1966	685	7.48	565	267	.2227		260	708	8.36	30.0	22.5	614	59	86
1967	645	7.46	274	275	.1972		262	746	7.94	28.0	21.0	549	42	76
1968	698	7.78	345	283	.1910		264	794	8.56	30.0	22.5	733	44	83
Mean	533	9.43		222	.2211			515	8.18	22.2			70	78

(continued)

TABLE A. 1, continued

	w^{min1}	w^{min2}	w^{min2}/y	τ^{sle}	τ^{exc}	τ^{def}	τ^{inf}	EO	η^{c}	Fed	Aut	State	Mun	Total
1948														
1949					10.9									
1950	64	46	124	8.0	10.9	1.7	1.7			135	38	154	19	346
1951	64	45	119	7.6	11.1	-.8	-.8			138	34	155	17	344
1952	64	46	115	8.4	11.3	-.3	-.3			152	30	167	15	364
1953	70	54	135	8.2	11.6	2.4	2.4			158	39	184	26	407
1954	78	59	136	8.2	11.6	-.3	-.3			141	33	174	22	370
1955	84	65	145	8.2	11.5	0.8	0.8			167	46	204	28	445
1956	88	67	147	8.4	11.4	1.6	1.6			256	60	195	23	534
1957	94	76	162	8.5	12.4	2.1	2.1			213	60	253	33	559
1958	97	81	166	9.2	13.3	0.2	0.2			222	66	257	30	575
1959	101	90	179	11.2	13.2	-.2	-.2			269	65	273	32	639
1960	85	80	149	11.2	14.9	0.1	2.8	-24.2	.116	275	66	297	37	675
1961	100	100	177	11.2	14.4	2.9	3.2	-3.4	.115	428	83	293	39	843
1962	86	86	148	11.5	14.9	4.8	3.3	-81.8	.133	383	101	368	44	896
1963	81	77	135	11.5	15.2	3.8	3.6	-42.8	.179	460	118	363	52	993
1964	75	71	124	12.4	18.0	4.4	3.2	-9.1	.166	451	145	379	49	1,024
1965	72	71	126	13.2	18.0	3.5	0.7	1,135.0	.165	311	122	482	62	977
1966	67	70	121	13.2	18.6	1.1	0.1	724.8	.162	363	123	476	69	1,031
1967	65	70	119	15.7	17.8	3.7	1.0	-241.2	.152	372	159	519	74	1,124
1968	64	67	108	16.8	21.0	1.0	0.9	-357.4	.183	331	155	566	91	1,143
Mean		70	139	10.7	14.3	1.7	1.4	122.2						698

Year	C^g Fed	C^g Aut	C^g State	C^g Mun	C^g Total	Tr Fed	Tr Aut	Tr State	Tr Mun	Tr Total	S^g Fed	S^g Aut	S^g State	S^g Mun	S^g Total
1948															
1949															
1950	115	38	96	38	287	77	38	58	19	192	19	77	19	19	134
1951	121	34	121	34	310	69	52	86	17	224	120	69	52	17	259
1952	91	45	152	45	333	76	61	91	15	243	106	75	15	15	212
1953	171	66	184	39	460	79	92	79	13	263	13	39	13	13	79
1954	141	76	185	43	445	76	87	98	11	272	163	43	21	22	250
1955	176	65	157	46	444	83	102	102	18	305	37	56	37	19	149
1956	180	60	158	52	450	113	105	113	15	346	-67	68	67	15	83
1957	153	80	160	60	453	113	140	113	20	386	-53	80	100	20	147
1958	137	72	222	72	503	138	144	102	24	408	114	143	114	30	401
1959	102	93	255	65	515	120	167	111	19	417	167	116	120	23	426
1960	205	92	319	62	678	132	172	113	22	439	128	172	153	22	476
1961	140	75	362	64	641	149	262	119	25	555	-273	229	174	19	149
1962	135	91	301	64	591	240	253	109	20	622	-361	124	187	21	-29
1963	122	73	333	72	600	158	249	102	24	533	-353	247	121	23	38
1964	142	82	238	68	530	160	274	119	23	576	-388	148	196	22	-22
1965	153	95	154	68	470	225	327	168	25	745	-240	248	49	51	108
1966	171	103	155	83	512	245	422	184	29	880	-169	343	152	64	390
1967	145	88	186	103	522	287	502	213	33	1,035	-417	359	33	98	73
1968	176	113	174	129	592	256	675	236	43	1,210	-29	175	177	157	480
Mean					491					608					435

(continued)

TABLE A. 1, continued

Year	I^g					E^g	H^g	D
	Fed	Aut	State	Mun	Total			
1948								
1949								
1950	96	57	77	19	249	96	38	115
1951	69	52	86	17	224	86	52	138
1952	61	30	76	15	182	106	61	136
1953	66	39	79	26	210	118	66	145
1954	76	43	87	22	228	109	65	141
1955	65	46	74	19	204	111	65	167
1956	67	37	75	23	202	113	60	195
1957	107	60	107	27	301	140	73	267
1958	108	102	168	30	408	162	84	246
1959	120	111	134	28	393	171	88	204
1960	106	150	194	26	476	198	99	201
1961	130	144	174	30	478	227	105	193
1962	121	160	212	38	531	267	110	203
1963	79	134	198	78	489	246	111	195
1964	93	112	235	73	513	279	99	207
1965	107	168	197	81	553	278	108	316
1966	97	158	188	84	527	275	128	286
1967	87	234	186	112	619	285	129	401
1968	86	190	181	166	673	378	143	391
Mean					388	192	89	216

APPENDIX B: REGRESSIONS FOR PUBLIC EXPENDITURES

Public expenditures are treated as predetermined instrument variables in this study. With a view to seeing the influence of certain exogenous variables (and public instruments) on the government's expenditure decisions over long periods, regressions were also run for this sector. As discussed in the text, these regressions may be viewed as the first-occurring block in an overall block-recursive model. In addition to the expenditure equations, three mongrel relations for T^{dir}, T^{ind}, and T are also given (see Table B. 1). A careful inspection of the coefficients of the mongrel equations indicates that the coefficients of these equations are not an approximate weighted sum of the coefficients of the individual tax equations. An aggregation bias is suggested, and the income elasticities of T^{dir} and T^{ind} appear to be biased upwards, that of T downwards, perhaps.

Finally, the trend rates of expansion of certain disaggregates of public expenditures are presented in Table B. 2. The results are analyzed in Chapter 4.

TABLE B. 1

The Expenditure and Some Mongrel Equations

Public Expenditures

(B1) $$W^g = -130 - .243\hat{C}^g + .027T^{disp}_{-1} + 17.7\tau^{inf} + .10Y'_{-1}$$
(-4.1) (-.22) (.58) (3.14) (20.3)
$\bar{R}^2 = .991$, $d = .99$

(B2) $$C^g = -79 + 47\tau^{inf} - 1.74\hat{W}^g + .19Y'_{-1}$$
(-.6) (4.36) (-3.3) (3.7)
$\bar{R}^2 = .753$, $d = 1.74$

(B3) $$E = -43 - .283\hat{H} - .14\hat{D} + .0169Y'_{-1} + .073T_{-1}$$
(-3.1) (-.46) (-1.38) (3.14) (3.09)
$\bar{R}^2 = .966$, $d = 1.76$

(B4) $$H = 7.6 + .0064\hat{D} + .021T_{-1} + .0045Y'_{-1}$$
(1.4) (.15) (2.42) (2.32)
$\bar{R}^2 = .947$, $d = 1.69$

(B5) $$D = -28.6 - .175E - 1.21(\Delta P/P) + .035Y'_{-1}$$
(-.5) (-.29) (-1.45) (1.81)
$\bar{R}^2 = .668$, $d = .99$

(B6) $$I^g = 2.74 + 8.11\tau^{def} + .194T_{-1} - .0164\hat{S}^g + .72t$$
(.07) (.82) (6.6) (-.93) (1.27)
$\bar{R}^2 = .892$, $d = 1.31$

(B7) $$G = 512 + 1.6T^{disp}_{-1} + 107\tau^{inf}$$
(2.5) (4.9) (1.97)
$\bar{R}^2 = .620$, $d = .91$

Some Mongrel Relations

$$T^{dir} = -260 + .107\hat{Y} - 23.5(Y^{na}/Y^a - 8.62(\Delta\hat{P}/P)$$
(-1.4) (5.36) (-.278) (-6.0)

$$- 31.4\tau^{pi}_{-1} + 28.5\tau^{c}_{-1}$$
(-1.99) (3.15)
$\bar{R}^2 = .956$, $d = 2.04$ (2SLS)

$$T^{ind} = -998 + .274\hat{Y}' + 89.4(Y^{na}/Y^a) - 92.2\tau^{def}$$
(-3.6) (7.5) (.63) (-3.9)
$\bar{R}^2 = .963$, $d = 1.22$

$$T = 1,915 + .142\hat{Y} + 448(Y^{na}/Y^a) + 49\tau^{pi}_{-1}$$
(3.55) (3.66) (2.05)

$$- 38.7\tau^{c}_{-1} + 87.6\tau^{exc} + 41.5\tau^{sle}$$
(-2.52) (2.65) (1.19)
$\bar{R}^2 = .980$, $d = 2.09$

TABLE B. 2

Exponential Trend Rates of Public Expenditures, 1950-68

Expenditure (1)	Geometric Trend Rate (Percent) (2)	Correlation (3)
Wages and Salaries		
Federal	2.97	.792
Autarquia	4.17	.939
State	3.28	.980
Municipal	3.85	.943
Goods and Services		
Federal	0.53	.111
Autarquia	2.09	.708
State	1.22	.199
Municipal	2.42	.863
Transfers		
Federal	3.49	.917
Autarquia	5.98	.975
State	2.32	.776
Public Saving		
Federal	-24.2	.449
Autarquia	4.45	.731
State	4.62	.422
Municipal	4.01	.605
Public Investment		
Federal	0.83	.211
Autarquia	4.46	.799
State	2.96	.773
Municipal	5.07	.852
Economic Classification		
Education	3.45	.951
Health	2.61	.921
Defense	2.40	.802

Note: The relation used is $G_i = a_i e^{b_i t}$, where G_i is ith public expenditure, t stands for serial number of years, and b gives the required exponential rate of annual changes.

APPENDIX C: DOCUMENTATION FOR COMPUTER PROGRAM FOR OPTIMAL-CONTROL-THEORY SOLUTIONS (SIGMA 7, VANDERBILT UNIVERSITY)

This program is intended to find the minimal-cost solution to a system described by a first-order linear difference equation:

$$Y_{i+1} = Y_i + AY_i + B\tau_i + CZ_i \tag{C1}$$

in which Y is a vector of state variables (n dimensional), τ is a vector of control variables (r dimensional), and Z is a vector of exogenous variables (s dimensional). A, B, and C are arrays (of dimension $n \times n$, $n \times r$, and $n \times s$) of coefficients. The subscripts refer to the times (or "periods") at which the variables are evaluated. The program produces both a set of control variable values for each period from period 0 to period N and the resulting state variable solutions that minimize the cost function:

$$\text{Cost} = \sum_{i=0}^{N} [(Y_i - Y_i^*)' \Omega (Y_i - Y_i^*) + (\tau_i - \tau_i^*)' \Theta (\tau_i - \tau_i^*)]/2 \tag{C2}$$

where Y_i^* and τ_i^* are "targeted" values of the state and control variables.

The first task is to convert the multidegree difference equations describing a given problem into the form C1 above that is expected by the system. As an illustration of what is involved, consider the following one-variable, one-control-variable problem:

$$P = 2P_{-1} - 3P_{-2} + P_{-3} + 7T - 2T_{-1} \tag{E3}$$

Two difficulties are illustrated: the difference equation is of third order and the control variable, T, is lagged. The reduction to form E1 is accomplished by the introduction of new "state" variables so that the state variable array is given by:

$$X_i = \begin{bmatrix} P_0 \\ P_{-1} \\ P_{-2} \\ T_0 \end{bmatrix}_i$$

and the control variable array $\tau_i = (T_{+1})_i$, where $(P_0)_i$ means P evaluated for period i, $(P_{-1})_i$ means P evaluated for period (i-1), etc. In terms of these variables Equation C3 is written:

$$(P_0)_{i+1} = 2(P_0)_i - 3(P_{-1})_i + (P_{-2})_i + 7(T_{+1})_i - 2(T_0)_i \qquad \text{(E4)}$$

so that everything on the left is for period i+1 and everything on the right is for period i. This equation determines element P_0 of the state variable vector. Equations for the other three elements must be written. But these are very simple because $(P_{-1})_{i+1}$ and $(P_0)_i$ are just different ways of saying "the value of P for period i." Therefore:

$$(P_{-1})_{i+1} = (P_0)_i \qquad \text{(E5)}$$

determines P_{-1}. Likewise, we have:

$$(P_{-2})_{i+1} = (P_{-1})_i \qquad \text{(E6)}$$

$$(T_0)_{i+1} = (T_{+1})_i \qquad \text{(E7)}$$

Casting these equations into the matrix form E1 we have:

$$\begin{bmatrix} P_0 \\ P_{-1} \\ P_{-2} \\ T_0 \end{bmatrix}_{i+1} = \begin{bmatrix} P_0 \\ P_{-1} \\ P_{-2} \\ T_0 \end{bmatrix}_i + \begin{bmatrix} 1 & -3 & 1 & -2 \\ 1 & -1 & 0 & 0 \\ 0 & -1 & -1 & 0 \\ 0 & 0 & 0 & -1 \end{bmatrix} \begin{bmatrix} P_0 \\ P_{-1} \\ P_{-2} \\ T_0 \end{bmatrix}_i + \begin{bmatrix} 7 \\ 0 \\ 0 \\ 1 \end{bmatrix} (T_{+1})_i + \begin{bmatrix} 0 \\ 0 \\ 0 \\ 0 \end{bmatrix} (1)$$

where the last (zero) matrix has been included to illustrate the fact that the C matrix and the exogenous variable must be present but can be nullified easily.

This illustration can be generalized with little effort to more elaborate cases. The results illustrate that one must introduce one new "state" variable for each state variable that appears in the original equation as a second-order difference; two new "state" variables must be introduced for each state variable that appears originally as a third-order difference, etc.; and one new "state" variable must be introduced for each difference order of a control variable.

DATA PREPARATION

To operate the program, input data must be prepared as described below. Note that in each case in which several input data are to appear on the same card any quantities in excess of 8 should be placed on a second card, any in excess of 16 on a third card, etc. That is, they should be punched 8 per card. Note also that the placement of the items on a card is not material: the quantities are to be punched in order and separated by one or more blanks or by a comma and zero or more blanks.

First. The number of periods to be projected, N of Equation E2, followed by the dimension of the state variable vector, n, the dimension of the control-variable vector, r, and the dimension of the exogenous-variable, s, all punched on one card

Second. The array A of Equation 1, punched by rows, each row beginning a new card

Third. The B array of Equation 1, punched by rows with each row beginning a new card

Fourth. The C array of Equation 1, punched by rows, each row beginning a new card

Fifth. A card containing any characters to be printed as an output heading

Sixth. The state variable-cost matrix Ω of Equation E2: input only the lower triangle of the matrix and input it by rows, beginning each row on a new card. Notice that this matrix must be positive semidefinite.

Seventh. The control variable-cost matrix Θ of Equation E2: input only the lower triangle by rows with each row beginning a new card. This matrix must be positive definite.

Eighth. The targeted state, control, and exogenous variables for each period, from period 0 to period N (N+1 periods): beginning a new card, input the state variables for period 0; then, beginning another new card, the control variables for period 0; then, beginning another new card, the exogenous variables for period 0; repeat for period 1, then for period 2, etc. For the last (Nth) period do not input exogenous variables, since they are not needed for the last period.

Ninth. If additional runs are to be made changing only the cost matrixes repeat steps 5, 6, and 7 (Fifth, Sixth, and Seventh) as many times as necessary.

OUTPUT

The program outputs as documentation each of the arrays, A, B, C, Ω, and Θ, as they are input. The heading-card data and resulting solution follow the Ω and Θ arrays for each run. For each period the actual state variables are output with the corresponding targeted values immediately below each element. Then follows the actual control variables immediately above the corresponding targeted values. After the last period output the value of the cost function resulting from these data is output.

ERROR MESSAGES

Three error messages originate from the program. They are as follows:

ERROR 1 *** CODE: n

If the value of the code n is -1, the cost matrix R is singular. If the code is positive, R is very nearly singular and the program cannot proceed.

ERROR 2 *** CODE: n

An array obtained as an intermediate result is singular (code value -1) or too nearly singular (code value positive) to proceed. The probable cause is failure of the cost matrix Q to be positive semidefinite.

ERROR 3 *** CODE: n

The problem described (by the first data card) is too large for the program. Notice that the program capacity can be changed by changing the size of the common array X and the value of the variable size that appear on cards numbered 0016 and 0017 in the source-program deck.

CARD DECK SET-UP

For infrequent use the card deck should be set up as follows:

```
!JOB ...........
!RES CORE=20, TIME=3
!FORTRAN

     program deck

!OLAY (GO),(UNSAT, (SCILIB))
!ASSIGN F:1, (FILE, XX), (OUTIN)
!ASSIGN F:2, (FILE, YY), (OUTIN)
!ASSIGN F:5, (DEVICE, SI)
!ASSIGN F:6, (DEVICE, LO)
!RUN
!DATA
```

data deck

!FIN

If the program is to be run frequently it is more efficient to save the program on the machine disk. This is done by the following deck set-up:

```
!JOB .........
!RES CORE=20, TIME=3
!FORTRAN

    program deck

!ASSIGN F:1, (FILE, XX), (OUTIN)
!ASSIGN F:2, (FILE, YY ), (OUTIN)
!ASSIGN F:5, (DEVICE, SI)
!ASSIGN F:6, (DEVICE, LO)
!OLAY(GO), (UNSAT, (SCILIB)), (LMN, name), (PERM)
!SAVE (ALL)
!FIN
```

where "name" is the name chosen for the program. This has currently been done using the name SAHOTA.

After this is done the program can be run with the following deck set-up:

```
!JOB .........
!RES CORE=20, TIME=3
!ASSIGN F:1, (FILE, XX), (OUTIN)
!ASSIGN F:2, (FILE, YY), (OUTIN)
!ASSIGN F:5, (DEVICE, SI)
!ASSIGN F:6, (DEVICE, LO)
!RUN (LMN, name)
!DATA

    data deck

!FIN
```

Here "name" is the name chosen for the program when saved on the disk. Currently, the name SAHOTA will work.

BIBLIOGRAPHY

Aaron, Henry. "Estimates of the Distributional Impact of Brazilian Taxes and Expenditures," Council for International Progress in Management (United States), Inc., report no. PIO/T 512-80336, Task Order 51 (July-August 1968), pp. 23, viii.

Abouchar, Alan. "Public Investment Allocation and Pricing for Transportation." In The Economy of Brazil, edited by Howard Ellis, pp. 345-75. Berkeley: University of California Press, 1969.

Aitken, A.C. "On Least Squares and Linear Combination of Observations," Proceedings of the Royal Society of Edinburgh 55 (1934-35): 42-48.

Allen, R.G.D. Mathematical Economics. London: Macmillan, 1956.

Anderson, Arthur, and Co. Income Tax Series. 294 Washington St., Boston, Mass., 1968.

Ando, Albert, and Stephen M. Goldfeld. "An Econometric Model for Evaluating Stabilization Policies." In Studies in Economic Stabilization, edited by Albert Ando, E. Cary Brown, and Ann F. Friedlander, pp. 215-87. Washington, D.C.: The Brookings Institution, 1968.

Arrow, Kenneth J. "The Economic Implications of Learning by Doing." Review of Economic Studies 29, no. 2 (June 1962): 155-73.

——, and Mordecai Kurz. Public Investment, the Rate of Return, and Optimal Fiscal Policy. Baltimore: The Johns Hopkins Press, 1970.

Balopoulos, E.T. Fiscal Policy Models for the British Economy. Amsterdam: North-Holland, 1967.

Basmann, Robert L. "A Generalized Classical Method of Linear Estimation of Coefficients in a Structural Equation." Econometrica, 25, no. 1 (January 1957): 77-83.

Baumol, William J. Economic Dynamics: An Introduction. 2d ed. New York: Macmillan, 1959.

——. Economic Theory and Operations Analysis. New ed. Englewood Cliffs, N.J.: Prentice-Hall, 1965.

Behrman, Jere, and Lawrence R. Klein. "Econometric Growth Models for the Developing Economy." In Induction, Growth and Trade: Essays in Honour of Sir Roy Harrod, edited by W. A. Eltis, M. F. G. Scott, and J. N. Wolfe, pp. 167-87. Oxford: Clarendon Press, 1970.

Bellman, R. Applied Dynamic Programming. Princeton, N. J.: Princeton University Press, 1962.

Bergsman, Joel, and Arthur Candal. "Industrialization: Past Success and Future Problems." In The Economy of Brazil, edited by Henry Ellis, pp. 29-73. Berkeley: University of California Press, 1969.

Bergsman, Joel, and Pedro Malan. "The Structure of Industry Protection in Brazil." Revista Brasileira da Economia 24, no. 2 (April/June 1970).

Bird, R. M., and Lue Henry De Wulf. Taxation and Income Distribution in Latin America: A Critical Review of Empirical Studies. IMF: 1973.

Bogart, P. J. M. van den, and Henry Theil. "Macrodynamic Policy-Making: An Application of Strategy and Certainty Equivalence Concepts to the Economy of the United States, 1933-1936." Metroeconomica 11, no. 1-II (April/August 1959): 149-67.

Brazil, Government of. Balanço dos Investimentos Publicos. Rio de Janeiro: annual since 1966.

——, Instituto de Pesquisa Economica Aplicada (IPEA), Ministry of Planning and Coordination. "Diagnôstico Preliminar Setor de Finanças Públicas." December 1965. Mimeographed.

——, Ministry of Finance. Revista de Finanças Publicas. Annual.

Bronfenbrenner, Martin, and T. Meyers. "Liquidity Functions in the American Economy." Econometrica 28, no. 3 (October 1960): 810-34.

Brown, Cary, and Richard J. Kruizenga. "Income Sensitivity of a Simple Personal Income Tax." Review of Economics and Statistics 41, no. 3 (August 1959): 260-69.

Cagan, Phillip. "The Monetary Dynamics of Hyperinflation." In Studies in the Quantity Theory of Money, edited by Milton Friedman et al., pp. 25-117. Chicago: The University of Chicago Press, 1956.

Chalmers, James A., and William A. Fischel. "An Analysis of Automatic Stabilizers in a Small Econometric Model." National Tax Journal 20, no. 4 (December 1967): 432-43.

Chow, Gregory C. "Optimal Stochastic Control of Linear Economic Systems." Journal of Money, Credit, and Banking 2, no. 3 (August 1970).

———. "Optimal Control of Linear Econometric Systems with Finite Time Horizons." International Economic Review 13, no. 1 (February 1972a): 16-25.

———. "How Much Could be Gained by Optimal Stochastic Control Policies." Annals of Economic and Social Measurement 1, no. 4 (October 1972b): 391-405.

———. "Problems of Economic Policy from the Viewpoint of Optimal Control." American Economic Review 43, no. 5 (December 1973): 825-37.

Christ, Carl F. Econometric Models and Methods. New York: Wiley, 1966.

Cochrane, D., and G. H. Orcutt. "Application of Least Squares Regression to Relationships Containing Autocorrelated Errors Terms." Journal of the American Statistical Association 44, no. 2 (March 1949): 32-61.

Cohen, L. "A More Recent Measure of the Built-In Flexibility of the Individual Income Tax." National Tax Journal 13, no. 2 (June 1960): 122-26.

Deaver, J. "The Chilean Inflation and the Demand for Money." In Varieties of Monetary Experience, edited by David Meiselman. Chicago: University of Chicago Press, 1970.

Diamond, Walter H. Digest on Brazilian Taxes and Business Organization. New York: Overseas Press, 1958.

Diz, A. "Money and Prices in Argentina 1935-1962." In Varieties of Monetary Experience, edited by David Meiselman. Chicago: University of Chicago Press, 1970.

Durbin, James, and G. S. Watson. "Testing for Serial Correlation in Least Squares Regression." Biometrika 37, no. 4 (December 1950): 409-28.

Eckstein, Otto. "Comparison of European and United States Tax Structure and Growth Implications." In National Bureau of Economic Research, U.S., The Role of Direct and Indirect Taxes in the Federal Revenue System. Princeton, N.J.: Princeton University Press, 1964.

Eijk, C.J. van, and J. Sandee. "Quantitative Determination of an Optimum Economic Policy." Econometrica 27, no. 1 (January 1959): 1-13.

Ellis, Howard S. "Corrective Inflation in Brazil, 1964-1966." In The Economy of Brazil, edited by Howard S. Ellis, pp. 177-212. Berkeley: University of California Press, 1969.

Friedman, Milton. The Theory of Consumption Function. National Bureau of Economic Research, Princeton, N.J.: Princeton University Press, 1957.

——. "The Demand for Money — Some Theoretical and Empirical Results Journal of Political Economy 67, no. 3 (June 1959): 327-51.

——. "Economic Miracles." Newsweek (January 21, 1974), p. 80.

Froomkin, J.M., and H.P. Wald, eds. Papers and Proceedings of the Conference on Agricultural Taxation and Economic Development. Cambridge, Mass.: Harvard Law School International Program in Taxation, 1954.

Fundaçāo Getúlio Vargas (FGV). Revista Brasileira da Economia 16, no. 1 (March 1962), and 20, no. 1 (March 1967).

——. Contas Nacionas. Annual.

——. Anúario Estatistico do Brazil. Annual.

——. Conjuntura Económica. Monthly.

——. O. Setor Publica Brasileiro 1947-1966. 1968.

——. Instituto Brasileiro de Economia, Pesquisa Sobre Orcamentos Familiares. Various cities and towns (processed), various rural areas (still unprocessed). Surveys for 1961-62 for cities and for 1962-63 for towns and rural areas. Rio de Janeiro, Brazil: FGV, date of processing not stated, but understood to be 1967.

Ghai, Dharam. Taxation for Development: A Case Study of Uganda. Nairobi, Kenya: East African Publishing House, 1968.

Gillespie, W. Irwin. "Effects of Public Expenditures on the Distribution of Income." In Essays in Fiscal Federalism, edited by Richard A. Musgrave, pp. 122-86. Washington, D.C.: The Brookings Institution, 1965.

Goldberger, Arthur S. Impact Multipliers and Dynamic Properties of the Klein-Goldberger Model. Amsterdam: North-Holland, 1959.

——, and Lawrence R. Klein. An Econometric Model of the United States, 1929-1952. Amsterdam: North-Holland, 1955.

Goldfeld, Stephen M. Commercial Bank Behavior and Economic Activity. Amsterdam: North-Holland, 1966.

Goode, Richard. The Individual Income Tax. Washington, D.C.: The Brookings Institution, 1964.

Gordon, Lincoln, and Engelbert L. Grommers. United States Manufacturing Investment in Brazil: The Impact of Brazilian Government Policies 1946-1960. Cambridge, Mass.: Harvard University Press, 1962.

Griliches, Zvi. "Specification Bias in Estimates of Production Function." Journal of Farm Economics 39, no. 1 (1957): 8-17.

——. "A Note on Serial Correlation Bias in Estimates of Distributed Lags." Econometrica 29, no. 1 (January 1961): 65-73.

——. "An Analysis of Specification Errors; Notes for a Course in Econometrics." Unpublished notes, Department of Economics, University of Chicago, 1963.

——. "Production Functions in Manufacturing: Some Preliminary Results." Paper presented at the Conference on Research in Income and Wealth, National Bureau of Economic Research, Inc., New York, December 15, 1965.

——. "Distributed Lags: A Survey." Econometrica 35, no. 1 (January 1967): 16-49.

——. "The Brookings Model Volume: A Review Article." The Review of Economics and Statistics 50, no. 2 (May 1968): 215-34.

Grumpel, Henry J., and R.G. de Souza. Taxation in Brazil. World Tax Series, Harvard Law School, International Program in Taxation. Cambridge, Mass.: The Riverside Press, 1957.

Grunfeld, Yahuda. "The Determinants of Corporate Investment." In The Demand for Durable Goods, edited by Arnold C. Harberger, pp. 211-66. Chicago: University of Chicago Press, 1960.

Gudin, Eugenio. "The Chief Characteristics of the Postwar Economic Development of Brazil." In The Economy of Brazil, edited by Howard S. Ellis, pp. 3-25. Berkeley: University of California Press, 1969.

Haavelmo, Trygve. "Multiplier Effects of a Balanced Budget." Econometrica 13, no. 4 (December 1945): 311-18.

Hall, R. E., and Dale W. Jorgenson. "The Quantitative Impact of Tax Policy on Investment Expenditure." In Tax Incentives and Investment Spending, edited by Gary Fromm. Washington, D. C.: The Brookings Institution, 1970.

Hall, R. L., and C. J. Hitch. "Price Theory and Business Behavior." Oxford Economic Papers, Old Series, no. 2 (May 1939): 12-45.

Hansen, Alvin H., and H. S. Perloff. State and Local Finances in the National Economy. New York: Norton, 1944.

Harberger, Arnold C. "The Incidence of the Corporation Income Tax." Journal of Political Economy 70, no. 3 (June 1962): 215-40.

——. "The Dynamics of Inflation in Chile." In Measurement in Economics: Essays in Honor of Yahuda Grunfeld, edited by Carl Christ et al., pp. 219-50. Stanford: Stanford University Press, 1963.

Heare, G. E. Brazil: Information for United States Business. Washington, D. C.: U. S. Department of Documents, 1961.

Heuss, F. G. Fiscal Policy for Growth without Inflation. Baltimore: Johns Hopkins Press, 1963.

Hickman, Bert G., ed. Quantitative Planning of Economic Policy. Washington, D. C.: The Brookings Institution, 1965.

Hoch, Irving. "Estimation of Production Function Parameters Combining Time-Series and Cross-Section Data." Econometrica 30, no. 1 (January 1962): 34-53.

Hockman, Harold M., and James D. Rodgers, "Pareto Optimal Redistribution." American Economic Review 59, no. 4, pt. I (September 1969): 542-57.

Holbrook, R. S. "Optimal Economic Policy and the Problem of Instrument Instability." American Economic Review 62, no. 1 (March 1972): 57-65.

Houthakker, H. S., and Lester D. Taylor. Consumer Demand in the United States, 1929-1970. Cambridge, Mass: Harvard University Press, 1966.

India, Government of. Statistical Abstract of India. New Delhi: Government of India Press. Annual.

International Monetary Fund (IMF). International Financial Statistics. Annual.

Japan Tax Association. Asian Taxation 1969. Tokyo: Japan Tax Association, 1970.

Johansen, Lief. Public Economics. Amsterdam: North-Holland, 1966.

Johnston, John. Econometric Methods. New York: McGraw-Hill, 1963.

Joint Tax Program. Problems of Tax Administration in Latin America. Papers and Proceedings of a Conference Held in Buenos Aires, Argentina, October 1961. Baltimore: Johns Hopkins Press, 1965.

Jorgenson, Dale W. "Anticipations and Investment Behavior." In The Brookings Quarterly Econometric Model of the United States, edited by J. S. Duesenberry et al., pp. 35-92. Chicago: Rand-McNally, 1969.

——. "A Comparison of Alternative Theories of Corporate Investment Behavior." The American Economic Review 58, no. 4 (September 1968): 681-713.

Kilpatrick, Robert W. "The Short-Run Forward Shifting of the Corporation Tax." Yale Economic Essays, vol. v (fall, 1965), pp. 355-420.

Klein, Lawrence R. "Single Equation vs. Equation System Methods of Estimation in Econometrics." Econometrica 28, no. 3 (October 1960): 866-71.

——. An Introduction to Econometrics. Englewood Cliffs, N. J.: Prentice-Hall, 1962.

——, and A. S. Goldberger. An Econometric Model of the United States 1929-1952. Amsterdam: North-Holland, 1955.

Klein, Lawrence R., and P. Taubman. "Impact of Accelerated Depreciation and Investment Tax Credit on Investment and General Economic Activity. In Tax Incentives and Investment Spending, edited by Gary Fromm. Washington, D.C.: The Brookings Institution, 1970.

Korff, Eurico. Imposto de Renda e Investimento. Edicao Financeiras, S.A. Rio de Janeiro, 1965.

Koyck, L.M. Distributed Lags and Investment Analysis, Contributions to Economic Analysis no. 4. Amsterdam: North-Holland, 1954.

Krischen, Etienne S., et al. Economic Policy in Our Times. Amsterdam: North-Holland, 1964.

Krischen, Etienne S., and Lucien Morissens. "The Objectives and Instruments of Economic Policy." In Quantitative Planning of Economic Policy, edited by Bert G. Hickman, pp. 111-33. Washington, D.C.: The Brookings Institution, 1965.

Krzyaniak, Marian, and Richard A. Musgrave. The Shifting of the Corporation Income Tax: An Empirical Study of Its Shortrun Effects upon the Rate of Return. Baltimore: Johns Hopkins Press, 1963.

Langoni, Carlos Geraldo. Distribucao da Renda e Desenvolvimento Economico do Brasil. Rio de Janeiro: Editora Expressao e Cultura, 1973.

Lewis, Stephen R., Jr. "Sources of Changes in Revenue from Import Taxes 1954-1963 (Pakistan)." Pakistan Institute of Development Economics, January 1965. Mimeographed.

——. "Agricultural Taxation in a Developing Economy." In Agricultural Development and Economic Growth, edited by Herman M. Southworth and Bruce F. Johnston, pp. 453-96. Ithaca, N.Y.: Cornell University Press, 1967.

LEX, Editora Proprietaria, Impreso Grafic Lex Ltda., São Paulo, LEX 15, no. 22 (1951, 1961, 1965).

Malinvaud, Edmond. Statistical Methods in Econometrics. Chicago: Rand-McNally, 1966.

Maneschi, Andrea. "The Brazilian Public Sector during the Sixties." In Brazil in the Sixties, edited by Riordan Roett. Nashville, Tenn.: Vanderbilt University Press, 1971.

Mann, H. B., and A. Wald. "On the Statistical Treatment of Linear Stochastic Difference Equations." Econometrica 11, no. 2 (June 1943): 173-220.

Margolis, J., and H. Guitton, eds. Public Economics: An Analysis of Public Production and Consumption and Their Relations to the Private Sector. New York: St. Martin's Press, 1969.

Maxwell, James A. "Counter-Cyclical Fiscal Policy of State and Local Governments." National Tax Journal 11, no. 4 (December 1958): 366-72.

May, Sydney. "Dynamic Multipliers and Their Use for Decision-Making." In Conference on Stabilization Policies convened by the Economic Council of Canada at the University of Western Ontario, London, August 30-September 1, 1965, pp. 155-96. Ottawa: Queen's Printer, 1966.

McNamara, Robert. "Address to the Board of Governors." IBRD/IMF Annual Meeting, Washington, D.C., September 25, 1972.

Meiselman, David, ed. Varieties of Monetary Experience. Chicago: University of Chicago Press, 1970.

Meyer, John R., and Edwin Kuh. The Investment Decision: An Empirical Study. Cambridge, Mass.: Harvard University Press, 1957.

Modigliani, Franco. "Liquidity Preference and the Theory of Interest and Money." Econometrica 12, no. 1 (January 1944): 45-236.

Morley, Samuel A. "Import Demand and Import Substitution in Brazil." In The Economy of Brazil, edited by Howard S. Ellis, pp. 283-316. Berkeley: University of California Press, 1969.

Musgrave, Richard A. The Theory of Public Finance: A Study in Public Economy. New York: McGraw-Hill, 1959.

——. "Estimating the Distribution of the Tax Burden." In Problems of Tax Administration in Latin America, pp. 31-111, Joint Tax Program, Baltimore: Johns Hopkins Press, 1965. Also printed in Colin Clark and G. Stuvel, eds. Income Distribution and the Statistical Foundations of Economic Policy. London: Bowes and Bowes, 1964.

——. "Provision for Social Goods." In Public Economics: An Analysis of Public Production and Consumption and Their Relations to the Private Sector, edited by J. Margolis and H. Guitton, pp. 124-44. New York: St. Martin's Press, 1969.

Musgrave, Richard A. "Cost-Benefit Analysis and the Theory of Public Finance." Journal of Economic Literature 7, no. 3 (September 1969): pp. 797-806.

Myrdal, Gunnar. Economic Theory and Underdeveloped Regions. London: Duckworth, 1959.

NBER and The Brookings Institution, Inc. The Role of Direct and Indirect Taxes in the Federal Revenue System, conference report. Princeton, N.J.: Princeton University Press, 1964.

Nerlove, Marc. "Distributed Lags and Estimation of Long-Run Supply and Demand Elasticities: Theoretical Considerations." Journal of Farm Economics 40, no. 3 (August 1958): 301-10.

——. Distributed Lags and Demand Analysis for Agricultural and Other Commodities, Agricultural Handbook no. 14. Washington, D.C.: U.S. Government Printing Office, 1958.

Neto, Antonio Delfim. "Tentativa de Explicacao das Causas que Determinaram a Expansao dos Meios de Pagamento en 1964-1965." In A Economia Brasileira e Suas Perspectivas, V, APEC Editoria, Rio de Janeiro, 1966.

——. "Nota Sobre A Estimação de Relação Producto/Capital Com Os Dados da Contabilidade Nacional." In Revista de Administração, University of São Paulo, no. 32 (July 1965), pp. 1-12.

——. Alguns Aspectos da Inflação Brasileira, Estudos Anpes no. 1. São Paulo: Editora Grafica Piratininga, 1965.

——, et al. Aspectos da Inflacao Brasileira e Suas Perspectivas para 1965. São Paulo: Associacao Nacional de Programacao Economica e Social, 1965.

Oldman, Oliver. "Tax Reform in El Salvador." Inter-American Law Review 6, no. 2 (July-December 1964): pp. 379-420.

——. "Controlling Income Tax Evasion." In Problems of Tax Administration in Latin America, Joint Tax Program, pp. 296-344. Baltimore: Johns Hopkins Press, 1965.

Paiva, Ray Miller. "Bases de uma politica para a melhoria tecnica da agricultura brasileira." Revista Brasileira de Economia (Rio de Janeiro) 21, no. 2 (July 1967).

Parks, R. W. "Efficient Estimation of a System of Regression Equations when Disturbances are Both Serially and Contemporaneously Correlated." Journal of the American Statistical Association 62, no. 320 (December 1967): pp. 500-9.

Pastore, Affonso Celso. "Inflação e Politica Monetaria No Brasil." Revista Brasileiro de Economia, no. 1 (1969): 92-123.

Pechman, Joseph A. "Yield on Individual Income Tax during Recession." National Tax Journal 7, no. 1 (March 1954). Also reprinted (with minor changes) in Conference of the Universities — National Bureau Committee for Economic Research. Policies to Combat Depression. Princeton, N.J.: Princeton University Press, 1956.

——. "Comment." In Problems of Tax Administration in Latin America, Joint Tax Program, pp. 90-102. Baltimore: Johns Hopkins Press, 1965.

Pigou, A. C. Public Finance (London: Kelley, 1928).

Pindyck, Robert S. Optimal Planning for Economic Stabilization — Contributions to Economic Analysis #81. Amsterdam: North-Holland/American Elsevier, 1973. Abridged version published in Pindyck. "Optimal Policies for Economic Stabilization." Econometrica 41, no. 3 (May 1973): 529-61.

Pontryagin, L., V. Boltyanskii, R. Gamkrelidge, and E. Mischenko. The Mathematical Theory of Optimal Processes. New York: Interscience, 1962.

Poole, William. "Alternative Paths to a Stable Full Employment Economy." Brookings Papers on Economic Activity 3. Washington, D.C.: The Brookings Institution, 1971, pp. 579-614.

Prest, A. R. "The Sensitivity of the Yield of Personal Income Tax in the United States." Economic Journal 71, no. 1 (1961).

——. "The Budget and Interpersonal Distribution." In The Budget and the Distribution of National Income, Institut International de Finances Publiques, Congres de Prague, pp. 80-98. Saarbrucken: Institute International de Finances Publiques, 1968.

——. Public Finance in Theory and Practice. 4th ed. London: Weidenfeld and Nicolson, 1970.

Price Waterhouse and Co. Information Guide for Doing Business in Brazil. Boston: Price Waterhouse, 1962, 1964, 1965, 1968.

Rafuse, Robert W., Jr. "Cyclical Behavior of State-Local Finances." In Essays in Fiscal Federalism, edited by Richard A. Musgrave, pp. 63-121. Washington, D.C.: The Brookings Institution, 1965.

Rao, V.K.R.V. "Investment, Income and the Multiplier in an Underdeveloped Economy." The Indian Economic Review (February 1942). Also printed in The Economics of Underdevelopment, edited by A.N.A. Aggarwala and S.P. Singh, pp. 205-18. Oxford: Oxford University Press, 1958.

Reviglio, Franco. "Social Security: A Means of Savings Mobilization for Economic Development." IMF Staff Papers 16, no. 2 (July 1967): 324-68.

Rijckeghem, Willy Van. "The 1959 Input-Output Table for Brazil." In The Economy of Brazil, edited by Howard S. Ellis, pp. 376-400. Berkeley: University of California Press, 1969.

Sahota, Gian S. Indian Tax Structure and Economic Development. Bombay: Asia Publishing House/Taplinger Publishers, New York, 1961.

———. "An Economic Analysis of Internal Migration in Brazil." Journal of Political Economy 76, no. 2 (March/April, 1968): 218-45.

———. "The Distribution of Tax Burden Amongst Education Classes in Brazil." Economic Development and Cultural Change 19, no. 2 (April 1971): 438-60.

———. "Public Expenditure and Income Distribution in Panama." Planning Commission, Government of Panama and USAID Panama, August 1972.

———. "Income Distribution, Marginal Utility of Income, and Tax Progressivity." Economic Times Annual, Bombay, special issue (summer 1974).

Samuelson, Paul A. "Interactions between the Multiplier Analysis and the Principle of Acceleration." Review of Economic Statistics 21 (1939): 75-78.

———. "The New Look in Tax and Fiscal Policy." Sub-Committee on Tax Policy, Joint Committee on the Economic Report, 84th Congress, Federal Tax Policy for Economic Growth and Stability. Washington, D.C., 1955, pp. 229-34.

Samuelson, Paul A. "Pure Theory of Public Expenditure and Taxation." In Public Economics: An Analysis of Public Production and Consumption and Their Relations to the Private Sector, edited by J. Margolis and H. Guitton, pp. 98-123. New York: St. Martin's Press, 1969.

Sandee, J., and J. H. Van de Pas. "The Effects of Flucturations in Public Expenditure and Taxation on Economic Growth." In Public Economics: An Analysis of Public Production and Consumption and Their Relations to the Private Sector, edited by J. Margolis and H. Guitton, pp. 388-405. New York: St. Martin's Press, 1969.

Sargan, J. D. "The Estimation of Economic Relationships Using Instrumental Variables." Econometrica 26, no. 3 (1958): 393-415.

Schipper, Lewis, Consumer Discretionary Behavior: A Comparative Study in Alternative Methods of Empirical Research. Amsterdam: North-Holland, 1964.

Sharp, Ansel M. "The Behavior of Selected State and Local Government Fiscal Variables." In Proceedings of the Fifty-Eighth Annual Conference on Taxation, edited by Walter J. Kress, pp. 599-614. Harrisburg, Pa.: National Tax Association, 1966.

Shoup, Carl S. Public Finance. Chicago: Aldine, 1969.

——. The Tax System of Brazil. Rio de Janeiro: FGV, 1965.

——, et al. The Fiscal System of Venezuela: A Report. Baltimore: Johns Hopkins Press, 1959.

Silva, Adroaldo M. da. "Money, Interest Rates and Inflation in Brazil." Ph. D. dissertation, Department of Economics, University of Chicago (1973).

Silva, Neyde Ramos da. "Functional Breakdown of State Finances" (in Portuguese), memorandum, Ministry of Finance, Government of Brazil. Rio de Janeiro, December 21, 1967.

Simonsen, Mario Henrique. "Inflation and the Money and Capital Markets of Brazil." In The Economy of Brazil, edited by Howard S. Ellis, pp. 133-61. Berkeley: University of California Press, 1969.

Slitor, Richard. "The Measurement of Progressivity and Built-In Flexibility." Quarterly Journal of Economics 62 (February 1948): 309-12.

Slitor, Richard. "The Enigma of Corporate Tax Incidence." Public Finance 18, no. 3-4 (1963): 323-32.

Smith, Gordon W. "Brazilian Agricultural Policy, 1950-1967." In The Economy of Brazil, edited by Howard S. Ellis, pp. 213-65. Berkeley: University of California Press, 1969.

Smith, Paul E. "A Note on the Built-In Flexibility of the Individual Income Tax." Econometrica 31, no. 4 (October 1963): 704-11.

Southworth, Herman M., and Bruce F. Johnston, eds. Agricultural Development and Economic Growth. Ithaca, N.Y.: Cornell University Press, 1967.

Spencer, Bryon G. "Shifting of the Corporation Income Tax: The Canadian Experience." Ph.D. dissertation, Department of Economics, Rice University, Houston, Texas, 1967.

Stigler, George J. "Director's Law of Public Income Distribution." The Journal of Law and Economics 13, no. 1 (April 1970): 1-10.

Suits, Daniel B. "Use of Dummy Variables in Regression Equations." Journal of American Statistical Association 52, no. 4 (December 1957): 548-51.

Taylor, L.D., and T.A. Wilson. "Three-Pass Least Squares: A Method for Estimating Models with a Lagged Dependent Variable." Review of Economics and Statistics 46, no. 4 (November 1964): 329-46.

Theil, Henri. "On the Theory of Economic Policy." American Economic Review 44, no. 2 (May 1956): 360-66.

——. Economic Forecasts and Policy. 2d rev. ed. Amsterdam: North-Holland, 1961.

——. "Linear Decision Rules for Macrodynamic Policy Problems. In Quantitative Planning of Economic Policy, edited by Bert G. Hickman, pp. 18-42. Washington, D.C.: The Brookings Institution, 1965.

Tinbergen, Jan. On the Theory of Economic Policy. Amsterdam: North-Holland, 1952.

——. Economic Policy: Principles and Design. Amsterdam: North-Holland, 1956.

Tinbergen, Jan. Development Planning. New York: McGraw-Hill, 1967.

Toukan, Abdul Rehman S. "The Implications of Achieving Fiscal Independence for Jordan." Ph.D. dissertation, Department of Economics, Vanderbilt University, 1966.

United Kingdom, Government of, Inland Revenue Service. Inland Revenue Statistics 1970. London: Her Majesty's Stationery Office, 1970.

United Nations. Per Capita National Product of Fifty Five Countries 1952-1954, Statistical Paper Series E, no. 4, 1957.

——. Statistical Yearbook. New York: UN, annual.

——. Yearbook of National Accounts Statistics. New York: UN, annual.

Wallich, Henry C. "Money and Growth — A Country Cross-Section Analysis," Yale University Economic Growth Center Paper no. 141. New Haven: Yale University Economic Growth Center, 1970.

Wallis, Kenneth F. "Lagged Dependent Variables and Serially Correlated Errors: A Reappraisal of Three-Pass Least Squares." Review of Economics and Statistics 49, no. 4 (November 1967): 555-67.

Watson, G.S., and E.J. Hannon. "Serial Correlation in Regression Analysis II." Biometrika 43, no. 4 (December 1956): 445-57.

Wogart, Jan Peter. "Secondary Expansion of Money Supply and Its Underlying Variables in Post-war Brazil." A Economia Brasileira e Suas Perspectivas, vol. 8. Rio de Janeiro: APEC Editoria, 1969.

Wulf, Lue Henry De, and R.M. Bird. "Fiscal Incidence Studies in Developing Countries: Survey and Critique." IMF, 1974.

Yaqub, Muhammad. "The Tax Structure of a Developing Country: A Case Study of Pakistan." Ph.D. dissertation, Department of Economics, Princeton University, 1966.

Yunus, Mohammad. "Optimal Allocation of Multiple Purpose Reservoir Water: A Dynamic Programming Model." Ph.D. dissertation, Vanderbilt University, 1971.

Zellner, A. "An Efficient Method of Estimating Seemingly Unrelated Regressions and Tests for Aggregation Bias." Journal of the American Statistical Association 57, no. 298 (June 1962): 348-68.

———, D. S. Huang, and L. C. Chan. "Further Analysis of the Short-Run Consumption Function with Emphasis on the Role of Liquid Assets." Econometrica 33, no. 3 (July 1965): 571-81.

Zellner, A., and H. Theil. "Three Stage Least Squares: Simultaneous Estimation of Simultaneous Equations." Econometrica 30, no. 1 (January 1962): 54-78.

NAME INDEX

SUBJECT INDEX

ABOUT THE AUTHOR

GIAN SINGH SAHOTA is Professor of Economics at Vanderbilt University. He has been a faculty member of Institute of Economic Growth and Punjab University and has had visiting assignments at Harvard, Manchester University, London School of Economics, University of São Paulo, Colorado University, and Institute of Economic Growth in India. He has served as economic advisor to the governments of Brazil, Panama, and Singapore and as consultant to the Ford Foundation, the River Basin Commission of Brazil, and the Urban Observatory of Metropolitan Nashville and Universities Center.

Professor Sahota has published a number of articles in such periodicals as the Journal of Political Economy, International Economic Review, Review of Economics and Statistics, American Journal of Agricultural Economics, Indian Economic Journal, Revista Brasileira da Economia, and Economic Development and Cultural Change. He is the author of Fertilizer in Economic Development (Praeger, 1969) and Indian Tax Structure and Economic Development (Institute of Economic Growth, Delhi, 1961).

Dr. Sahota earned his M.A. in Economics at Punjab and Leeds Universities and his Ph.D. in Economics at the University of Chicago.

RELATED TITLES
Published by
Praeger Special Studies

CHRONIC INFLATION IN LATIN AMERICA
Felipe Pazos

CONTEMPORARY BRAZIL: ISSUES IN ECONOMIC AND POLITICAL DEVELOPMENT
edited by H. Jon Rosenbaum and William G. Tyler

EXPROPRIATION OF U. S. INVESTMENTS IN CUBA, MEXICO, AND CHILE
Eric N. Baklanoff

EXPROPRIATION OF U. S. PROPERTY IN SOUTH AMERICA: Nationalization of Oil and Copper Companies in Peru, Bolivia, and Chile
George M. Ingram

THE INTERNATIONAL COFFEE AGREEMENT: A Study in Coffee Diplomacy
Bart S. Fisher